Exoplanets

Detection, Formation, Properties, Habitability

John W. Mason (Editor)

Exoplanets

Detection, Formation, Properties, Habitability

Springer

Published in association with
Praxis Publishing
Chichester, UK

PRAXIS

Editor
Dr John W. Mason
Olympus Mons
51 Orchard Way
Barnham
West Sussex PO22 0HX
UK

Front cover illustrations: (Main image) A computer-generated simulation of a Jupiter-sized exoplanet shown crossing in front of the disk of its parent star. Image courtesy Jeffery Hall, Lowell Observatory. (Smaller image) Artist's concept of a planetary disk around a brown dwarf. Image courtesy NASA/JPL-Caltech/T.Pyle (SSC).

Back cover illustrations: (Top) Artist's impression of a Saturn-mass planet orbiting the sun-like star HD149026, with atmosphere based on models by James Cho. Image courtesy Greg Laughlin, University of California, Santa Cruz. (Bottom) Artist's impression of the Jupiter-sized planet discovered transiting a star 500 light-years from Earth. Image courtesy Jeffery Hall, Lowell Observatory.

SPRINGER–PRAXIS BOOKS IN ASTRONOMY AND PLANETARY SCIENCES
SUBJECT *ADVISORY EDITORS*: Philippe Blondel, C.Geol., F.G.S., Ph.D., M.Sc., Senior Scientist, Department of Physics, University of Bath, UK; John Mason, B.Sc., M.Sc., Ph.D.

ISBN 978-3-540-74007-0 Springer Berlin Heidelberg New York

Springer is part of Springer-Science + Business Media (springer.com)

Library of Congress Control Number: 2007935198

Cover design: Jim Wilkie
Author-generated LaTex, processed by EDV-Beratung, Germany

Printed on acid-free paper

Contents

Editor's Preface

An extrasolar planet or exoplanet is a planet orbiting a star (or remnant of a star) beyond our Solar System. As of autumn 2007, about 250 exoplanets had been discovered around 220 different stars, including nearly two dozen multiple planet systems. No less than five exoplanets have been discovered orbiting the star 55 Cancri; one of the planets has nearly four times the mass of Jupiter, another is comparable with Jupiter in mass, two are slightly less massive than Saturn, while the innermost planet has a mass similar to that of Uranus.

Around 2300 years ago, the Greek philosopher Epicurius reflected on the existence of planets around other stars, and of life on those planets:

> *"There are infinite worlds both like and unlike this world of ours... We must believe that in all worlds there are living creatures and plants and other things we see in this world."*

And in the 16th Century, the medieval scholar Giordano Bruno, in his work *De l'infinito, universo e mondi,* speculated:

> *"There are countless suns and countless Earths all rotating around their suns in exactly the same way as the seven planets of our system. We see only the suns because they are the largest bodies and are luminous, but their planets remain invisible to us because they are smaller and non-luminous. The countless worlds in the universe are no worse and no less inhabited than our Earth."*

Extrasolar planets became a subject of scientific investigation in the mid-19th Century, and although there were some unsubstantiated claims as to their discovery, it was not known how common they were, how similar they were to the planets of the Solar System, or indeed how typical the make up of our Solar System was in comparison with planetary systems around other stars. There was also the question of the habitability of such planets. Were there Earth-like planets orbiting other stars and, if so, could they have the necessary surface conditions to support some form of life?

What actually constitutes a planet? In February 2003, the Working Group on Extrasolar Planets (WGESP) of the International Astronomical Union produced a reasonable working definition of a "planet", agreeing to revise the definition as and when necessary, and as our knowledge improves. The WGESP considered that objects with true masses below the limiting mass for thermonuclear fusion of deuterium (currently calculated to be ~13 Jupiter masses (~13 M_J) for objects of solar metallicity) that orbit stars or stellar remnants are "planets", no matter how they formed. As it happens, this deuterium-burning limit at ~13 M_J resides near the upper-end of the observed exoplanet mass distribution.

The WGESP also decided that the minimum mass/size required for an extrasolar object to be considered a "planet" should be the same as that used in our Solar System. Here, of course, there has been very considerable deliberation and debate arising out of the resolutions passed at the IAU General Assembly in Prague in August 2006, mainly in relation to the status of "dwarf" bodies such as Pluto, Eris and Ceres within our own Solar System. As far as detected exoplanets are concerned, the minimum mass object detected to date is the 0.00007 M_J object (40 per cent the mass of Mercury) orbiting the pulsar PSR 1257+12, but the lowest mass companions to ordinary stars which have been discovered to date are Gl 876 d, which has a minimum mass of 0.0185 M_J (about 5.9 Earth masses), OGLE-05-390L b, which has an estimated mass of 0.017 M_J (about 5.4 Earth masses) and Gl 581 c, which has a minimum mass of 0.0158 M_J (about 5 Earth masses).

The WGESP also decided that substellar objects with true masses above the limiting mass for thermonuclear fusion of deuterium are "brown dwarfs", no matter how they formed nor where they are located. Furthermore, free-floating objects in young star clusters with masses below the limiting mass for thermonuclear fusion of deuterium are not "planets", but are "sub-brown dwarfs" (or whatever name is most appropriate).

The first confirmed detections of exoplanets were made in early 1992, by the radio astronomers Aleksander Wolszczan and Dale Frail, but rather surprisingly these were not found around an ordinary star, but a pulsar – the superdense remnant of a massive star that has exploded as a supernova. The first definitive detection of an exoplanet orbiting an ordinary main-sequence star came in October 1995 with the announcement, by Michel Mayor and Didier Queloz of the University of Geneva, of an exoplanet orbiting the star 51 Pegasi. This discovery ushered in the modern era of exoplanet discovery, and since 2000 about 20–30 exoplanets have been discovered every year, with the most detections, by far, during 2007.

New discoveries and significant developments in exoplanet research continue at a frenetic pace, and it is difficult to keep up with progress in this exciting field. This multi-author volume comprises a collection of eleven topical reviews, each presented as a separate chapter, and covering an important aspect of exoplanet studies. The contributions have been written by scientists at the forefront of research in the selected areas, in a style which, we hope, will be accessible not only to advanced undergraduate students and beginning graduate students, but also to professional astronomers working in the field.

Although the direct imaging of exoplanets is extremely difficult at the present time, a variety of indirect detection methods are available. In Chapter 1, Patrick Irwin provides an overview of exoplanet detection techniques. The most successful take advantage of the fact that a planet orbiting a distant star can make its presence known through small, regular variations in the radial velocity or position of its parent star. However, exoplanets are increasingly being detected by observing the minute decrease in the light of the host star if an exoplanet happens to pass in front of it (in transit), or through techniques such as gravitational microlensing. So many exoplanets have now been found that it is possible to consider the statistics of the mass and orbital parameter distributions, and Chapter 1 includes a collection of plots showing the exoplanet mass distribution, their orbital period and orbital radius distribution, distributions of mass and radius and of eccentricity and radius for known exoplanets, and the distribution of host star metallicity. Chapter 1 concludes with a discussion of selection effects for different exoplanet detection programmes, and a look ahead to planned transit surveys and the techniques being developed for direct optical detection.

In Chapter 2, Jian Ge takes a detailed look at the most successful method employed to date for exoplanet detection, that of Doppler planet surveys. Of the roughly 250 exoplanets discovered to date, over 90 per cent have been detected by single object Doppler techniques. This chapter outlines the theory of the two principal Doppler methods: one using high resolution cross dispersed echelle spectrographs (the echelle method) and the other using dispersed fixed-delay interferometers (the DFDI method). Both methods have been successfully used for detecting new exoplanets. The main results of Doppler planet surveys over the past decade are then summarised, together with early results in the development of new Doppler techniques, especially multiple object techniques. Chapter 2 presents the scientific motivation for the next generation large-scale multi-object Doppler planet surveys and possible new science which will be addressed. Past experience has shown that the ability to move from single-object to multi-object observations has facilitated large-scale astronomical surveys (e.g. the Sloan Digital Sky Survey), and has consistently led to dramatic new discoveries. It is anticipated that similar advances will result from multi-object Doppler planet surveys in the next decade.

Another important exoplanet detection technique, that of gravitational microlensing, is reviewed by David Bennett in Chapter 3. This method relies upon chance alignments between background source stars and foreground stars which may host planetary systems. The background source stars serve as light sources that are used to probe the gravitational field of the foreground stars and any planets that they might host. The author explains how the microlensing method is unique among exoplanet detection methods in a number of respects, particularly in its ability to find low-mass planets at separations of a few AU. The basic physics of the microlensing method is reviewed together with typical planetary microlensing events. The author shows how such microlensing events may be used to enable the measurement of planetary orbital parameters, and he reviews early observational results highlighting the exoplanets discovered by microlensing to date. Finally, the author demonstrates that a low-cost, space-based microlensing survey can provide a comprehensive statistical census of extrasolar planetary systems with sensitivity down to 0.1 Earth-masses at separations ranging from 0.5 AU to infinity.

As George Rieke explains in Chapter 4, exoplanets move within tenuous disks of dust (and early-on, gas) that are relatively easy to detect. The dust intercepts energy from the parent star more efficiently than a planet can, and thus scatters and reradiates energy in far larger amounts than a planet could. In the process, it imposes its own signatures on this output. We know of hundreds of planetary systems through observation of circumstellar disks of dust, and we can learn indirectly about them if we can read these signatures. The author discusses the formation and evolution of protoplanetary disks in the context of terrestrial planet formation. He shows that although there is a well-defined overall pattern of protoplanetary disk characteristics, there is a wide range of starting conditions, e.g. disk masses, along with some variation in evolutionary timescales. Such differences presumably translate into a wide range of properties for the planetary systems that develop within these disks. The process of terrestrial planet formation continues well beyond the protoplanetary stage, and produces disks of debris from the planetesimal collisions. The observed behaviour of these debris disks can test many hypotheses regarding the evolution of the Solar System. Debris disks also enable astronomers to probe many different examples of how planetary systems evolve, since there are $\sim$150 known examples within 50pc.

The interesting connection between brown dwarfs and exoplanets is explored by I. Neill Reid and Stanimir Metchev in Chapter 5. Brown dwarfs form like ordinary stars but, with masses below 0.075 solar masses, or 1.5×10^{29} kg, they fail to ignite core hydrogen fusion. Lacking a central energy source, they cool and fade on timescales that are rapid by astronomical standards. Consequently, the observed characteristics of old, cold brown dwarfs provide insight into the expected properties of gas-giant exoplanets. The chapter focusses on brown dwarfs as companions to main-sequence and evolved stars. Following a brief introduction to the intrinsic properties of brown dwarfs, including their observed characteristics and classification, the authors examine the different observational techniques used to identify very low mass companions of stars and review the advantages and challenges associated with each method. The authors summarise the results of various observational programs, particularly those regarding companion frequency as a function of mass and separation, and discuss the so-called 'brown dwarf desert'. The implications of these results for brown dwarf and planetary formation mechanisms are considered. The chapter concludes with a discussion of future surveys for low mass companions, particularly direct imaging programs that will have sufficient sensitivity to detect objects of planetary mass.

The detection of the first exoplanet around the G2V star 51 Pegasi in 1995 was a landmark discovery. The presence of this Jupiter mass planet in a very close 4.2-day orbit around the host star was quickly confirmed, and corroborated by Doppler evidence for more of these close-orbiting Jupiter mass planets (dubbed 'hot Jupiters') around a number of other nearby stars. Developments in experimental capabilities have meant that so called 'hot Saturns' and 'hot Neptunes' have also been discovered, and these close-orbiting planetary systems are discussed in detail by Hugh Jones, James Jenkins and John Barnes in Chapter 6. As the authors explain, although 51 Pegasi-like objects dominated early discoveries, other types of planets are considerably more common. The 51 Pegasi class were found first because they were the easiest to detect by the radial velocity method. In addition to being favoured by radial velocity surveys, the bias is even stronger in transit surveys. All known transiting exoplanets have periods less than a week. Although our overall knowledge of exoplanets has been fuelled by the growth in the sheer number and also by the broad range of parameter space now populated, close-orbiting planets characterised with a combination of precise radial velocity measurements and transit photometry have played a key role. In these close-orbiting systems it is possible to determine the mass and radius of the planet, which in turn yields constraints on its physical structure and bulk composition. The transiting geometry also permits the study of the planetary atmosphere without the need to spatially isolate the light from the planet from that of the star. This technique (known as transit spectroscopy or occultation spectroscopy) has enabled photometric and spectroscopic measurements of exoplanets to be made. As the authors of Chapter 6 make clear, the wide range of properties of close-orbiting planets has stimulated a plethora of physical models to explain their properties. They provide the sharpest test for theories of formation, e.g., gravitational instability versus core-accretion, the role of stellar metallicity in determining planetary core mass and how an irradiating star influences planetary contraction and migration, e.g., type I, type II and delayed migration. With the continuous development of experimental techniques, close-orbiting terrestrial-mass exoplanets are the exciting new frontier in astrophysics and will test a wide range of theoretical predictions.

The dynamical properties of multiple planet systems are reviewed by Rory Barnes in Chapter 7. As the author explains, the study of exoplanet dynamics is severely hampered by observational uncertainties. Although the detections themselves are robust, the orbital elements have significant uncertainties. The most problematic aspect of the Doppler technique is the mass-inclination degeneracy. If the inclination, the angle between the plane of the orbit and a reference plane, can be determined by a complementary method, such as astrometry or transits, this degeneracy may be broken, and the planetary masses and full three dimensional orbits identified. The mass-inclination degeneracy therefore makes many simulations, analyses, and hypotheses unreliable. Generally, in the dynamical studies discussed in Chapter 6, the masses are assumed to be the "minimum mass" – the mass if the orbit was exactly edge-on. Statistically, this choice is expected to be reasonably accurate. The Doppler technique also limits the ranges of planetary masses and orbital radii that may be observed, and so the observed planets may not be all the planets in a system. Consequently, the conclusions presented in Chapter 6 are subject to revision as additional planets may exist in each system that are either low-mass or orbit at large distances, and these unseen companions may significantly alter the best-fit orbits of the known planets. The author describes how the orbits of planets evolve due to tidal, resonant, and/or secular (long-term) effects. Basic analytical and numerical techniques can describe these interactions, and the author reviews orbital theory and analytical methods (secular theory and resonant interactions), and shows how N-body integrations are used to determine the evolution of a system. Multiple planet systems may also evolve chaotically, and some principles of chaos theory are described. Finally, the author discusses the current distributions of dynamical properties of known multiple exoplanetary systems, possible origins of these distributions, and compares exoplanetary systems with the Solar System.

There is increasing evidence that planets are ubiquitous, and may form around stars over a wide range in stellar masses. After a star dies, the planets may remain, and in some circumstances there may be a new epoch of planet formation after the main sequence. In Chapter 8, Steinn Sigurdsson discusses scenarios for the retention and formation of planets after the death of the parent star, and the prospects for detection, including current known post-main sequence systems. Planets in the so-called 'stellar graveyard' are, in many cases, easier observational targets than planets around main sequence stars, and different detection techniques may also be brought to bear, in some cases with much higher sensitivity, allowing the detection of low mass planets. This is particularly true in the case of the three exoplanets detected around the millisecond pulsar PSR 1257+12, which at 0.00007, 0.13 and 0.12 Jupiter masses are the lowest mass exoplanets discovered to date. The author discusses theories as to the origin of planets around pulsars, including the pulsar planet in the globular cluster Messier 4, before turning his attention to the detection of planets around white dwarfs. He also describes the recent exciting discovery of a giant planet around the extreme horizontal branch star V391 Pegasi. This is a well known pulsating subdwarf, a star that has terminated core hydrogen fusion on the stellar main sequence and evolved through a red giant branch phase. The planet must originally have been closer to the star, but moved outwards as the star lost mass, avoiding being swallowed by the red giant envelope as the star expanded. As the author explains, planets detected in the stellar graveyard reflect the 'live' population of planets, and in some cases provide potentially strong constraints on planet formation processes, and the general planet population.

A survey of currently known planet-hosting stars indicates that approximately 25 per cent of extrasolar planetary systems are within dual-star environments. Several of these systems contain stellar companions on moderately close orbits, and the existence of exoplanets in such binary systems has confronted dynamicists with many new challenges, as Nader Haghighipour explains in Chapter 9. Questions such as how are these planets formed, whether binary-planetary systems host terrestrial and/or habitable planets, how habitable planets form in such dynamically complex environments, and how such planets acquire the ingredients necessary for life, are among major topics of research in this area. Chapter 9 begins with a review of the dynamics of a planet in a binary star system, and in particular whether the orbit of a planet around its host star would be stable. The author then examines the formation of planets in binary star systems. In spite of the observational evidence that indicates the majority of main and pre-main sequence stars are formed in binaries or clusters, and in spite of the detection of potentially planet-forming environments in and around binary stars, planet formation theories are still unclear in explaining how planets may form in multi-star environments. The author then discusses the formation of giant and terrestrial planets in moderately close binary-planetary systems, and reviews the current status of planet formation theories in this area. The habitability of a binary system is then examined. Models of habitable planet formation in and around binary systems are presented, and their connections to models of terrestrial planet formation and water-delivery around single stars are discussed. Chapter 9 ends with a discussion of the future prospects for research in the field of planets in binary star systems.

The theme of the habitability of planets and the search for life beyond the Solar System is explored in detail by Victoria Meadows in Chapter 10. In its most conservative definition, a 'habitable world' is a solid-surfaced world, either a planet or moon, which can maintain liquid water on its surface. This definition is based on the fact that water is the one common constituent used by an enormous array of life forms on the Earth. Life may also be present in the atmospheres of planets, or in subsurface water tables or oceans, even in our own Solar System. However, as the author explains, when searching for life beyond our Solar System, we adopt the more conservative definition of the presence of surface water, because this definition also has the advantage of describing worlds that would be more detectable as habitable, even over enormous distances. After introducing the concept of habitable zones around stars which may harbour planets, the author explains how even a conservative definition of habitability still encompasses a vast array of potential worlds that could be considered habitable, without being similar to the present-day Earth. The techniques and space missions which will enable the direct detection of Earth-sized planets are then described, and aspects of the remote detection of planetary characteristics are outlined. Although characterising a planet for the ability to support life is an exciting first step, it is a precursor to the search for any indications that the planet already harbours life. Such signs of life, either past or present, when inferred from very distant measurements are called 'remote-sensing biosignatures'. As the author carefully explains, the search for these is based on the premise that widespread life will modify the atmosphere and surface of its planet, and that such modifications will be detectable on a global scale. The chapter concludes with a look at how such biosignatures might be detected.

There is good reason to hypothesise that giant exoplanets will be attended by significant moon systems. Moon systems exhibit diverse characteristics, and present unique environments – possibly even suitable habitats for life. As Caleb Scharf outlines in the final chapter, Chapter 11, such exomoons may share many characteristics with those in our own Solar System, as well as represent alternatives - possibly including temperate Mars- or Earth-sized bodies. In our own Solar System the majority of giant planet moons harbour substantial water ice mantles. The inferred internal structure and observed activity of many suggests the potential for extensive subsurface liquid water, both currently and in the past. A well known example of this is Jupiter's icy moon, Europa. Liquid water is vital for all forms of terrestrial life, through its integrated roles in biochemistry and geophysics. By contrast, the thick atmosphere and rich, low-temperature, hydrocarbon chemistry of Saturn's largest moon, Titan, points towards a highly complex surface environment paralleling some of the conditions on the early Earth, and conceivably offering alternative pathways for complex phenomena such as life. As the author concludes, detecting the presence of moons in exoplanetary systems is rapidly approaching feasibility, and will open a new window on such objects and their potential habitability.

This book has benefited from the support and assistance of a large number of people. I would like to offer my sincere thanks to all of the contributing authors for their considerable efforts, perseverance and enthusiasm for this project. I am indebted to Frank Herweg of Springer, Heidelberg for his invaluable support and advice in the preparation of the LaTeX files for this book, including his work on a number of the illustrations prior to publication. I am also most grateful to my wife Jane Mason for her assistance in the preparation of the Index, and to John and Margaret Dowling for help with proof reading. Finally, I am indebted to Imogen Millard, Sue Peterkin and Romy Blott of Praxis Publishing for their very considerable assistance at all stages in the organization and coordination of this project, and to Clive Horwood, Publisher, for his encouragement, advice and patience throughout.

Barnham, November 2007 John W. Mason

List of Contributors

John R. Barnes
Centre for Astrophysics Research
University of Hertfordshire
College Lane
Hatfield
Hertfordshire AL10 9AB
England
j.r.barnes@herts.ac.uk

Rory Barnes
Lunar and Planetary Laboratory
University of Arizona
1629 E. University Blvd.
Tucson
Arizona AZ 85721
USA
rory@lpl.arizona.edu

David P. Bennett
Research Associate Professor
Astrophysics and Cosmology
University of Notre Dame
225 Nieuwland Science Hall
Notre Dame
Indiana IN 46556-5670
USA
bennett@nd.edu

Jian Ge
Professor, Department of Astronomy
211 Bryant Space Science Center
University of Florida
Gainesville
Florida FL 32611
USA
jge@astro.ufl.edu

Nader Haghighipour
Institute for Astronomy and NASA
Astrobiology Institute
University of Hawaii-Manoa
2680 Woodlawn Drive
Honolulu
Hawaii HI 96822
USA
nader@ifa.hawaii.edu

Patrick G. J. Irwin
Atmospheric, Oceanic and Planetary
Physics
Clarendon Laboratory
Department of Physics
University of Oxford
Parks Road
Oxford OX1 3PU
England
irwin@atm.ox.ac.uk

James S. Jenkins
Centre for Astrophysics Research
University of Hertfordshire
College Lane
Hatfield
Hertfordshire AL10 9AB
England
j.s.jenkins@herts.ac.uk

Hugh R.A. Jones
Professor, Centre for Astrophysics Research
University of Hertfordshire
College Lane
Hatfield
Hertfordshire AL10 9AB
England
h.r.a.jones@herts.ac.uk

Victoria S. Meadows
Professor, University of Washington
Department of Astronomy
Box 351580
Seattle
Washington WA 98195
vsm@astro.washington.edu

Stanimir Metchev
Department of Physics & Astronomy
University of California
430 Portola Plaza
Los Angeles
California CA 90095-1547
metchev@astro.ucla.edu

I. Neill Reid
Space Telescope Science Institute
3700 San Martin Drive
Baltimore,
Maryland MD 21218
USA
inr@stsci.edu

George H. Rieke
Regents Professor of Astronomy and Planetary Sciences
Steward Observatory
933 N. Cherry St.
The University of Arizona
Tucson
Arizona AZ 85721
USA
grieke@as.arizona.edu

Caleb A. Scharf
Columbia Astrobiology Center
Columbia Astrophysics Laboratory
Columbia University
550 West 120th Street, MC5247
New York NY 10027
USA
caleb@astro.columbia.edu

Steinn Sigurdsson
Professor, Department of Astronomy & Astrophysics
525 Davey Laboratory
The Pennsylvania State University
University Park
Pennsylvania PA 16802
USA
steinn@astro.psu.edu

1 Detection Methods and Properties of Known Exoplanets

Patrick G. J. Irwin

Summary. Following the historic discovery of the first extrasolar planet, 51 Pegasi b, in 1995 (Mayor and Queloz, 1995) more than 200 planets orbiting other stars have now been catalogued. The vast majority of these planets have been detected with the radial velocity technique, which is biased towards heavy, close-orbiting planets. However, the number of lighter, more distantly orbiting known exoplanets is increasing steadily and, in addition, a growing fraction of exoplanets have now been discovered using other detection methods that may be more successful in detecting terrestrial-type planets. In this chapter we will review the main physical properties of the exoplanets (and their parent stars) discovered to date (28 February 2007) and will review the expectations of forthcoming observations.

1.1 Introduction

The question of just how unique our Solar System is has intrigued philosophers and scientists for centuries. While it has generally been assumed that there are almost certainly other planets orbiting other stars, it was not until the historic discovery of 51 Pegasi b in 1995 by Mayor and Queloz (1995) that the first conclusive proof of the non-uniqueness of the Solar System was obtained. The planet that was discovered though, and most of those discovered since with the same radial velocity technique (Sect. 1.2.1), is very different from the planets of our Solar System. 51 Peg b (the exoplanetary naming convention is to list the star name followed by 'b', 'c' ... in order of the planet's discovery) has a mass greater than or equal to 0.46 M_J, (where M_J is the mass of Jupiter) and orbits at a distance of only 0.05 AU in a period of just 4.2 days! The surface temperature of the planet, so close to its star, is calculated to be enormous ($\sim$ 1400 K) and the planet has been dubbed a 'hot Jupiter'.

1.2 Detection of Extrasolar Planets

Directly observing extrasolar planets is extremely difficult given the large brightness contrast between a star and its planets and also the small angular separation. For example, if our own Solar System were observed at a distance of, say, 5 parsecs, the greatest angular separation of the Sun and Jupiter would be just 1 arcsecond with the Sun appearing 10^9 times brighter at visible wavelengths. Under these conditions it would be impossible to pick Jupiter out from the Sun's glare (Lewis, 2004). One possible solution to this problem is to search for planets around dimmer stars such as white and brown dwarfs. Searches for extrasolar planets around white dwarfs have so far been unsuccessful (e.g. Burleigh et al., 2003; Friedrich et al., 2006), but four planets/brown dwarfs (Sect. 1.4.2) have now been directly imaged about brown dwarfs, the first being imaged by the Very Large Telescope (VLT) orbiting a brown dwarf, situated 200 light years away, at a distance of $\sim$ 60 AU (Chauvin et al., 2005a). Another strategy is to attempt to detect the planet at wavelengths near the peak of the planet's Planck function. Observing at 50 μm rather than 0.6 μm reduces the flux ratio to 10^4 for the Sun-Jupiter system, but at these longer wavelengths the diffraction-limited angular resolution of any achievable telescope would be insufficient.

Although direct optical detection of extrasolar planets initially appeared very difficult, it was realised that it might be possible to indirectly detect them through their influence on the motion of the central star. There are two ways of doing this: 1) by observing the *radial velocity* of the star as the planetary system rotates about its centre-of-mass and 2) by observing the actual reflex motion[1] of the star against the heavens (*astrometry*). In addition, it also came to be realised that there was a chance that an extrasolar planet could be detected if it transited in front of its star, while other detection methods, such as gravitational lensing, revealed themselves serendipitously. There are now numerous methods of detecting extrasolar planets, which will be briefly summarised.

1.2.1 Radial Velocity Detections

For a planet of mass M_p in a circular orbit of radius a about a star of mass M_*, the star and planet will both orbit about their centre-of-mass, situated at a distance $2aM_p/(M_p + M_*)$ from the star. Equating the gravitational force with the centripetal force acting on the star, and assuming that $M_* \gg M_p$, the maximum velocity of the star v in the line of sight of an observer may be shown to satisfy $v^2 = G(M_p \sin i)^2/2M_* a$, where i is the inclination of the planet's orbit with respect to the observer, i.e. the angle between the normal to the orbital plane of the planet and the line from the star to the observer on the Earth. The radial velocity method can determine both $M_p \sin i$ and also, from the shape of the variation of v with time, the eccentricity, e, of the planet's orbit. It is worth noting that unless the inclination can be determined from other methods such as astrometry (Sect. 1.2.2), this method only provides a lower limit on the planet's mass. The technique is most effective for larger mass planets orbiting close to the lower mass stars (i.e. G and K type) since this gives the largest line-of-sight stellar velocity and it is crucial to

[1] The reflex motion of the star is caused by both it and the planet orbiting their common centre of mass.

be able to distinguish the radial velocity of the star due to the orbit of a planet from the naturally occurring turbulent velocities present in a stellar photosphere. An example of a measured radial velocity curve for the star GJ 446 (Butler et al., 2004) is given in Fig. 1.1.

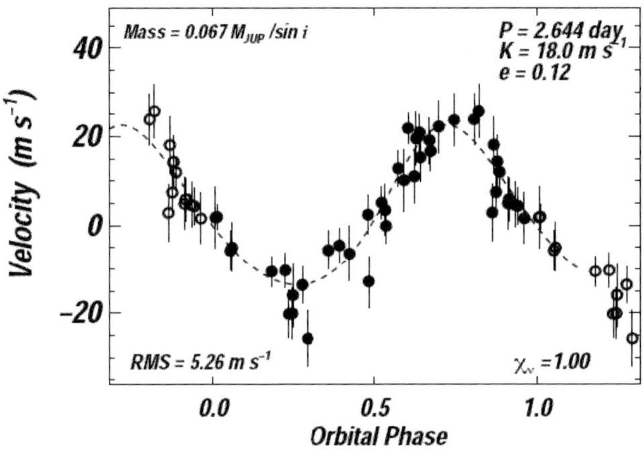

Fig. 1.1. Measured velocities vs orbital phase for GJ 436 (Butler et al., 2004). The dotted line is the radial velocity curve from the best-fit solution: $P = 2.644$ days, $e = 0.12$, $M \sin i = 0.067 M_J$

Since the discovery of 51 Peg b there have been detections (almost all by the radial velocity technique) of over 200 extrasolar planets. Indeed it is now estimated that more than 6% of sun-like stars have a detectable 'wobble' due to the orbit of at least one Jupiter-mass planet. At the time of writing (28 February 2007), the total number of planets listed in the Extrasolar Planets Encyclopedia (http://www.obspm.fr/planets) was 215 in 185 planetary systems (including 21 multiple planet systems). Most of the recent radial velocity planet searches have been able to detect velocity variations as small as 10 m/s (Marcy et al., 2003) and so a Sun-Jupiter system (for which the Sun's radial velocity is 13.2 m/s) should have been just about detectable and, indeed, such planets are now regularly being found. For example (Wittenmyer et al., 2007) report the discovery of 47 UMa c, a planet with mass 1.34 M_J, low eccentricity and an orbital radius $a = 7.73$ AU. Recent improvements have meant that current observations can now achieve even greater accuracies of 3 m/s and thus the number of planets detectable by this technique is steadily increasing. In addition, the current data sets only last for $\sim$ 10 years. As measurements continue, and the sensitivity improves, the discovery of more Jupiter-like planets orbiting far from their star with longer periods is expected. At the time of writing 26 exoplanets have now been catalogued with an orbital distance greater than 3 AU.

1.2.2 Astrometry

Given a sequence of observations of a star's position of sufficiently high accuracy relative to the celestial sphere, the reflex motion of the star caused by the orbit of a planet around it can be detected. This can be used to determine both the absolute mass and orbital inclination of a planet. Considering the motion of the star and planet about their common centre of mass we can see that the reflex amplitude of the star is $a_* = a_p M_p / M_*$, where a_* and a_p are the distances from the centre-of-mass to the star and planet respectively. Thus, this method is most effective for large mass planets orbiting at some distance from their parent stars. In addition, since what is actually measured is the angular position of the star, the method is clearly best for planetary systems within a few parsecs of the Earth.

The accurate measurement of a star's position over a number of years is a challenging task. Current optical systems have an absolute accuracy of a few milliarcseconds. However this precision can be improved through the use of long-baseline interferometry. The VLT and Keck currently have programmes to do this and are expected to achieve accuracies of 30 μas (microarcseconds), which should be sufficient to observe the reflex motion of the stars of several extrasolar giant planets already discovered. In addition, there are two space missions planned to exploit this technique. The NASA SIM (Space Interferometry Mission) is due for launch sometime betwee 2009 and 2015 and will be able to achieve 1 μas accuracy, while the ESA GAIA spacecraft, which is a follow-up to ESA's Hipparcos mission, is due to launch in 2011. Although not an interferometric instrument, GAIA aims to observe 1 billion stars with magnitude brighter than 20, with an accuracy of 10–20 μas at magnitude 15.

1.2.3 Transit Detections

For extrasolar planets, there is a small, but finite, chance that the orbital inclination i will be very close to 90° and thus that a planet will periodically pass between the planet's star and the Earth. If the planet is sufficiently large, then the drop of intensity of the starlight can be detected and used to determine both i and also the radius of the planet.

The first published detection of a planetary transit (using the STARE transit camera (Charbonneau et al., 2000)), was of the planet HD 209458 b, which orbits its star at a distance of 0.046 AU in a period of 3.5 days (Henry et al., 2000). The transit was observed the next year with the Hubble Space Telescope (HST) (Fig. 1.2) and Brown et al. (2001) concluded, from the transit depth, that the planet had a radius of 1.35 R_J (where R_J is the radius of Jupiter). This figure has recently been revised to 1.32 R_J (Knutson et al., 2007).

Assuming HD 209458 b to be typical, and until more transits of this type are observed there is no reason to think otherwise, these observations showed that the massive, close-orbiting planets discovered by the radial velocity survey were not just rocky cores, but large Jupiter-sized objects. The radius observed is considerably larger than that expected from a planet cooling in isolation and Burrows et al.

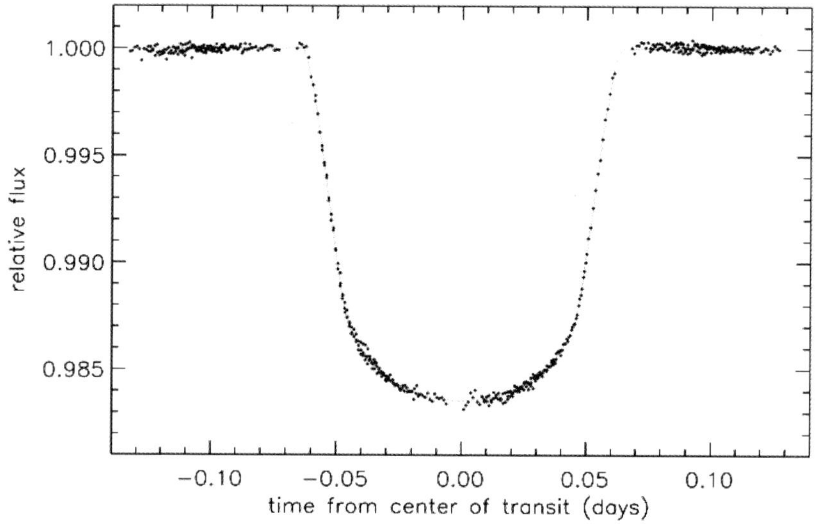

Fig. 1.2. HST observation of transit of HD 209458 b (Brown et al., 2001)

(2000) proposed that irradiation from the star inhibits convection and thus cooling/contraction. This idea was developed by Bodenheimer et al. (2001) and Guillot and Showman (2002).

A number of other extrasolar planetary transits have been observed since 1999, using projects such as OGLE (Sect. 1.2.4). Most lead to a dip in intensity of the order of 1%, and at these levels care must be taken to ensure that phenomena such as sunspot variations or isolated or blended eclipsing binary systems are not mistaken for planet detections (e.g. Mandushev et al., 2005; O'Donovan et al., 2006b, 2007).

Transit Spectroscopy

Soon after the first transit of HD 209458 b was observed, it was realised that observations at a number of different wavelengths might be used to infer the atmospheric transmission of the planet's atmosphere, since a planet's effective cross-sectional area will be larger at wavelengths where its atmosphere is more strongly absorbing than at others. Just such a study is reported by Charbonneau et al. (2002) who used HST observations near 600 nm to search for the atmospheric sodium absorption lines predicted for 'hot Jupiters' by radiative transfer models such as Sudarsky et al. (2003). The absorption line was duly detected, the first ever detection of an exoplanetary atmosphere, although the magnitude of the absorption was found to be less than that predicted by cloud-free radiative transfer models suggesting that clouds high in the atmosphere of this planet reduce the absorption band depth. Brown et al. (2002), Richardson et al. (2003a) and Richardson et al. (2003b) extended this campaign to the infrared, searching for CO, H_2O and CH_4 absorption, and recently Deming et al. (2005) detected a weak absorption due to CO at 4325 cm^{-1} and also suggested the presence of a high level cloud at, or above, 3.3 mbar.

In addition to direct detection of atmospheric absorption during transits, a gas giant orbiting as close to its star as HD 209458 b will get very hot in its upper atmosphere leading possibly to exospheric loss. Vidal-Madjar et al. (2003) report HST observations of atomic hydrogen absorption of starlight during several transits of HD 209458 b. They interpret this observation as being due to absorption by hydrogen atoms that have exospherically escaped the planet's atmosphere and are now beyond the Hill radius[2] of the planet. They further conclude that if the timescale for this evaporation is comparable to the lifetime of the stellar system then it may explain why so few 'hot Jupiters' are found orbiting with periods less than $\sim$ 3 days. More recent HST observations by Vidal-Madjar et al. (2004) have also detected exospherically escaping carbon and oxygen atoms. Such atoms should be too heavy to escape by the Jean's mechanism, responsible for the hydrogen escape, and instead Vidal-Madjar et al. (2004) suggest that hydrodynamic escape (or 'blow-off') is responsible, whereby the outward flow of exospherically escaping hydrogen atoms carry with them heavier atoms such as carbon and oxygen.

1.2.4 Microlensing

For several years now there have been campaigns to observe galactic bulge microlensing events, with a view to searching for dark matter and extrasolar planets. In this technique, light from a distant (source) star is observed as another star at intermediate distance (the lens star) passes close to, or in front of it. Light from the source star is gravitationally bent around the lens star and thus its apparent magnitude changes during the event. Two such campaigns are OGLE (Udalski, 2003) and MOA (Bond et al., 2001). In addition to lensing events, such programmes are also sensitive to planetary transits and to date, OGLE has detected the transits of five previously unknown extrasolar planets.

In 2003, both observatories observed a remarkable microlensing event shown in Fig. 1.3 where, in addition to the central peak in source star brightness due to the gravitational lensing of the lens star, two additional sharp peaks were observed which are interpreted as being due to the microlensing of a planetary companion to the lens star. Bond et al. (2004) conclude, assuming the lens star to be a main sequence M dwarf, that the planet has a mass of 1.5 M_J, and orbits the lens star at a distance of approximately 3 AU.

OGLE has now detected three further planets through gravitational microlensing events. For future observations we will see later in Sect. 1.4 that gravitational lensing is the only detection method that is capable of sensing terrestrial planets orbiting some distance from their stars (dubbed 'cool Earths'). In addition to the continuation of the OGLE and MOA campaigns, other ground-based campaigns include PLANET, which is a collaboration of telescopes in the southern hemisphere observing since 1995. The sensitivity of microlensing campaigns to 'cool Earths' would be further advanced by placing the telescope in space and proposed mis-

[2]The Hill radius gives the limit of the gravitational sphere of influence of a body in orbit about another heavier body, in this case the central star.

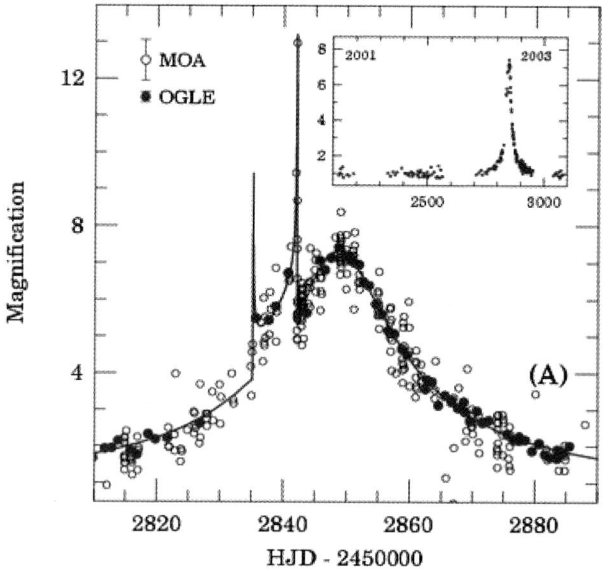

Fig. 1.3. Observation of gravitational microlensing by a planet by OGLE (Bond et al., 2004). Inset panel shows all OGLE data from 2001 to 2003, while the main figure shows a close-up of the data for 2003 for both OGLE and MOA.

sions include GEST (Galactic Exoplanet Survey Telescope) and Microlensing Planet Finder (MPF).

1.3 Properties of Observed Extrasolar Planets

So many planets have now been found that it is possible to consider the statistics of the mass and orbital parameter distributions, as has been done by Collier Cameron (2002), and Marcy et al. (2003). Radial velocity measurements can only provide information on the distribution of $M_p \sin i$. However, it can be shown (Jorissen et al., 2001) that for a random distribution of planetary systems, the distribution of $M_p \sin i$ is very close to the distribution of M_p and thus statistical conclusions on the overall mass distribution can be inferred from the distribution of $M_p \sin i$ for known exoplanets, shown in Fig. 1.4.

Considering the selection effects of radial velocity measurements, a predominance of heavy planets might be expected. However, most of the planets discovered so far have $M_p \sin i < 10 M_J$, and the distribution of planets rises rapidly for smaller masses. A power law fit to the distribution is also plotted in Fig. 1.4, where the number of planets N has been assumed to vary with planetary mass as $N = \alpha (M_p \sin i)^{\beta}$. Fitting only to the well sampled distribution where $M_p \sin i < 4 M_J$, values of $\alpha = 44.98$ and $\beta = -0.95$ are derived, which are found to reasonably well ap-

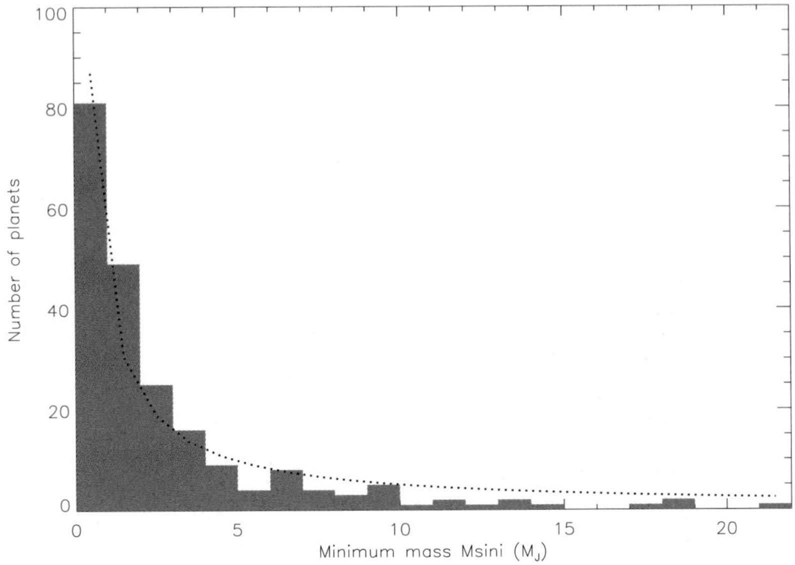

Fig. 1.4. Distribution of $M_p \sin i$ of currently known exoplanets. Also plotted is the curve $N = \alpha(M_p \sin i)^\beta$, where $\alpha = 44.98$ and $\beta = -0.95$, which is described in the text

proxinate the rest of the distribution. Hence, to first order it would appear that the number of planets falls approximately linearly with the planetary mass.

The smallest exoplanets discovered to date are OGLE-05-390L b and GJ 876 d (Rivera et al., 2005) which have estimated masses of only $\sim 5.5M_{Earth}$ and $\sim 7.5M_{Earth}$, respectively. In contrast, there is an apparent absence of heavy extrasolar planets with mass above the deuterium-burning limit for brown dwarfs of $\sim 13.6M_J$ (Lewis, 2004). This apparent absence of very large mass planets has become known as the 'Brown Dwarf Desert' and it has been suggested that brown dwarfs might be formed by a different process from planets, leading to them orbiting at much greater distances than is currently detectable with the radial velocity technique. However, very recently a few heavy mass exoplanets have been discovered, the heaviest being GQ Lup b and HD 41004 B b which have an estimated $M_p \sin i$ of $21.5M_J$ and $18.4M_J$ (Zucker et al., 2004) respectively. Hence, the 'Brown Dwarf Desert' may prove not to be quite so barren as has been previously thought, supporting the suggestion of Jorissen et al. (2001) that there is no reason to ascribe the transition between giant planets and brown dwarfs to the threshold mass of deuterium ignition.

The distribution of exoplanet orbital periods is shown in Fig. 1.5, which appears to have a slight bimodal distribution, with peaks at 3 days and 500 days.

The distribution of exoplanet orbit radii is shown in Fig. 1.6 and it is found that a large fraction of known exoplanets orbit within 1 AU. However, given that planets with larger orbital distances take longer to orbit and current observation programmes have only been running for 10 years or so and are becoming more

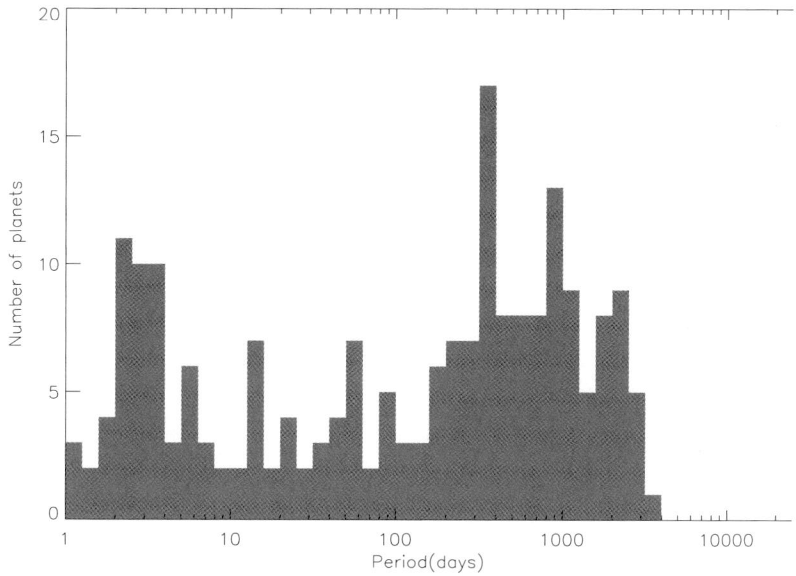

Fig. 1.5. Orbital period distribution of known exoplanets.

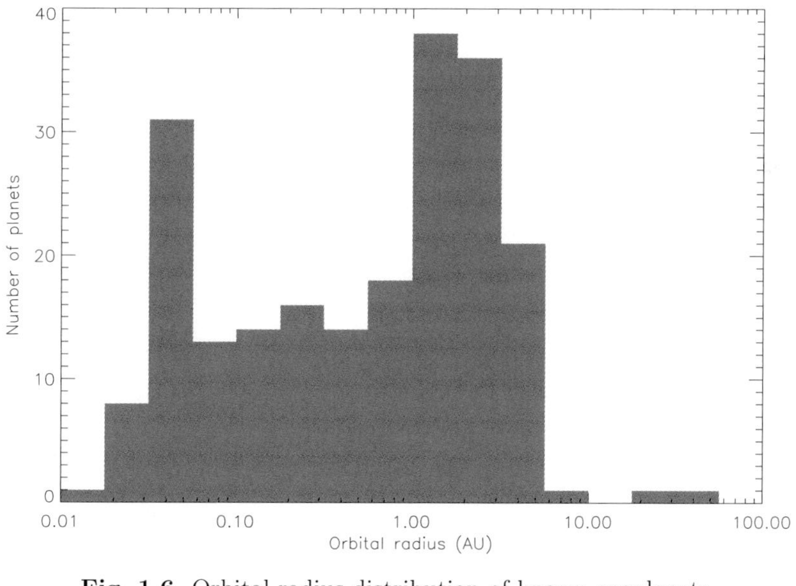

Fig. 1.6. Orbital radius distribution of known exoplanets.

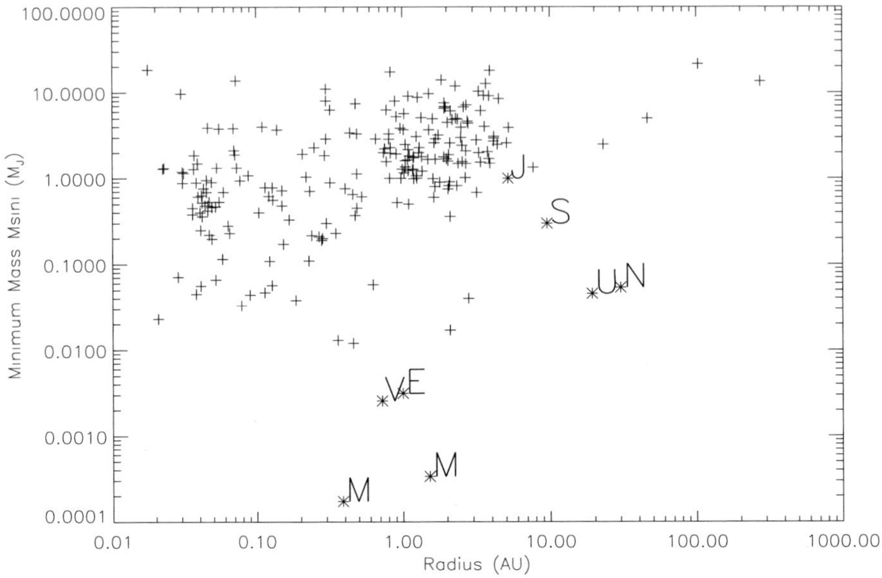

Fig. 1.7. Distribution of mass and radius for known exoplanets. Solar System planets are indicated by letter.

precise all the time, there is good reason to suspect that there is a large population of planets orbiting beyond 3 AU (Marcy et al., 2003) which will soon be detected.

Fig. 1.7 shows $M_p \sin i$ for known exoplanets plotted against their orbital distance and there can be seen to be a general decrease in the number of massive planets ($M_p > 4M_J$) orbiting within 0.3 AU. Such planets would be eminently detectable using the radial velocity method so we can be confident that they are really not there. A possible explanation for this is that the migration mechanism of massive planets is either inefficient within 0.3 AU or too efficient and thus that massive planets straying within 1 AU fall all the way into the star (Marcy et al., 2003). Alternatively, as discussed in Sect. 1.2.3 it may be that planets closer than this quickly evaporate (Vidal-Madjar et al., 2003).

There is a massive and uniform spread in the eccentricities of exoplanets between 0 and 0.9 (Fig. 1.8), which suggests that there is a common mechanism for pumping the eccentricity of extrasolar planets. It can also be seen from Fig. 1.8 that the eccentricity distribution for planets in multiple-planet systems is indistinguishable from that for single planet systems. For the multiple planet systems known, eccentricity pumping may result from planets migrating in their circumstellar disc, leading to occasional mutual capture and resonance. Subsequent close encounters may lead to scattering and ejection of planets. This scenario explains the orbital resonances commonly seen in multiple-planet systems and also the occurrence of 'hierarchical' systems (ones with only a few, widely separated planets), where some of the planets have presumably been ejected. Single planet systems may be the end result of such interactions, where all other giant planets have been lost through ejec-

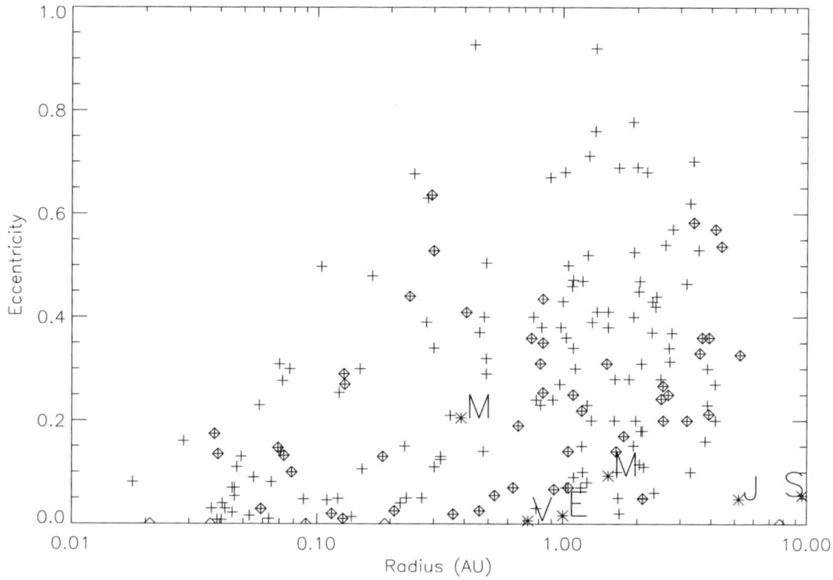

Fig. 1.8. Distribution of eccentricity and radius for known exoplanets. In this plot Solar System planets are indicated by letter and planets in multi-planet systems are indicated by diamonds.

tion. Alternatively it could just be that single planet systems actually have other planets which have just not been detected yet.

An intriguing discovery is of a multiple planet system around the star HD 69830 which comprises three Neptune mass planets (Lovis et al., 2006) and possibly also an asteroid belt (Beichman et al., 2005).

It has been pointed out by Charbonneau (2006) that the precision achieved by Lovis et al. (2006) means that it is now more likely that terrestrial-type planets may be detected by the radial-velocity method, since the Sun is unusually hot and massive compared to other nearby stars. The 'habitable zone' of other stars is likely to be closer to the star and coupled with their lower mass the 'wobble' introduced by a terrestrial planet's mass may now be just about detectable.

The analysis of the metallicity of stars which have planetary companions is very revealing (Fig. 1.9). The [Fe/H] ratio is defined as the abundance of iron in a star to that found in the Sun, expressed on a logarithmic scale. Thus a star with [Fe/H]=1 has 10 times the abundance of iron (and other metals) as the Sun. From Fig. 1.9 it can be seen that, as found by Fischer and Valenti (2003) and Santos et al. (2004), the distribution rises rapidly at the high metallicity end and thus the great majority of known exoplanets orbit stars with a metallicity equal to, or greater than that of our Sun (Sudarsky et al., 2003). These observations strongly suggest that the presence of dust in proto-stellar nebulas is very important for the formation of planets and thus favours the core-accretion model of planetary formation (Pollack et al., 1996).

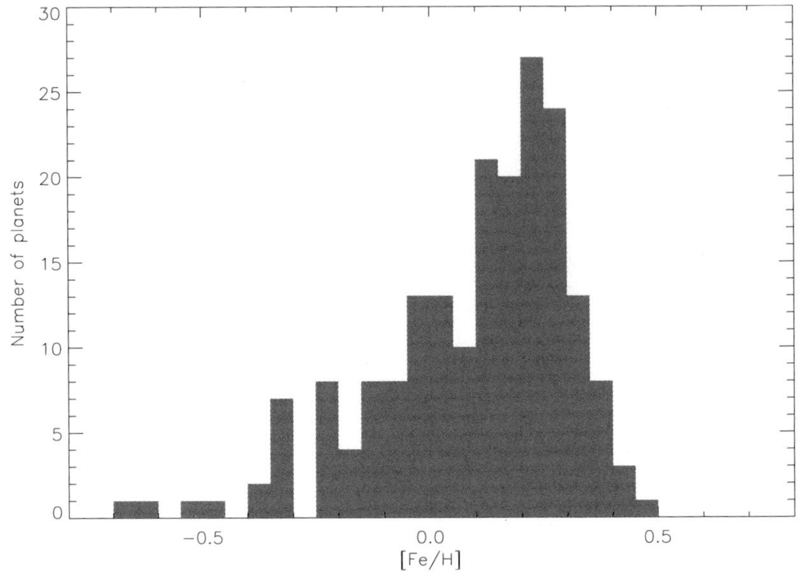

Fig. 1.9. Distribution of star metallicity for known exoplanetary systems.

1.4 Sensitivity and Future Methods for Detection of Extrasolar Planets

We have seen that there are a number of ways of detecting the existence of extrasolar planets, most indirect. All the techniques have their own advantages and disadvantages and the different selection effects of these detection methods are summarised in Fig. 1.10, on which are plotted the mass and orbital radii of known exoplanets, together with characteristics of the Solar System planets.

Currently employed detection methods are biased towards close-orbiting heavy planets and thus very few lighter terrestrial-like planets have so far been found, with the lowest mass for planet orbiting an active star so far being estimated as $5.5M_{Earth}$ (Sect. 1.3). Three earth-mass extrasolar planets have actually been discovered, but these do not orbit a main sequence star, but instead have been observed orbiting the pulsar PSR 1257+12 (Wolszczan and Frail, 1992; Wolszczan, 1994). Although no terrestrial planets have so far been discovered, there is no reason to think that they are not present and as measurement techniques improve, it is widely hoped that terrestrial planets may soon start being detected.

As can be seen, the radial velocity technique is best for detecting heavy, close orbiting planets, and thus the planets found so far are clustered in the top left corner of Fig. 1.10. The limit of detectability of existing measurements is shown, together with the expected improvement due to ever increasing sensitivity and longer observation runs. Radial velocity programmes currently under way include the Anglo-Australian Planet Search (e.g. Carter et al., 2003), the California and Carnegie Planet Search,

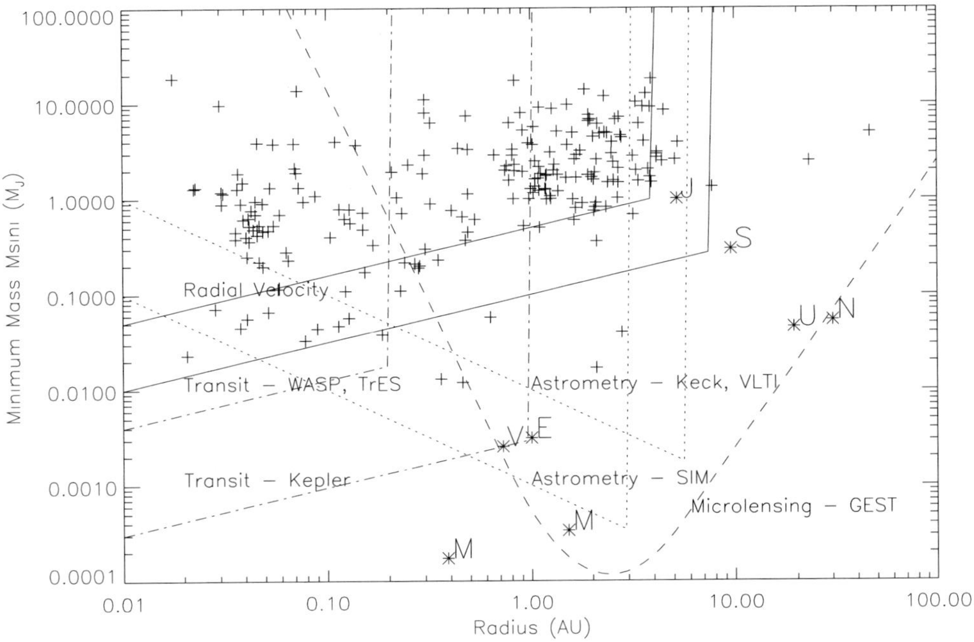

Fig. 1.10. Selection effects of different exoplanet detection programmes. Solar System planets are indicated by letter.

ELODIE (Naef et al., 2004) and CORALIE (Mayor et al., 2004). Analysis methods are rapidly becoming more sophisticated and Charbonneau (2006) note that the radial velocity method may soon start turning up terrestial-like planets (Sect. 1.3). Transit observations also favour shorter periods, but can also detect lighter planets, especially the planned space-based missions (Sect. 1.4.1). The selection limits of astrometric observations (Sect. 1.2.2), both for ground-based programmes such as at the Keck Observatory and the VLT, and for space-based missions such as the forthcoming Space Interferometry Mission (SIM), are expected to start probing into the terrestrial planet region of the diagram. This technique is complemented by the microlensing technique (Sect. 1.2.4) which similarly is more sensitive for space-based proposals (such as GEST), than terrestrial ones. We will now consider future observations using these methods and what may reasonably be expected of them.

1.4.1 Transit Programmes

A number of transit surveys have been planned for the next few years – both ground- and space-based. They may be conveniently split into two categories: 1) Deep surveys, which have small pixel size, can see faint stars, but do not cover a wide area of sky; and 2) Wide surveys, which cover a wide area of the sky with large pixel size, but cannot see fainter stars. The *a priori* probability of transit detection is the ratio of the diameter of a star to the diameter of a planet's orbit (Lewis,

2004). For the Sun-Jupiter system this is $1.4 \times 10^{11}/5.2 \times 10^{13} = 1.8 \times 10^{-3}$. Hence, transit surveys need to observe lots of star systems to have sufficient probability of planetary detection and so both survey methods attempt to view many stars simultaneously, taking special care not to confuse real planetary transits with other phenomena (e.g. Mandushev et al., 2005).

One ground-based wide survey is TrES (Transatlantic Exoplanet Survey), a network of three 99-mm aperture field-flattened Schmidt telescopes based at Palomar Observatory, Lowell Observatory and the STARE transit camera at the Observatorio del Teide, Canary islands. STARE made the first observation of a planetary transit (HD 209458 b) (Sect. 1.2.3) and the TrES network has now discovered two further planets, TrES-1 (Alonso et al., 2004) and TrES-2 (O'Donovan et al., 2006a). The infrared radiation of TrES-1 has since been observed directly with the Spitzer telescope by Charbonneau et al. (2005) who report a surface temperature of over 1000K. Another ground-based wide-survey is SuperWASP, which has two facilities: one based in the Canary Islands and another soon to be operating in South Africa. Both instruments are comprised of five 11-cm aperture wide-angle cameras and are developments of the WASP0 prototype instrument (Kane et al., 2004; Pollacco et al., 2006). Both SuperWASPs began operations in 2004 and are able to monitor 10,000 stars simultaneously over a $15° \times 15°$ field of view. So far, two planets have been discovered, WASP-1b and WASP-2b (Collier Cameron et al., 2007).

Space-based wide-survey observations are predicted to be both more sensitive and less prone to false identifications of planetary transits. One such project was the Sagittarius Window Eclipsing Extrasolar Planet Search (SWEEPS), which used the Hubble Space Telescope in 2004 to search for transit events (Sahu et al., 2006). A current mission is COROT, which is a French-led project to place a small 0.27-m telescope into orbit to study astroseismology and also detect extrasolar planet transits. COROT was launched on 27 December 2006 and started its first observing run on 8 February 2007. It will observe an area of the sky of size $2.8° \times 2.8°$ for $2\frac{1}{2}$ years. The US-led Kepler mission is a Schmidt telescope with 1.4-m primary mirror and a 0.95-m aperture, due for launch in October 2008. Kepler will continuously and simultaneously monitor the brightness of approximately 100,000 A–K dwarf (main-sequence) stars brighter than 14th magnitude in the Cygnus-Lyra region along the Orion arm, for a period of 4 years.

Deep transit surveys will also be conducted by microlensing programmes, which are outlined in Sect. 1.2.4.

1.4.2 Direct Optical Detection

Almost all of the currently known exoplanets have been discovered through indirect methods. However, four extrasolar planetary-mass objects have now been directly imaged about brown dwarfs, although their masses are towards the top end and in some cases exceed what might really be classified as planets and instead might be better described as brown dwarfs ($\sim 13M_J$). The objects are: 2M1207 b ($5 - 8M_J$) (Song et al., 2006; Chauvin et al., 2005a), GQ Lup b ($10 - 40M_J$) (McElwain et al., 2007), AB Pic b ($13 - 14M_J$) (Chauvin et al., 2005b) and SCR 1845 b ($9 - 65M_J$)

(Biller et al., 2006). All objects except SCR 1845 b orbit at great distance from their parent stars. In addition, we have also seen that methods have successfully been developed to study the spectra of exoplanets though differencing methods (Sect. 1.2.3). In this section we will look at other methods of directly detecting the reflected starlight or thermal emission of exoplanets about nearby stars.

Doppler Spectral Separation

As a planet orbits a star, part of the starlight will be reflected by the planet towards the observer. This component will be Doppler-shifted by an amount depending on the planet's orbital velocity ($\sim$ 100 km/s), rather than the star's ($\sim$ 10 m/s) and can be extracted using very high-resolution ground-based spectroscopy and correlation techniques. Collier Cameron and Leigh (2004) review the current status of a number of direct planetary detections achieved by this technique. Assuming the planetary radius is known, these observations may be used to estimate the visible planetary albedo. For example, the albedo of τ Bootis b (Collier Cameron et al., 1999; Charbonneau et al., 1999) is estimated by Leigh et al. (2003a) to be less than 0.39. Similarly, the albedoes of ν Andromeda b and HD 75289 b are estimated to be less than 0.3 (Collier Cameron et al., 2002) and 0.14 (Leigh et al., 2003b) respectively. These albedoes are much less than Jupiter's (0.5).

Differential Direct Detection

Models of the expected spectra of extrasolar giant planets (e.g. Sudarsky et al., 2003) show that the reflected sunlight from such planets will be significantly affected by absorption of atmospheric constituents such as sodium and carbon monoxide, whereas the stellar spectrum is expected to be smoothly varying. Hence, these absorption features provide a possible means of discriminating between the light reflected by a planet and the direct stellar light. Wiedemann et al. (2001) report just such a detection of the 3-μm methane absorption of τ Bootes b. There are other programmes in development, such as TRIDENT (Marois et al., 2005) on the 3.6m Canada-France-Hawaii-Telescope (CFHT), which observes the edge of a methane absorption band between 1.5 and 1.8 μm.

Interferometric Imaging

Using two telescopes, separated by a long baseline of precisely controlled optical length D, the beams may be combined with a phase difference of π to completely eliminate the light from the central star. Constructive interference will then occur at a number of angles θ where $D \sin\theta = (2n+1)\,\lambda/2$ and n is an integer. By varying the baseline D (assuming fixed wavelength λ), a range of constructive interference angles can be examined to attempt to detect either the weak stellar reflection or thermal emission of an extrasolar planet. Both the Keck Observatory and VLT have long baseline interferometric observation programmes in development. Another interesting project is the Large Binocular Telescope Interferometer (LBTI) in Arizona, which achieved 'First Light' in October 2005. LBTI uses adaptive optics and a beam combiner including a dielectric material to correct for colour dependence of light interference. LBTI will operate in the infrared (3-5 μm) and should be able to detect planets further than 0.03" from their stars. Nulling interferometry is also the planned mode of the proposed ESA Darwin space mission, and a possible mode of the proposed NASA Terrestrial Planet Finder (TPF-IR). In these mission plans, a fleet of large telescopes would fly in formation and the light combined in a central hub using precisely controlled phase delays. Due to their very long baselines, low temperatures and no atmospheric absorption, these missions will be able to not only directly detect extrasolar planets in the infrared, but also measure their emission spectra, allowing the composition of their atmospheres to be determined.

Coronagraphic Imaging

Finally, in this technique, which is only suitable for space missions, light from the central star is eliminated using a mask in the focal plane. The method is used in solar studies to study the corona and prominences of the Sun's atmosphere, from which its name is derived. The technique may be used by a version of the proposed NASA Terrestrial Planet Finder (TPF-C). In addition, the new James Webb Space Telescope (scheduled for launch in 2013) will house the NIRCam instrument, which includes a coronagraphic module, operating from 2-5 μm. This system will be capable of 10^8–10^9 high contrast imaging for separations >0.1 arcsecond. In addition, tunable narrow-band filters will allow the measurement of spectra from 2.5–4.5 μm at low resolution.

1.5 Conclusions

The discovery of extrasolar giant planets has been one of the most exciting astronomical discoveries in the last twelve years. It has proved beyond doubt that planetary formation is a common by-product of star formation, although some of the systems discovered so far appear peculiarly exotic compared to our own. Analysing the characteristics of these systems is already giving us a new insight into how our own Solar System formed and has placed very important constraints on planetary formation theories. With $\sim$ 215 giant extrasolar planets discovered, attention is starting to switch to the detection of more Earth-like terrestial planets.

A number of programmes proposed for the next 20–30 years will be able to address these questions. Of particular interest are the NASA TPF-IR and ESA Darwin missions, which through their infrared nulling interferometry approach will be able to measure the thermal emission spectra of any planets discovered. The Earth's atmosphere has been substantially modified by the presence of life, which produces high levels of oxygen, which is then photolysed to form ozone in the upper atmosphere. Ozone has a very clear signature in the thermal infrared and the detection of significant quantities of this gas in the atmosphere of an extrasolar planet would give a strong indication of the occurence of life-like processes. Hence, it is hoped that within our lifetime, it may be possible to indirectly detect the presence of life in another Solar System, which will have profound implications on our view of ourselves and on our place in the universe.

References

Alonso, R. et al. 2004, TrES-1: The transiting planet of a bright K0 V star, *ApJ*, **613**, L153

Beichman, C.A. et al. 2005, An excess due to small grains around the nearby K0 V star HD 69830: Asteroid or cometary debris?, *ApJ*, **626**, 1061

Biller, B.A., L.M. Close, W. Brandner and S. Kellner 2006, Discovery of a Brown Dwarf very close to the Sun: A methane-rich brown dwarf companion to the low-mass star SCR 1845-6357, *ApJ*, **641**, L141

Bodenheimer, P., D.N. Lin & R.A. Mardling 2001, On the Tidal Inflation of Short-Period Extrasolar Planets, *ApJ*, **548**, 466

Bond, I.A. et al. 2001, Real-time difference imaging analysis of MOA galactic bulge observations during 2001, MNRAS, **327**, 868

Bond, I.A. et al. 2004, OGLE-BLG-235/MOA 2003-BLG-53: A planetary microlensing event, *ApJL*, **606**, 155

Brown, T.M. et al. 2001, Hubble Space Telescope time-series photometry of the transiting planet of HD 209458, *ApJ*, **552**, 699

Brown, T.M., K.G. Libbrecht & D. Charbonneau 2002, A Search for CO Absorption in the Transmission Spectrum of HD 209458b, Publications of the Astronomical Society of the Pacific, **114**, 826

Burleigh, M., F. Clarke & S. Hodgkin 2003, Imaging extrasolar planets around nearby white dwarves. In: *Scientific Frontiers in Research on Extrasolar planets*, vol 294, ed by D. Deming & S. Seager (ASP Conference Series) pp 111-115.

Burrows, A. et al. 2000, On the Radii of Close-in Giant Planets, *ApJL*, **534**, 97

Butler, R.P. et al. 2004, A Neptune-Mass Planet Orbiting the Nearby M Dwarf GJ 436, *ApJ*, **617**, 580

Carter, B.D. et al. 2003, A planet in a circular orbit with a 6 year period, *ApJL*, **593**, 43

Charbonneau, D. et al. 1999, An upper limit on the reflected light from the planet orbiting the star tau Bootis, *ApJL*, **522**, 145

Charbonneau, D. et al. 2000, Detection of Planetary Transits Across a Sun-like Star, *ApJL*, **529**, 45

Charbonneau, D. et al. 2002, Detection of an extrasolar planet atmosphere, *ApJ*, **568**, 377

Charbonneau, D. et al. 2005, Detection of thermal emission from an extrasolar planet, *ApJ*, **626**, 523

Charbonneau, D. 2006, A neptunian triplet, *Nature*, **441**, 292

Chauvin, G. et al. 2005a, Giant planet companion to 2MASSW J1207334-393254, *Astron. & Astrophys.*, **438**, L25

Chauvin, G. et al. 2005b, A companion to AB Pic at the planet/brown dwarf boundary, *Astron. & Astrophys.*, **438**, L29

Collier Cameron, A. et al. 1999, Probable detection of starlight reflected from the giant planet orbiting tau Bootis, *Nature*, **402**, 751

Collier Cameron, A. et al. 2002, A search for starlight reflected from nu And's innermost planet, MNRAS, **330**, 187

Collier Cameron, A. 2002, What are hot Jupiters made of? *Astron. & Geophys.*, **43**, 4.21

Collier Cameron, A. & C. Leigh 2004, Tomographic studies of exoplanet atmospheres, *Astron. Nachr.*, **325**, 252

Collier Cameron, A. et al. 2007, WASP-1b and WASP-2b: two new transiting exoplanets discovered with SuperWASP and SOPHIE, MNRAS, **375**, 951

Deming, D. et al. 2005, A new search for carbon monoxide absorption in the transmission spectrum of the extrasolar planet HD 209458b, *ApJ*, **622**, 1149

Fischer, D.A. & J.A. Valenti 2003, Metallicities of stars with extrasolar planets. In: *Scientific Frontiers in Research on Extrasolar planets*, vol 294, ed by D. Deming & S. Seager (ASP Conference Series), pp 117-128

Friedrich, S. et al. 2006, Search for giant planets around white dwarfs with HST, Spitzer and VLT, 18th European Workshop on White Dwarfs. ASP Conference Series, **999**

Guillot, T. & A.P. Showman 2002, Evolution of "51 Pegasus b-like" planets, *A&A*, **385**, 156

Henry, G.W. et al. 2000, A transiting "51-Peg-like" planet, *ApJL*, **529**, 41

Jorissen, A., M. Mayor & S. Udry 2001, The distribution of exoplanet masses, *A&A*, **379**, 992

Kane, S.R. et al. 2004, Results from the Wide-Angle Search for Planets prototype (WASP0) – I. Analysis of the Pegasus field, MNRAS, **353**, 689

Knutson, H.A. et al. 2007, Using stellar limb-darkening to refine the properties of HD 209458b, *ApJ*, **655**, 564

Leigh, C. et al. 2003a, A new upper limit on the reflected starlight from tau Bootis b, MNRAS, **344**, 1271

Leigh, C. et al. 2003b. A search for starlight reflected from HD 75289b, MNRAS, **346**, L16

Lewis, J.S. 2004, *Physics and Chemistry of the Solar System*, 2nd edition, Elsevier Academic Press.

Lovis, C. et al. 2006, An extrasolar planetary system with three Neptune-mass planets, *Nature*, **441**, 305

Mandushev, G. et al. 2005, The challenge of wide-field transit surveys: The case of GSC 09144-02289, *ApJ*, **621**, 1061

Marcy, G.W. et al. 2003, Properties of extrasolar planets. In: *Scientific Frontiers in Research on Extrasolar planets*, vol 294, ed by D. Deming & S. Seager (ASP Conference Series), pp 1-16.

Marois, C. et al. 2005, TRIDENT: An infrared differential imaging camera optimized for tge detection of methanated substellar companions, *Publ. Astr. Soc. Pacific*, **117**, 745.

Mayor, M. & D. Queloz 1995, A Jupiter-Mass Companion to a Solar-Type Star, *Nature*, **378**, 355

Mayor, M. et al. 2004, The CORALIE survey for southern extra-solar planets. XII. Orbital solutions for 16 extra-solar planets discovered with CORALIE, *A&A*, **415**, 391

McElwain, M.W. et al. 2007, First high-contrast science with an Integral Field Spectrograph: The substeallar companion to GQ Lupi, *ApJ*, **656**, 505

Naef, D. et al. 2004, The ELODIE survey for northern extra-solar planets. III. Three planetary candidates detected with ELODIE, *A&A*, **414**, 351

O'Donovan, F.T. et al. 2006a, TrES-2: The first transiting planet in the Kepler field, *ApJ*, **651**, L61

O'Donovan, F.T. et al. 2006b, Rejecting astrophysical false positives from the TrES transiting planet survey: The example of GSC 03885-00829, *ApJ*, **644**, 1237

O'Donovan, F.T. et al. 2007, Outcome of six candidate transiting planets from a TrES field in Andromeda, *ApJ* (in press).

Pollacco D.L. et al. 2006, The WASP project and the SuperWASP cameras, *Publ. Astro. Soc. Pacific*, **118**, 1407

Pollack, J.B. et al. 1996, Formation of the Giant Planets by concurrent accretion of solids and gas, Icarus, **124**, 62

Richardson, L.J., D. Deming, S. Seager 2003a, Infrared Observations during the Secondary Eclipse of HD 209458b. II. Strong Limits on the Infrared Spectrum Near 2.2 mum, *ApJ*, **597**, 581

Richardson, L.J. et al. 2003b, Infrared Observations during the Secondary Eclipse of HD 209458b. I. 3.6 Micron Occultation Spectroscopy Using the Very Large Telescope, *ApJ*, **584**, 1053

Rivera, E.J. et al. 2005, A $\sim 7.5 M_{Earth}$ Planet orbiting the nearby star, GJ 876, *ApJ*, **634**, 625

Sahu, K.C. et al. 2006, Transiting extrasolar planetary candidates on the galactic bulge, *Nature*, **443**, 535

Santos, N.C., G. Israelian & M. Mayor 2004, Spectroscopic [Fe/H] for 98 extra-solar planet-host stars, *A&A*, **415**, 1153

Song, I., G. Schneider, B. Zuckerman, J. Farihi, E.E. Becklin, M.S. Bessell, P. Lowrance and B.A. Macintosh 2006, HST NICMOS Imaging of the Planetary-mass Companion to the Young Brown Dwarf 2MASSW J1207334-393254, *ApJ*, **652**, 724

Sudarsky, D., A. Burrows & I. Hubeny 2003, Theoretical Spectra and Atmospheres of Extrasolar Giant Planets, *ApJ*, **588**, 1121

Udalski, A. 2003, The Optical Gravitational Lensing Experiment. Real time data analysis systems in the OGLE-III survey, *Acta Astron.*, **53**, 291

Vidal-Madjar, A. et al. 2003, An extended upper atmosphere around the extrasolar planet HD 209458 b, *Nature*, **422**, 143

Vidal-Madjar, A. et al. 2004, Detection of oxygen and carbon in the hydrodynamically escaping atmosphere of the extrasolar planet HD 209458 b, *ApJL*, **604**, 69

Wiedemann, G., D. Deming & G. Bjoraker 2001, A Sensitive Search for Methane in the Infrared Spectrum of tau Bootis, *ApJ*, **546**, 1068

Wittenmyer, R.A., M. Endl and W.D. Cochran 2007, Long-period objects in the extrasolar planetary systems 47 Ursae Majoris and 14 Herculis, *ApJ*, **654**, 625

Wolszczan, A. & D.A. Frail 1992, A planetary system around the millisecond pulsar PSR1257+12, *Nature*, **355**, 145

Wolszczan, A. 1994, Confirmation of earth mass planets orbiting the millisecond pulsar PSR B1257+12, *Science*, **264**, 538

Zucker, S., T. Mazeh, N.C. Santos, S. Udry and M. Mayor 2004, Multi-order TODCOR: Application to observations taken with the CORALIE echelle spectrograph, *Astron. & Astrophys.*, **426**, 695

2 Doppler Exoplanet Surveys: From Single Object to Multiple Objects

Jian Ge

Summary. Doppler planet surveys are the major tool for discovering new exoplanets. Of 200-plus known exoplanets discovered to date, about 90% were discovered by single object Doppler techniques. This chapter summarizes the results of Doppler planet surveys in the past decade, and new progress and early results in the development of new Doppler techniques, especially multiple object techniques. It also presents the scientific motivation for the next generation large-scale multi-object Doppler planet surveys and possible new science to be addressed. In the history of astronomy, the ability to move from single-object to multi-object observations has enabled large-scale astronomical surveys (e.g., the Sloan Digital Sky Survey) and consistently led to dramatic new discoveries. We anticipate similar advances will also occur with multi-object Doppler planet surveys in the next decade.

2.1 Introduction

One of the most surprising astronomical developments of the last 15 years has been the discovery of an abundant population of extra-solar planets. Surveys to date have detected over 230 extrasolar planets (see the exoplanet website at http://exoplanet.eu; Butler et al. 2006 and Udry et al. 2007), of which approximately 90% were found by detecting the reflex motion in the host star from precise measurements of the star's radial velocity (RV). In this chapter, we will describe the Doppler technique, primary results to date using single object Doppler instruments, and introduce future large-scale RV surveys using multi-object Doppler instruments.

2.2 Description of the Doppler Method

Currently, there are two major Doppler methods: one using high resolution cross-dispersed echelle spectrographs (the echelle method) and the other using dispersed fixed-delay interferometers (the DFDI method). Both methods have been successfully used for detecting new planets (e.g., Butler et al. 2006 for a summary of exoplanets detected by the Doppler techniques). Here we briefly describe both methods.

2.2.1 The High Resolution Cross-Dispersed Echelle Method

The RV method using high resolution optical spectrographs was proposed for detecting extrasolar planets in the 1950's by monitoring spectral line shifts of the target star caused by the gravitational pull of the planets (Struve 1952). However, it was difficult to obtain better than several tenths of a kilometer per second precision before the 1970's due to the use of inadequate instrument calibration methods, in which the reference beam does not follow the same optical path as the stellar beam (Griffin 1967). In order to detect Jupiter like planets, a Doppler precision of ~ 10 m/s is required (e.g., the velocity semi-amplitude of the Sun caused by the gravitational pull of Jupiter is about 12.3 m/s over 11.86 years).

In 1973, Griffin and Griffin proposed to use telluric absorption lines as a reference for RV measurements to eliminate the differential motion between the reference and stellar beams in order to achieve high RV precision (Griffin & Griffin 1973). This method was further developed by Campbell & Walker in the late 1970's (Campbell & Walker 1979). Instead of using telluric absorption lines, which vary and also shift slightly due to line saturation and atmospheric winds, they used a toxic hydrogen fluoide (HF) gas cell which produces the R branch of the 3-0 vibration rotation band in the wavelength region of 8650-8800 Å (Campbell & Walker 1979). A Doppler precision of $\sim$15 m/s was achieved. Walker et al. monitored a total of 21 bright solar-type stars using this method and a Coudé spectrograph on the Canada-France-Hawaii 3.6-m telescope (CFHT) over 12 years. Although they did not detect any exoplanets, their results indicate that less than 5% of solar-type stars have planets larger than 2 Jupiter masses within 5 AU (Walker et al. 1995). This result is consistent with the recent conclusion based on the California-Carnegie planet survey (Marcy et al. 2005).

In the late 1980s, molecular iodine was chosen to use as a reference instead of HF. Unlike HF, molecular iodine is non-toxic and has thousands of absorption lines in the wavelength region of 5000-6200 Å, which can be used for tracking the instrument velocity drifts and also instrument response changes. Another calibration method using a stabilized Fabry-Perot etalon interferometer was also developed at the same time and has reached a precision of ~ 8 m/s over ~ 5 years (McMillan et al. 1993). The iodine absorption calibration method was used to achieve $\sim$3 m/s Doppler precision with the Lick Hamilton echelle spectrograph in the 1990s (Butler et al. 1996). Several other groups have also achieved similar Doppler precision using the iodine calibration method (Cochran & Hatzes 1993; Brown et al. 1994). The iodine calibration method has become popular for high precision RV measurements since then.

In the early 1990s, the ThAr separate beam calibration method reached a milestone, delivering ~ 10 m/s Doppler precision and long term stability using fibers for feeding the star and calibration light to the spectrograph (Baranne et al. 1996). The first extrasolar planet around a solar-type star, 51 Peg, was detected by this method (Mayor & Queloz 1995). Fiber feeding has played a key role in allowing the instrument to be installed in an isothermal environment to increase the instrument thermal and mechanical stability.

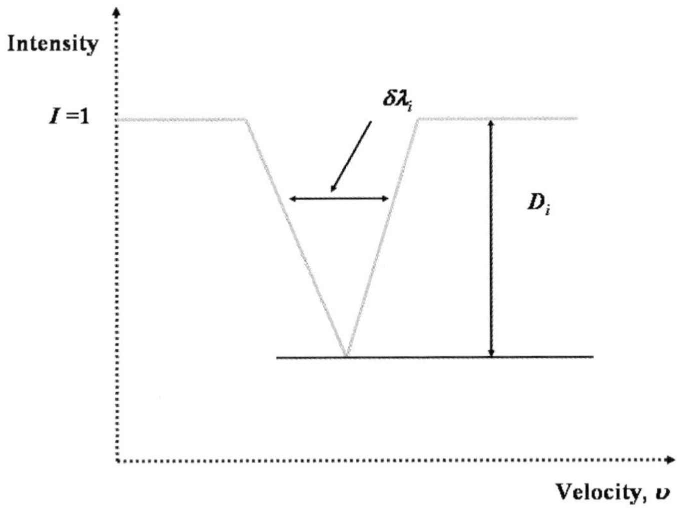

Fig. 2.1. Echelle working principle for Doppler RV measurements. $\delta\lambda_i$ is the intrinsic line width of the absorption line, D_i is the absorption line depth.

In the cross-dispersed echelle method, the Doppler precision is fundamentally limited by the total number of photons collected by the spectrograph. Figure 2.1 shows the principle for RV measurements using a high resolution echelle spectrograph. The photon-limited Doppler precision of an echelle instrument can be described as

$$\sigma_{RV} = \frac{1}{\sqrt{(\Sigma_{i=1,N}(1/\sigma_{e,i}^2))}}, \tag{2.1}$$

$$\sigma_{e,i} = \frac{\epsilon_I}{dI/dV} \tag{2.2}$$

where N is the total number of pixels, ϵ_I is the uncertainty in the intensity at pixel i, and dI/dV is the local slope of the absorption line (Butler et al. 1996).

For a fully resolved absorption line, the derived intrinsic Doppler measurement error due to the photon noise is:

$$\sigma_{e,i} = \frac{c\delta\lambda_i}{\lambda D_i \sqrt{F_i}}, \tag{2.3}$$

where the total photon number collected by each line is $F = \eta A \delta\lambda_i \Delta t S$, η is the total detection efficiency from the telescope to the detector, A is the telescope photon collecting area, Δt is the total exposure time and S is the stellar flux in photons cm^{-2} s^{-1} Å^{-1}. For an echelle with a resolving power of $R = \lambda/\delta\lambda_o$, where $\delta\lambda_o$ is the spectral purity of the spectrograph, the total measured photon-noise limited Doppler error for an absorption line can be approximately described as

$$\sigma_{e,o} \approx (\frac{\delta\lambda_o}{\delta\lambda_i})^{3/2}\sigma_{e,i}, \tag{2.4}$$

when $\delta\lambda_o > \delta\lambda_i$. Therefore, the total Doppler error over a wave band of $\Delta\lambda$ can be approximated as

$$\sigma_{RV} \propto \delta\lambda_i S^{-0.5} \Delta\lambda^{-0.5} R^{-1.5} D^{-1}, \tag{2.5}$$

(also see Hatzes & Cochran, 1992). This formula shows that the echelle Doppler precision strongly depends on spectral resolution ($-3/2$ power of the spectral resolution) and is also related to the wavelength coverage and stellar flux ($-1/2$ power of the wavelength coverage and stellar flux). This is the main reason that most of the planet hunting echelle spectrographs have a spectral resolution R >60,000 at optical wavelengths since the typical width of stellar absorption lines for a solar-type star is a few km/s. The Doppler precision also depends on stellar properties: a star with deep and narrow lines (such as late type stars) tends to produce a higher Doppler precision than a star with shallow and broad lines (such as early type stars) using the same spectrograph with the same exposure time. This is one of the main reasons that current optical Doppler planet surveys are mainly focused on late type stars.

Currently, two major calibration methods are used in the cross-dispersed echelle instruments for measuring the instrument velocity drifts: the iodine absorption cell and the ThAr emission lamp. These two methods have their own advantages and disadvantages. Some advantages of iodine cell calibration are that (1) thousands of iodine absorption lines are superimposed on top of the stellar absorption lines; (2) both the starlight and iodine absorption share a common optical path, so the iodine absorption lines simultaneously track changes in the instrument point spread function due to the same physical effects causing instrumental drifts affecting the stellar absorption lines. Therefore, iodine cell calibration enables reaching photon-noise limited Doppler precision. However, the major limitation for the iodine method is that iodine has absorption line bands clustered in the visible (5000-6200 Å) and also the absorption cell absorbs about 30% of the incoming photons. These limit the application of the iodine method for mainly observing relatively bright solar-type stars which have peak fluxes around the visible. The method becomes less efficient for late type stars such as M dwarfs which have peak fluxes at wavelengths much longer than the visible.

The main advantage of using the ThAr calibration method is that the ThAr lamp has hundreds of strong emission lines over the entire optical and near-IR wavelength range (Palmer & Engleman 1983; Hinkle et al. 2001) so it can be used for RV measurements over a wavelength band much broader than the iodine calibration technique. For instance, the High-Accuracy Radial velocity Planetary Searcher (HARPS) uses the entire 380-690 nm for RV measurements. In addition, since the calibration beam is separated from the stellar beam, no stellar photons are absorbed by the calibration optics, increasing the RV measurement throughput by ~30%. No overhead time is required to take the star and iodine templates during observation, increasing the observation efficiency. However, the main drawback for the ThAr method is that the entire instrument needs to be installed in an isothermal and mechanically stable environment, possibly even in a vacuum chamber such as HARPS, in order to minimize the differential movement between the reference beam and stellar beam. Also, fibers must be used in order to minimize the differential motion between the incoming stellar and ThAr beams. In HARPS, a fiber mode scrambler is applied to further reduce the fiber illumination variation caused by the seeing and the fiber guiding changes in order to reach sub m/s RV precision (Pepe et al. 2000; Rupprecht et al. 2004). Therefore, the instrument design becomes more complicated than that using the iodine calibration method.

2.2.2 The Dispersed Fixed-Delay Interferometer Method

The RV method using a Michelson type interferometer was proposed in the late 1970s (Gorskii & Lebedev 1977, and Beckers & Brown 1978). The Doppler shifts of the incoming spectral lines are measured through monitoring the interference fringe phase shifts. For a Michelson interferometer with an optical path difference d between the two interferometer arms, the fringe order m is determined by

$$m\lambda = d. \tag{2.6}$$

For a small wavelength shift, $\delta\lambda$, the Doppler shift, ΔV, can be derived as

$$\Delta V = \frac{c\lambda}{2\pi d}\Delta\phi, \tag{2.7}$$

where $\phi = 2\pi m$ is the phase of the interferences (Ge et al. 2002).

This kind of interferometer with a narrow band pass has been successfully used for very high precision Doppler measurements of the Sun (e.g., sub m/s precision for the Global Oscillation Network Group (GONG) measurements, Harvey 2002 private communications; Harvey et al. 1996). This narrow-band Michelson type interferometer with a fixed delay is suitable for observing bright sources such as the Sun, but not for faint targets such as other stars.

In 1997, David Erskine of Lawrence Livermore National Lab proposed to use a combination of a Michelson type interferometer with a moderate resolution spectrograph for stellar RV measurements. The addition of the spectrograph separates the interference fringes at different wavelengths to increase the fringe visibility (or contrast) for each absorption line and the wavelength band in order to obtain high precision RV measurements for faint sources such as stars. Figure 2.2 shows a schematic layout of this kind of instrument concept (Erskine & Ge 2000; Ge 2002; Ge et al. 2002). The fringe visibility is defined as

$$\gamma = \frac{I_{max} - I_{min}}{I_{max} + I_{min}}, \tag{2.8}$$

where I is the fringe intensity. High-precision RV measurements can be achieved by summing independent RV measurements over many different spectral lines within the instrument wavelength coverage. This approach is called the dispersed fixed delay inferferometer (DFDI) method. The initial lab experiments and telescope observing with a prototype at the Lick 1 m telescope demonstrated its feasibility for stellar RV measurements (Erskine & Ge 2000; Ge et al. 2002).

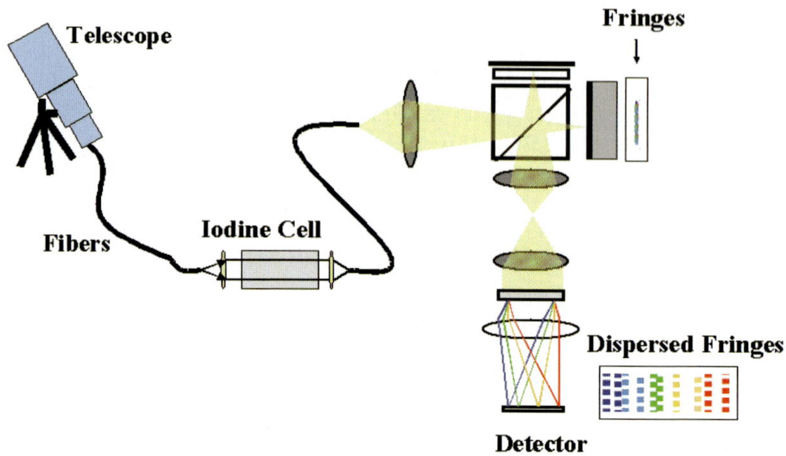

Fig. 2.2. Principle of a dispersed fixed delay ineterferometer, a combination of a Michelson interferometer with a moderate dispersion spectrograph. The spectrograph separates fringes from different wavelengths to allow high precision RV measurements using a broadband spectrum.

In this approach, the photon-limited Doppler precision is described as

$$\sigma_{RV} = \frac{1}{\sqrt{\Sigma_{i=1,N}(1/\sigma_{f,i}^2)}}, \tag{2.9}$$

$$\sigma_{f,i} = \frac{c\lambda}{\pi d\gamma_i\sqrt{2F_i}}, \tag{2.10}$$

where γ_i is the fringe visibility, and F_i is the photon flux in each of the N wavelength channels (Figure 2.3, Ge 2002; van Eyken et al. 2004). For a Gaussian-shaped absorption line, the intrinsic Doppler precision can be derived as

$$\sigma_{f,i} \approx \frac{c\delta\lambda_i}{\lambda D_i\sqrt{F_i}}, \tag{2.11}$$

which is the same as that for the echelle spectrograph (Ge 2002). This is not surprising since the intrinsic Doppler precision is totally determined by the spectral line intrinsic properties, irrelevant of measurement methods. However, when a spectrograph with a spectral resolution, $\delta\lambda_o > \delta\lambda_i$, is used to separate fringes from different wavelengths, the measured Doppler error per fringe becomes

$$\sigma_{f,o} = (\frac{\delta\lambda_o}{\delta\lambda_i})^{1/2}\sigma_{f,i}. \tag{2.12}$$

The total Doppler error over a wave band of $\Delta\lambda$ can be approximately described as

$$\sigma_{RV} \propto \delta\lambda_i S^{-0.5}\Delta\lambda^{-0.5}R^{-0.5}D^{-1}. \tag{2.13}$$

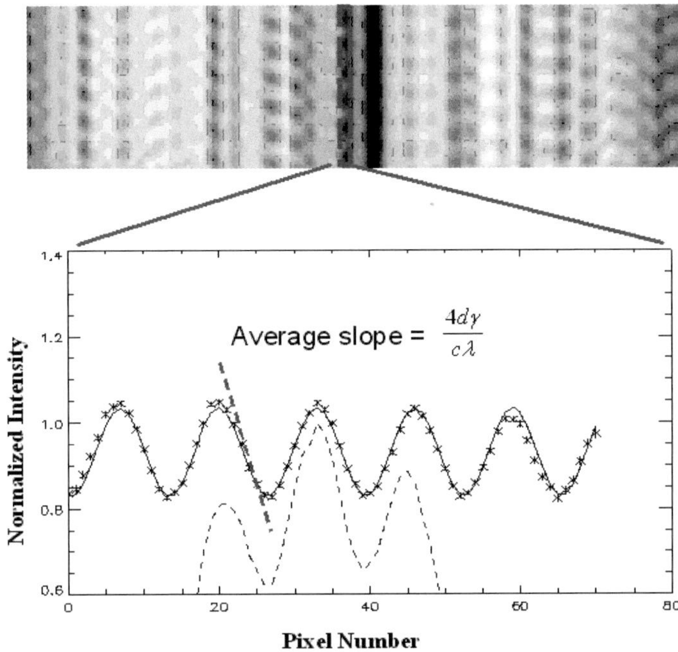

Fig. 2.3. An example of a fringing spectrum. The dashed line is the raw fringe. The cross points are the fringe after flat fielding and normalization, which can be fit with a sinusoidal function (the solid line) to extract the fringe phase information. The Doppler shift, ΔV is proportional to the phase shift, $\Delta\phi$.

This formula resembles that for the echelle (Eq. 2.5); the main difference is the dependence on the instrumental resolution (1/2 power for DFDI; 3/2 power for echelle). This allows DFDI instruments to use a medium resolution, high efficiency, first-order grating spectrometer for dispersing the fringes, producing a dramatically reduced size (and cost) of the instrument, while maintaining high precision for RV measurements. The fringing data for a single star requires only a small area in the detector plane for recording. This latter property enables simultaneous RV measurements of many objects using a reasonably sized detector (Ge 2002). The high throughput and multi-object capability are the main advantages of DFDI compared to the single-object echelle approach. The high throughput gain can offset the Doppler sensitivity loss due to the use of a much lower dispersion grating for a DFDI instrument than for an echelle instrument.

The simple instrument response, the sinusoidal function, created by the two-beam interference allows the DFDI method to be easily adopted to other wavelengths for maximizing the Doppler detection sensitivity for different stellar spectral types. The wavelengths other than the optical region include near UV and blue wavelength region (380-500 nm), red region (700-1000 nm) and near IR region (1.0-1.8 μm). Therefore, the DFDI survey instruments can be designed to target stars from late F type in the near UV to optical to late M types in the near IR. On the other hand, although the multi-object DFDI instrument interferometer is usually coupled with a first order grating spectrometer, it can also be designed to couple with a high efficiency cross-dispersed high resolution echellette or echelle spectrograph to gain additional Doppler sensivity by increasing the operation wavelength coverage and dispersion power. Since the Doppler sensitivity weakly depends on the spectrograph resolution, the spectrograph can still be designed to have moderate to high resolution (such as R$\sim$ 20,000) to reach high Doppler sensitivity. This kind of design can still leave sufficient detector resources to pack fringing spectra from tens of objects on a large size CCD detector (such as 4kx4k CCD) to allow multi-object high precision RV measurements.

2.3 Main Results from Single Object Doppler Planet Surveys

Since the first extrasolar planet was discovered around a main sequence (MS) solar-type star, 51 Peg, in 1995 (Mayor & Queloz 1995), a total of $\sim$ 200 new planets have been detected by the Doppler surveys using a dozen ground-based telescopes with various sizes, from 0.6 meters to 10 meters, and different Doppler instruments (e.g. Frink et al. 2002; Vogt et al. 2000; Butler et al. 2006; Mayor et al. 2003; Cochran et al. 2007; Sato et al. 2007). Except the planet HD 102195b, which was detected by the recently developed DFDI method (Ge et al. 2006), all of the RV planets were detected with single object high resolution echelle spectrographs. Due to limited telescope resources, the detection rate for RV surveys has reached a plateau of about 30 new planets per year in very recent years (Fig. 2.4). Here we summarize the main results from past RV planet surveys. Details for most of the conclusions can be found in recent review articles (Marcy et al. 2005; Udry, Fischer & Queloz, 2007).

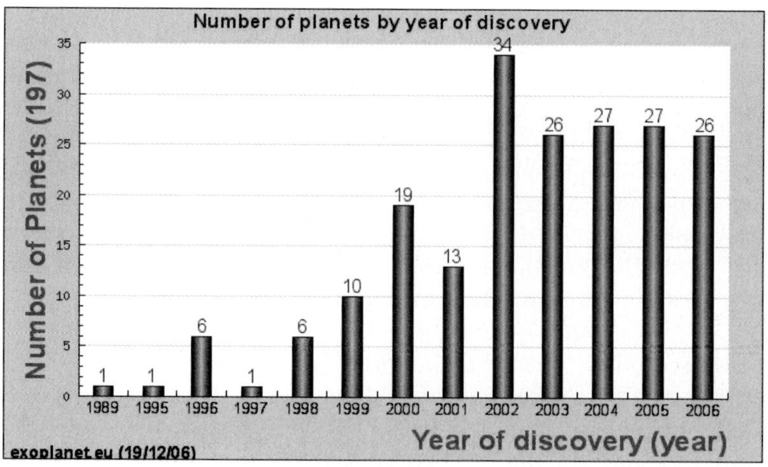

Fig. 2.4. Planets detected by the ground-based telescope in the last 15 years. The plot was created using the program in the exoplanet.eu website kindly provided by Jean Schneider.

2.3.1 Main Conclusions on Giant Planets

Despite the heterogeneity of the RV planet data sets obtained by many different groups, using different target samples, instruments, telescopes, and data reduction techniques, some basic conclusions based on the existing sample of giant planets can be drawn. For instance,

- The planet mass distribution for giant planets with ($M sini > 0.2$ M_J) has shown a clear power-law with $dN/dM \approx M^{-1.05}$ for the FGKM survey stars (Marcy et al. 2005).
- About 1% of FGK stars in both the Lick+Keck+AAT and CORALIE survey samples host hot Jupiters (orbital period, P≤10 d) (Marcy et al. 2005; Udry, Fischer & Queloz, 2007). It appears that single giant planets have a 3 day period pileup, while giant planets in multiple planet systems do not have the 3 day pileup (Wright et al. 2007).
- There are a total of 179 planets detected so far within 200 pc with RV methods and a total of 212 planets detected using Doppler techniques to date.
- About 5-7% (or higher percentage) of FGK stars in both the Lick+Keck+AAT and CORALIE survey samples have giant planets within 5 AU.
- Extrasolar giant planets have much larger eccentricities than the planets in the Solar System. The eccentricities range from 0 to 0.93. The median of the eccentricities of the extrasolar giant planets with orbit radii > 0.1 AU is about 0.25 (Marcy et al. 2005). The eccentricities in multiple planet systems are not higher than that in single planet systems (Wright et al. 2007).
- There appears to be a lack of giant planets with masses larger than $\sim 2M_{Jup}$ with intermediate and short periods (of <100 d) around single stars (Udry, Fischer & Queloz, 2007). Most of the massive planets with short and intermediate periods are found in multiple star systems (Zucker & Mazeh 2002). It appears that the maximum planet mass increases with the distance from the host star (Udry, Fischer & Queloz, 2007).
- There is a strong correlation between the giant planet occurrence and the planet host star metallicity. (e.g. Gonzalez 1997; Santos et al. 2001; Fischer & Valenti 2005). In the Lick-Keck-AAT survey sample, the occurrence of planets is consistent with being proportional to the square of the metallicity of the host stars (Fischer & Valenti 2005). However, the recent detection of giant planets around metal poor stars (Cochran et al. 2007) suggests that the occurrence of planets may be constant at low metallicity (e.g. [Fe/H]~ -0.5; also see Santos et al. 2004).
- Among stars with planets, at least 14% of them have multiple planets. Including those with RV trends, the percentage may be as high as 34% in the Lick+Keck+AAT sample (Wright et al. 2007). However, this high percentage may be partially due to detection bias. A large fraction of multiple planets are in resonant orbits.
- Evolved A type stars (subgiants with 1.1-2.0 $M_\odot$) have a higher giant planet percentage than solar-type stars (Johnson et al. 2007). G giants have more mas-

sive planets than solar-type stars (e.g. about 2% of G giants have planets with $>5M_J$ versus about 1% for solar-type stars (Sato et al. 2007)).
- Fewer than 1.3% of M dwarfs host giant planets (Endl et al. 2007). The occurrence of low mass planets (super-Earth masses) is much higher than giant planets around M dwarfs (Udry, Fischer & Queloz, 2007). Most of the planets around M dwarfs are in multiple planet systems (Mayor et al. 2007).

2.3.2 New Super-Earth Mass Planet Results

In the last 3 years, a new population of low mass planets in the 5-21 Earth mass range with a few day periods have been detected with high precision cross-dispersed echelle spectrographs (Santos et al 2004; McArthur et al. 2004; Butler et al. 2004; Rivera et al. 2005; Bonfils et al. 2005; Vogt et al. 2005; Udry et al. 2006; Lovis et al. 2006; Udry et al. 2007; Melo et al. 2007; Pepe et al. 2007). This becomes possible due to the improvement in Doppler sensitivity of planet survey instruments from 3-10 m/s before 2003 to 1-3 m/s post-2003. So far a total of 13 such planet systems orbiting nearby stars have been announced. Most of them are detected around M dwarfs, which may be due to observational biases.

Although the sample of super Earth mass and Neptune-mass planets is still relatively small, their properties show different trends than the known giant planets. Unlike the giant planets, whose frequency scales as the square of the host-star metallicity (Fischer & Valenti 2006), the Neptune-mass planet frequency seems to weakly depend on the host-star metallicity (see Udry, Fischer & Queloz 2007; Mayor et al. 2007). There also appears to be a lack of the 3-day orbital period pile up for these low-mass planets, unlike the giant planets (Marcy et al. 2005). In addition, the discovery of two low-mass planets with microlensing indicates that cool Neptune-mass planets may be common (Beaulieu et al. 2006; Gould et al. 2006).

All of these early results indicate that they may belong to a distinct planet population whose formation and evolution may be very different from that of the giant planets. For instance, they may be formed without accumulating a substantial amount of gaseous material, unlike the gas giant planets; i.e. they may be terrestrial rocky planets. Their lower mass may make them less capable of opening up a gap in the protoplanetary disk, so they may have undergone a very different migration history than the gas giants, and some may have been formed in situ.

2.4 Science Needs for Multiple Object Doppler Planet Surveys

Despite the fact that over 200 known exoplanets have provided important information about planet masses and orbital parameters, many more exoplanets are needed for statistical characterization of emerging classes of planets and tests of detailed theoretical models for planet formation and evolution. A larger planet sample could also be used for study of the diversity of exoplanets and correlations with stellar properties, measurement of planet mass and orbital functions and their correlations, and also discovery of new planet populations. On the other hand, the planet detection speed has substantially slowed down in recent years (Fig. 2.4) although more and more ground-based telescopes have been heavily used in planet detection using Doppler instruments. This slowdown is mainly caused by the limitation of the single object capability with echelle spectrographs. Furthermore, most of the bright suitable stars (e.g., FGKM dwarfs, subgiants, and G and K giants) with $V < 8$ have already been searched by previous RV surveys. In order to increase the survey sample, stars fainter than $V > 8$ will become the primary survey targets, which requires more telescope time in order to obtain a similar RV precision and maintain a similar planet detection speed to the previous surveys.

In an effort to significantly increase the sample of known extrasolar planets, several new RV surveys have recently been initiated. In particular the N2K (Fischer et al. 2005) and ELODIE metallicity-biased (Da Silva et al. 2006) surveys together target ~3000 stars, and deliberately bias their sample toward metal-rich old MS target stars since giant planet occurrence strongly depends on the host star metallicity (Fischer & Valenti 2005). However, these single object RV surveys will have very limited constraints on the planetary systems around metal-poor, late-type, giant, or active stars.

In recent years, a strong need has emerged to have a large homogeneous sample of giant planets. This large homogeneous sample is critical for testing various models of the planet formation, migration and dynamical evolution (e.g., Ida & Lin 2004a,b; Alibert 2005). These models can now provide quantitative predictions for the planet mass function and orbital parameter distributions, and their correlations with the properties of the host stars (i.e., metallicity, mass). Direct comparison between theory and observational data becomes a critical step to test the validity of the physical mechanisms or assumptions used in the models. To have this kind of comparison, it is essential to have a large sample of planetary systems discovered in surveys with well-characterized selection effects. Unfortunately, no such sample currently exists. Although the sample of currently known planets is relatively large, the data come from different groups using different target selection, observation techniques, cadence strategy, and candidate selection criteria. The planet data is quite heterogeneous and it is difficult to use them to do quantitative comparisons. A recent study by Armitage (2007) shows the size of statistically complete samples for testing planet theories is strikingly small. Armitage (2007) was only able to use a total of 22 planets from the the planet sample in the Fischer & Valenti (2005) paper. Due to the small size of the sample, Armitage (2007) could not distinguish different planet migration models. It is quite clear that there is a strong need for a large homogeneous sample of planets extracted from a large-scale RV survey using a clearly defined survey strategy and plan, and well characterized instruments. This requirement can only be achieved with new generation RV instruments with multi-object capabilities, which can substantially increase the survey sample size without introducing strong selection biases.

The recently-developed multi-object DFDI technique is well suited to conduct a large-scale next generation multi-object Doppler planet survey on wide field telescopes in the next decade. It is critical to have wide field telescopes for this application since they can cover hundreds of bright survey stars within their field of view (FOV) for simultaneous multi-object RV monitoring. There are quite a few existing and planned wide-field telescopes which are potentially suitable for the next generation large-scale multi-object RV planet survey; e.g., the Sloan Digital Sky Survey (SDSS) 2.5m telescope, with a 7 square degree FOV; the AAT 4m telescope, with a 3 square degree FOV; the LAMOST 4m telescope, with a 20 square degree FOV; the 4.2m Discovery Channel Telescope, with a 3 square degree FOV; and the 8m LSST telescope, with a 3 square degree FOV. A multi-object RV instrument has already demonstrated its feasibility at the SDSS telescope. The following sections summarize the progress, the early results and a planned Multi-object APO Radial-Velocity Exoplanet Large-area Survey (MARVELS, formerly ASEPS) at the SDSS telescope in 2008-2014.

2.5 Early Results from a Multi-Object Doppler Planet Survey

The first DFDI instrument used for planet searches was the Exoplanet Tracker (ET) that was commissioned at the KPNO 0.9-m Coude Feed/2.1-m telescope in late 2003. ET was used for a planet survey of ~150 solar-type stars with $V = 7.6$–9 in 2004–2006, and since Fall 2006 the instrument has been available to the public.

The instrument Doppler precision over a few hours' baseline was measured in April 2007; the results are shown in Fig. 2.5. The measurement errors are consistent with photon noise limits. The total measured detection efficiency, including the telescope, seeing, fiber, instrument and detector losses, is 18% under typical seeing conditions (1.5 arcsec) at the KPNO Coude Feed/2.1 m. This efficiency is about four times higher than that reached with the state-of-the-art echelle Doppler instrument HARPS on the ESO 3.6 meter telescope (Pepe et al 2002). One new planet, orbiting the $V = 8.05$ G8V star HD 102195, has been discovered by ET (Ge et al. 2006; see Fig. 2.6).

The first DFDI instrument designed for the MARVELS survey at the SDSS telescope was constructed at the University of Florida in 2005-2006 with support from the W.M. Keck Foundation and designated as the W.M. Keck Exoplanet Tracker, or Keck ET. It was commissioned at the SDSS telescope in spring 2006. Figure 2.7 shows ET fiber plugging in the SDSS fiber cartridge and the ET setup on an optical bench in a thermally controlled instrument enclosure. The Keck ET is based upon the design of the single-object KPNO ET. The Keck ET consists of eight subsystems: the multi-object fiber feed, the iodine cell, the fixed-delay interferometer system, the slit, the collimator, the grating, the camera, and the 4k × 4k CCD. The instrument contains four auxiliary subsystems for interferometer control, instrument calibration, photon flux monitoring, and thermal control. The instrument is fed with 60 fibers of 200 μm core diameters, which are coupled to 180 μm core diameter short fibers from the 2.5-m telescope; the latter measurement corresponds to $3''$ on the sky at $f/5$. The spectral resolution for the spectrograph is

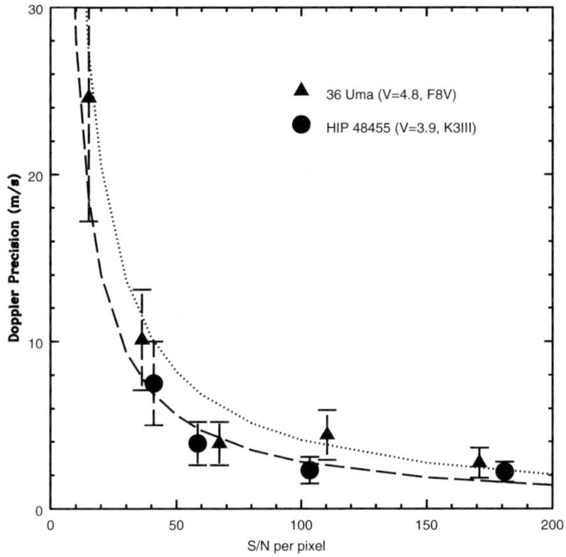

Fig. 2.5. Doppler precision measurements with KPNO ET in April 2007. The dotted and long dashed lines represent the photon noise limits for 36 UMa ($V = 4.8$, F8V) and HIP 48455 ($V = 3.9$, K2III), respectively. The RMS RV measurement errors are consistent with the photon noise limits. The best photon limiting precision is 1.4 m/s and 2.4 m/s for HIP 48455 (V = 3.9, K2III) and 36 UMa, respectively.

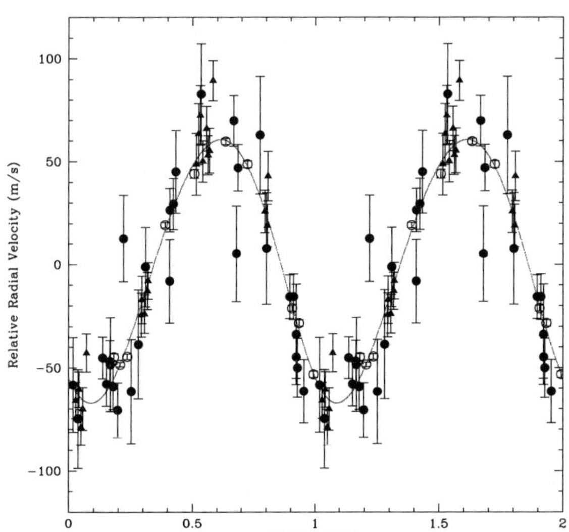

Fig. 2.6. The first planet discovered with a prototype of the MARVELS spectrograph, a single-fiber instrument operating on the Kitt Peak 2.1-m telescope (Ge et al. 2006a). Points show the measured radial velocity of the star against the orbital phase, with one repetition. The smooth curve shows the best-fit orbit, which is nearly circular with a period of 4.11 days.

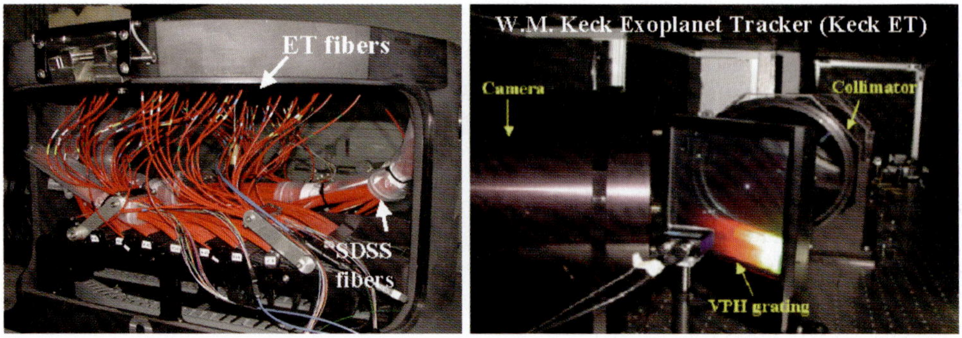

Fig. 2.7. *left:* An SDSS fiber cartridge with ET fibers plugged ready for multi-object ET observations. *right:* Part of the setup of the Keck ET multi-object Doppler instrument in April 2006.

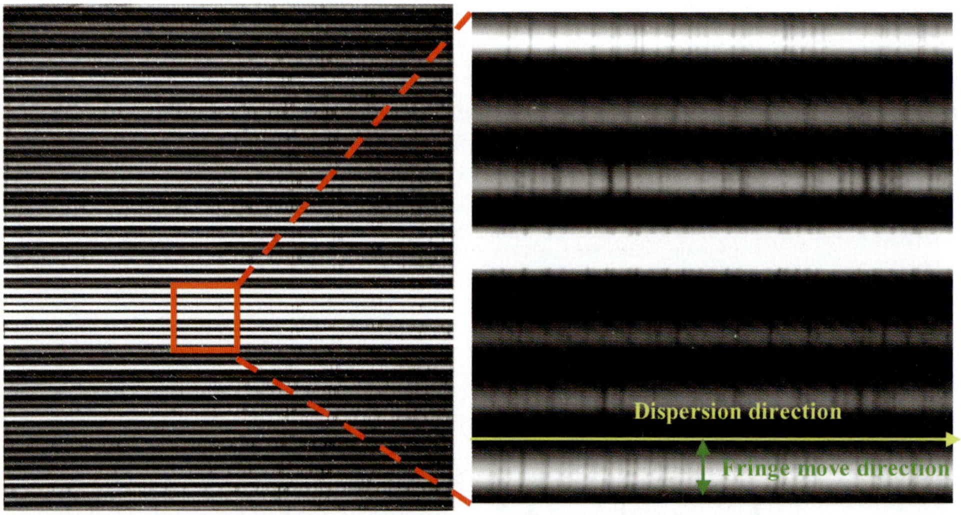

Fig. 2.8. *left:* 59 stellar fringes recorded by the 4kx4k CCD of Keck ET. The brightest star is $V = 8$ and the faintest star is $V = 12$. *right:* Expanded fringing spectra of the central region. Fringes can be clearly seen.

R=5,100 and the wavelength coverage is 900 Å, centered at $\sim$5400 Å. Details of the instrument design can be found in Ge et al. (2006b), Wan et al. (2006), and Zhao & Ge (2006). Figure 2.8 shows the spectral format on the 4kx4k CCD detector.

The current Keck ET has one spectrograph and one $4k \times 4k$ CCD camera to capture only one of the two interferometer outputs, and has a 5.5% detection efficiency from the telescope to the detector without the iodine cell under the typical seeing condition at Apache Point Observatory (APO). (The instrument will be upgraded to capture both interferometer outputs and have better throughput in Spring 2008 before the MARVELS survey starts in July 2008.) The instrument can record 59 objects in a single exposure (a slight modification planned this fall will increase this to 60 spectra). The instrument Doppler precision was measured with the day sky scattered light, which offers a stable, homogeneous RV source for simultaneously calibrating the instrument performance for all of the sky fibers. The sky spectra had an average signal-to-noise ratio of ~150 per pixel, or a total of 1.8×10^9 photons. The average RMS error over a few hours of RV measurements for the 59 fibers is 6.3±1.3 m/s in 2006 November; the corresponding average photon-limit error is 5.5±0.5 m/s. Figure 2.9 shows the RV accuracy of the sky measurements as a function of the recorded signal: the short term RMS errors are consistent with the photon-noise limit errors. Figure 2.10 shows a comfirmation of the planet HD 209458b using the Keck ET in fall 2006.

The instrument's long term precision has been measured using the sky scattered light during two extended periods: 45 days in fall 2006, and 150 days in winter/spring 2007. The RMS RV measurement errors for these periods, after photon noise errors are subtracted, are 11.7 ± 2.7 m/s and 11.3 ± 2.5 m/s, respectively. Figure 2.11 shows the RV measurements with one of the fibers in 2007. These measurement errors are mainly caused by inhomogeneous illumination of the slit, image aberration, and the interferometer comb aliasing (sampling on the detector).

The instrument's long term RV stability has been derived from sky observations between December 2006 to May 2007. The average RV offsets of later months compared to the first month data are all within ±2.0 m/s.

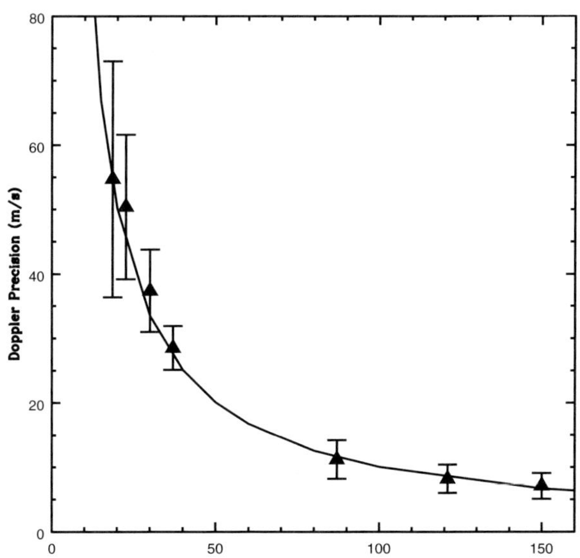

Fig. 2.9. Doppler precision measurements with Keck ET on 4 April 2007. The solid line is the photon noise limit. The triangle dots represent the average RMS error of the day sky RV measurements of the 59 fibers.

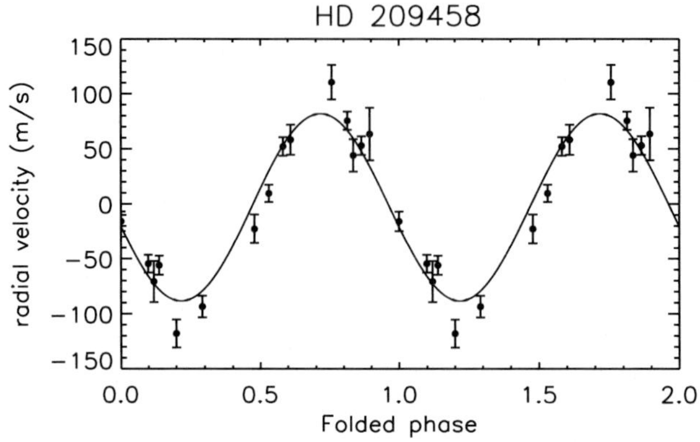

Fig. 2.10. Keck ET observations of the previously known transiting planet system HD 209458. Points show the measurements (with one repetition of the orbit), while the smooth curve is the prediction based on earlier observations from other telescopes (Henry et al. 2000).

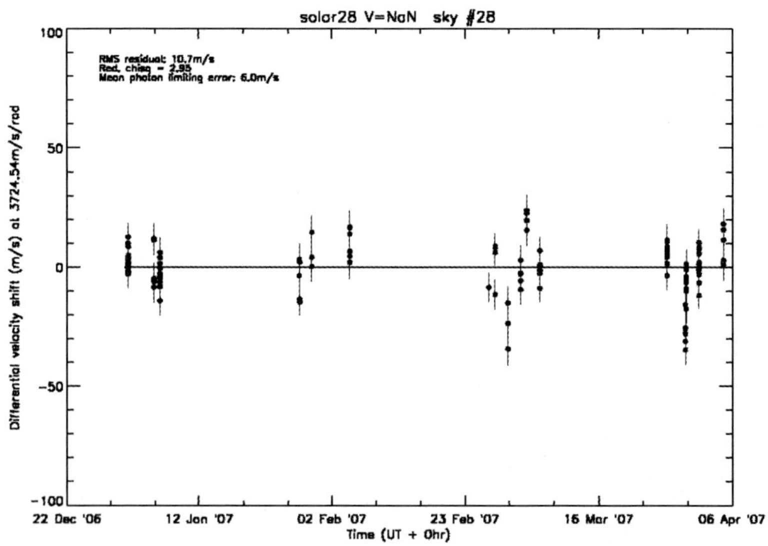

Fig. 2.11. The long term RV monitoring of the sky using the Keck ET in 2007.

The Keck ET is currently being used in a pilot survey of $\sim$ 2000 $V = 8$–12 solar-type stars in 30 different fields and will be completed by June 2008 before MARVELS starts in July 2008. The early results from the Keck ET pilot program have demonstrated that the multi-object DFDI instrument is capable of conducting a large-scale planet survey at the SDSS telescope.

2.6 New Planet Science to be Addressed by Next Generation Multi-Object RV Planet Surveys

The next generation multi-object RV planet survey is becoming the most efficient way to detect and charcterize hundreds and even thousands of new giant planets in the next decade. Within each of the SDSS telescope FOVs, there are over one hundred candidate stars with V<12. A coupling of the multi-object DFDI instruments with the SDSS telescope will allow simultaneous monitoring of over 100 stars for planet detection and characterization with a moderate to high RV precision (about 10 m/s Doppler precision for a V$\sim$11 solar-type star with a DFDI type instrument in an hour integration at the SDSS telescope). This kind of accuracy is sufficient to uncover most of the giant planets with a Jupiter mass and a year period since such a planet produces RV variations with a semi-amplitude of $\sim$30 m/s on a solar-type star.

Below we use the MARVELS survey as an example to illustrate important planet science to be addressed by the next generation multi-object RV planet surveys.

The baseline MARVELS survey plan is to monitor a total of V$\sim$8-12 10,000 MS and subgiant stars and 1,000 giant stars in 2008-2014 using two Keck ET instruments with 120 object capability with the precision and cadence needed to detect giant planets with orbital periods of a few days to $\sim$2 years in 2008-2014. The survey can be extended well beyond 2014 to detect and characterize additional long period planets.

The MARVELS survey has the capability to accommodate additional specialty instruments for exploring new planet parameter space, such as the low-mass star (mainly M dwarf) and super Earth mass planet regimes in 2008-2014. Figure 2.12 illustrates the parameter space that a full-scale MARVELS planet survey can cover in the next decade.

2.6.1 Giant Planet Science

The MARVELS optical survey with Keck-ET type instruments will be able to provide the comprehensive data set needed to characterize the population of giant planets with short and intermediate periods (e.g. P< 2 years), and to test theoretical models of the formation, migration, and dynamical evolution of giant planet systems. In detail:

- The MARVELS survey will explore a much larger range of the parameter space of stellar properties than any of the previous and current RV surveys as a single survey project. For instance, as shown in Fig. 2.12, stellar masses in the range 0.6-5 $M_\odot$ will be explored with the optical Keck ET type survey instruments (the 0.2-0.6 $M_\odot$ range can be explored with a red-sensitive survey instrument), while stars of a wide range of metallicity, activity level and age will be included in the target sample.
- The broad selection of target stars will make MARVELS ideal for studying the correlation of planet systems with stellar metallicity, mass, multiplicity, age, evolutionary stage, activity level, and rotation velocity.

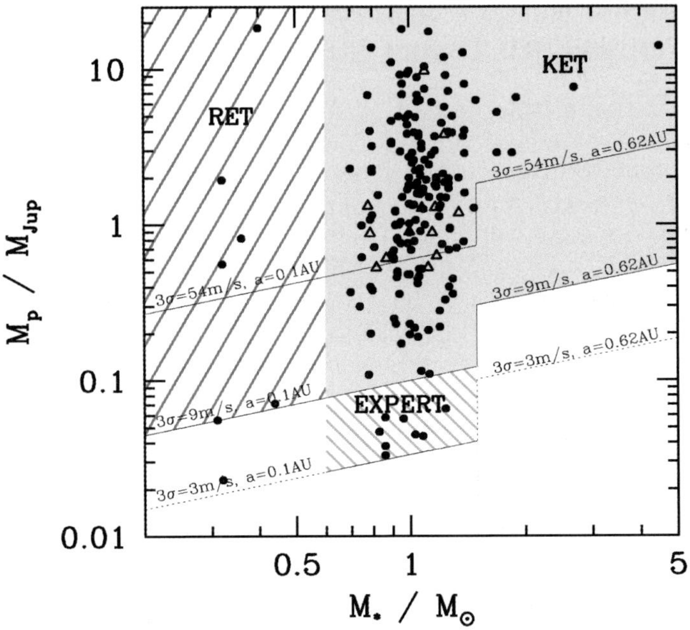

Fig. 2.12. The planet mass (Y-axis) versus stellar mass (X-axis) showing the exploration space for the MARVELS planet survey. The filled circles are the known planets detected by the RV method, the open triangles are the known planets detected by the transit method. The grey region is probed by the Keck ET instruments, while the striped regions marked with RET and EXPERT, would be accessible to the red-sensitive and high-precision, optical RV instruments, respectively.

- If planets around metal poor stars are detected by MARVELS, they will provide much stronger constraints on theoretical models of planet formation and evolution in metal-poor environments. Any detection of giant planets around stars with [Fe/H]<-0.5 would support the hybrid planet formation model for giant planets (i.e., giant planets form through disk instability in the metal poor disk environments (e.g., Boss 2002) vs. through core accretion in the metal-rich disk environment). On the other hand, a possible null detection of any giant planets around stars with [Fe/H]<-0.5 may place significant constraints on the disk instability model (Durisen et al. 2007).
- The detection of planets around giant stars will provide an important sample for a systematic study of planet formation and evolution around massive stars. The detection of new planets around active stars will offer the first significant sample of planets around these relatively young stars and allow direct comparison of their properties with the well established planet populations around older stars.
- A combination of the large number of new planets from a homogeneous survey sample with the survey efficiency calculations from the survey simulation models

can be used to quantitatively test the planet distributions predicted by current planet formation models (i.e., Ida & Lin 2004a,b; Alibert 2005)

- The large number of short and intermediate period planet systems probed by MARVELS are the key to understanding the migration of giant planets after their formation. The semi-major axis distribution of short and intermediate period giant planets depends on the protoplanet disk conditions, the planet migration rate and the migration stopping mechanisms. Therefore, the measured distribution can test competing models for giant planet migration (Mordasini, private communication; Ida and Lin 2004a,b; Armitage 2007; Kornet et al. 2005; Ford & Rasio 2006; Wu et al. 2007)
- The large sample of planets will enable us to confirm or reject the marginal trends currently seen in the distributions of planet parameters with much higher confidence levels, such as the shortage of massive planets with 10-100 day periods, the planet mass fucntion at the high mass end (the "brown dwarf desert" region).
- The MARVELS optical survey will identify rare and interesting planetary systems for follow-up observations. These could include massive hot Jupiters, rapidly interacting multiple planet systems, very-hot Jupiters, and/or planets with extremely high eccentricities. Furthermore, stars with short period planets are also promising targets for longer term high-precision radial velocity monitoring to detect long period companions at large orbital separations and/or multiple planet systems. Multiple planet systems are particularly valuable for providing clues to the orbital evolution that shapes planetary systems.
- Transiting planets to be detected by the MARVELS survey will be important for detailed studies of planet properties. RV detected transiting planets are not biased toward the largest (radius) objects at fixed mass compared to those detected by transit surveys which are biased toward bloated and short-period planets (e.g., Gaudi 2006). These selection effects can be very difficult to quantify, particularly for ground-based transit surveys.
- MARVELS will be the best survey to date for exploring the "brown dwarf desert," the apparent paucity of $\sim 15 - 80 M_J$ companions to solar-type stars. MARVELS may discover new planetary systems with properties that have not previously been seen and have not been anticipated by theory.

2.6.2 Comparison with Other Planet Surveys

Compared to other on-going and planned RV searches using single-object spectrographs, MARVELS will monitor relatively faint stars, but many more of them. This survey target selection makes MARVELS especially powerful for detecting relatively short and intermediate period giant planets (P<2 years) with large velocity amplitudes. The MARVELS survey is complementary to transiting planet surveys which are sensitive to short period planets, but only those edge-on systems. The MARVELS survey is also complementary to microlensing planet searches (Bennett et al. 2007; Gould et al. 2007) and to space-based astrometric searches such as Space

Interferometry Mission (SIM) PlanetQuest and GAIA, which are more sensitive to longer period planet systems (several years or longer).

Ground-based transit surveys are primarily sensitive to Jupiter-sized planets orbiting bright stars, and are strongly biased toward planets with periods less than a few days. Space based transit surveys, such as Kepler, can detect relatively long period planets. However, the total detection number is very small because the transit probability declines as $P^{-2/3}$. Based on the close-in giant planet frequencies derived by Gould et al. (2006), Kepler should find ~ 20 giant planets with periods between 5 days and two years, most of which will be detected around the $V \sim 14$ Kepler stars.

The proposed next-generation ground and spaced-based microlensing surveys (Bennett et al. 2007; Gould et al. 2007) are primarily sensitive to planets with separations of several AU and can only yield statistics for planets since the host stars of detected planets will generally be too distant and faint for follow-up.

SIM will primarily focus on a small number of stars (~ 100) to detect terrestrial like planets. SIM has the best sensitivity in discovering planets at orbital distances close to its 5 year mission lifetime, considerably greater than those found by MARVELS.

GAIA will survey 100,000 V<12 MS stars comparable to MARVELS over its five-year mission lifetime and can potentially detect ~1,000 Jupiter-mass planets around these stars (Lattanzi et al. 2000, Sozzetti et al. 2005). Like SIM, GAIA's sensitivity will peak near orbital periods close to the mission lifetime, and so most of the detected planets will have several AU orbits.

Therfore, for the next decade, MARVELS will provide a unique large homogeneous sample of giant planets with orbital periods of a few days to over one year for characterizing giant planets and studying planet formation and dynamical evolution.

2.6.3 Super-Earth Mass Planets

One property of the MARVELS survey (common to all multi-object RV surveys) is that most of the fibers are devoted to faint stars which require longer exposure times and have limited Doppler precision. However, there will be a handful of bright stars in each field for which there is the opportunity to carry out an RV survey at much higher precision in both optical and red wavelengths. This simply requires building an instrument with higher resolution, and thus higher precision, based on the same multi-object DFDI technology. With additional high precision optical and red sensitive multi-object RV instruments, MARVELS can become a powerful tool for detecting a large number of super-Earth mass planets.

A large sample of super-Earth mass planets can help to address the frequency of super-Earth mass planets around nearby stars, what fraction of them are in the habitable zone, their mass and orbital parameter distribution such as semi-major axis and eccentricity, and their correlations with stellar properties of their host stars. Furthermore, a larger sample of these planets will lead to the detection of possible transiting planets, which will be used to determine the typical planet density and mass-radius relation, to test the idea that these are indeed rocky planets.

2.7 Conclusions

The discovery of over 200 extrasolar giant planets in the last decade has led to many unexpected results. Now we know that most of the giant planets have much more eccentric orbits than the Solar System planets. Most of the known Jupiter-mass exoplanets orbit their host stars much closer than Jupiter. The Solar System is not typical compared to all of the known exoplanet systems. Various new theories about planet formation and evolution have been proposed to explain these new planet properties. These theories have also made many testable predictions. In order to understand the detailed physical processes that lead to planet formation, migration and evolution, a large sample of planets drawn from a well-characterized, homogeneous survey of stars with known properties is required to make quantitative comparisons to the theoretical predictions. However, this is an extremely challenging task for single-object RV planet surveys due to the limited instrument efficiency and worldwide telescope resources. New survey instruments with much greater observational efficiency than the single object RV survey instruments are urgently needed. This need is being met with the recently developed multi-object DFDI instruments with $\sim$100 object capability.

The next generation multi-object RV planet surveys based on these multi-object DFDI instruments will produce a large homogeneous and well characterized planet sample for understanding planet formation and evolution mechanisms in the next decade. Compared to the planned single object RV, transit, microlensing, and astrometric planet surveys in the next decade, the multi-object RV planet surveys will not only provide complementary information about planet mass and orbital distributions, but also play a unique and critical role in understanding planet populations and formation and evolution mechanisms, by providing a large and statistically well-defined sample of massive, short-to-intermediate period planets orbiting relatively bright host stars with a range of intrinsic properties to test various predictive planet formation, migration and dynamical evolution models.

Acknowledgement

The author is grateful to Brian Lee and Julian van Eyken who helped to make some of the plots in this chapter, to Scott Gaudi, Eric Ford and Eric Agol who shared some of the MARVELS science thoughts and to Scott Fleming who read the manuscript carefully and provided valuable comments. The author is also grateful to the rest of the MARVELS science and engineering team members who made many contributions to the science, hardware and software development that produced the results shared in this chapter. The Keck ET has been supported by the W.M. Keck Foundation. The MARVELS project has been supported by NSF, NASA and the University of Florida.

References

Alibert, Y., Mordasini, C., Benz, W., & Winisdoerffer, C. 2005, Models of giant planet formation with migration and disc evolution, *A&A*, **434**, 343

Armitage, P.J. 2007, Massive planet migration: Theoretical predictions and comparison with observations, *ApJ* in press/astro/ph-arXiv:0705.3039

Baranne, A. et al. 1996, ELODIE: A spectrograph for accurate radial velocity measurements, *A&AS*, **119**, 373

Beaulieu, J.-P. et al. 2006, Discovery of a cool planet of 5.5 Earth masses through gravitational microlensing, *Nature*, **439**, 437

Beckers, J.M., & Brown T.M., 1978, Solar circulation measurements: consideration and plans, *Osser. Mem. Astrophys. Obs. Arcetri.*, **No. 106**, 189

Bennett, D. P., et al. 2007, An Extrasolar Planet Census with a Space-based Microlensing Survey, ExoPTF White Paper (arXiv:0704.0454)

Bonfils, X. et al. 2005, The HARPS search for southern extra-solar planets. VI. A Neptune-mass planet around the nearby M dwarf Gl 581, *A&A*, **443**, L15

Boss, A. P. 2002, Stellar Metallicity and the Formation of Extrasolar Gas Giant Planets, *ApJ*, **567**, L149

Brown, T.M. et al. 1994, The AFOE: A spectrograph for precise Doppler studies, *PASP*, **106**, 1285

Butler, R.P. et al. 1996, Attaining Doppler Precision of 3 M s-1, *PASP*, **108**, 500

Butler, R.P., et al. 2004, A Neptune-Mass Planet Orbiting the Nearby M Dwarf GJ 436, *ApJ*, **617**, 580

Butler, R.P., et al. 2006, Catalog of Nearby Exoplanets, *ApJ*, **646**, 505

Campbell, B., & Walker, G.A.H. 1979, Precision radial velocities with an absorption cell, *PASP*, **91**, 540

Cochran, W.D., & Hatzes, A.P., 1993, McDonald Observatory Planetary Search - A high precision stellar radial velocity survey for other planetary systems, In: *Planets around pulsars; Proceedings of the Conference, California Inst. of Technology, Pasadena, Apr. 30-May 1, 1992 (A93-36426 14-90)*, pp 267-273.

Cochran, W.D. et al. 2007, A Planetary System Around HD 155358: The Lowest Metallicity Planet Host Star, astro/ph-arXiv:0705.3228

da Silva, R., et al. 2006, Elodie metallicity-biased search for transiting Hot Jupiters. I. Two Hot Jupiters orbiting the slightly evolved stars HD 118203 and HD 149143, *A&A*, **446**, 717

Durisen, R.H., Boss, A.P., Mayer, L., Nelson, A., Rice, K., & Quinn, T.R. 2007, Gravitational Instabilities in Gaseous Protoplanetary Disks and Implications for Giant Planet Formation, In: *Protostars and Planets V*, B. Reipurth, D. Jewitt, & K. Keil, eds. (Tucson: University of Arizona Press), pp 607

Endl, M. et al. 2007, New results from the McDonald Observatory and ESO/VLT planet surveys, In: *Extreme Solar Systems, Santorini, June 25-29, 2007*

Erskine D., & Ge, J. 2000, A Novel Interferometer Spectrometer for Sensitive Stellar Radial Velocimetry, In: *Proc. Imaging the Universe in Three Dimension*, Edited by W. van Breugel and J. Bland-Hawthorn, ASP Conference Series, **195**, pp 501

Fischer, D.A., et al. 2005, The N2K Consortium. I. A Hot Saturn Planet Orbiting HD 88133, *ApJ*, **620**, 481

Fischer, D.A., & Valenti, J., 2005, The Planet-Metallicity Correlation, *ApJ*, **622**, 1102

Frink, S. et al. 2002, Discovery of a Substellar Companion to the K2 III Giant Draconis, *ApJ*, **576**, 478

Ford, E. B., & Rasio, F. A. 2006, On the Relation between Hot Jupiters and the Roche Limit, *ApJ*, **638**, L45

Gaudi, B. S. 2006, Statistics and Simulations of Transit Surveys for Extrasolar Planets, In: *ASP Conf. Series: Transiting Extrasolar Planets Workshop*(arXiv:astro-ph/0612141)

Ge, J., 2002, Fixed Delay Interferometry for Doppler Extrasolar Planet Detection, *ApJ*, **571**, L165

Ge, J. et al. 2002, An Externally Dispersed Interferometer for Sensitive Doppler Extrasolar Planet Searches, *PASP*, **114**, 1016

Ge, J., et al. 2006a, The First Extrasolar Planet Discovered with a New-Generation High-Throughput Doppler Instrument, *ApJ*, **648**, 683

Ge, J., et al. 2006b, A new-generation multi-object high throughput Doppler instrument for a planet survey at the SDSS Telescope, In: *Proc.SPIE*, **6269**, pp 75

Gonzalez, G. 1997, The stellar metallicity-giant planet connection, *MNRAS*, **285**, 403

Gorskii, S.M., & Lebedev, V.P., 1977, Interference-phase method of measuring ray velocities in the solar atmosphere, *Izv. Krym. Astrofiz. Obs.* **57**, 228

Gould, A., Dorsher, S., Gaudi, B. S., & Udalski, A. 2006, Frequency of Hot Jupiters and Very Hot Jupiters from the OGLE-III Transit Surveys toward the Galactic Bulge and Carina, *Acta Astronomica*, **56**, 1

Gould, A., Gaudi, B. S., & Bennett, D. P. 2007, Ground-based Microlensing Surveys, ExoPTF White Paper (arXiv:0704.0767)

Griffin, R.F., 1967, A Photoelectric Radial-Velocity Spectrometer, *ApJ*, **148**, 465

Griffin, R., & Griffin, R. 1973, Accurate wavelengths of stellar and telluric absorption lines near lambda 7000 Angstroms, *MNRAS*, **162**, 255

Harvey, J.W. et al. 1996, The Global Oscillation Network Group (GONG) Project, *Science*, **272**, 1284

Hatzes, A.P., & Cochran, W.D., 1992, Spectrograph Requirements for Precise Radial Velocity Measurements, In: *ESO Workshop on High Resolution Spectroscopy with the VLT. Proceedings*, held in Garching, Germany, February 11-13, 1992. Editor, M.-H. Ulrich; Publisher, European Southern Observatory, Garching bei Munchen, Germany, pp 275

Henry, G.W., Marcy, G.W., Butler, R.P. & Vogt, S.S. 2000, A Transiting "51 Peg-like" Planet, *ApJ*, **539**, L41

Hinkle, H.H., Jorce, R.R., Hedden, A., Wallace, L., & Engleman, R. Jr. 2001, Wavelength Calibration of Near-Infrared Spectra, *PASP*, **113**, 548

Ida, S., & Lin, D. N. C. 2004a, Toward a Deterministic Model of Planetary Formation. I. A Desert in the Mass and Semimajor Axis Distributions of Extrasolar Planets, *ApJ*, **604**, 388

Ida, S., & Lin, D. N. C. 2004b, Toward a Deterministic Model of Planetary Formation. II. The Formation and Retention of Gas Giant Planets around Stars with a Range of Metallicities, *ApJ*, **616**, 567

Johnson, J. et al. 2007, Retired A stars and their planets, In: *Extreme Solar Systems, Santorini, June 25-29, 2007*

Juric, M., & Tremaine, S. 2007, The Eccentricity Distribution of Extrasolar Planets, *ApJ*, submitted (astro-ph/0703160)

Kornet, K. et al. 2005, Formation of giant planets in disks with different metallicities, *A&A*, **430**, 1133

Lattanzi, M. G., Spagna, A., Sozzetti, A., & Casertano, S. 2000, Space-borne global astrometric surveys: the hunt for extrasolar planets, *MNRAS*, **317**, 211

Lovis, C., et al. 2006, An extrasolar planetary system with three Neptune-mass planets, *Nature*, **441**, 305

Marcy, G., Butler, R.P., Fischer, D., Vogt, S., Wright, J.T., Tinney, C.G., Jones, H.R.A., 2005, Observed Properties of Exoplanets: Masses, Orbits, and Metallicities, *PThPS*, **158**, 24

Mayor, M. & Quolez, D. 1995, A Jupiter-Mass Companion to a Solar-Type Star, *Nature*, **378**, 355

Mayor, M., et al. 2003, Setting New Standards with HARPS, *The Messenger (ISSN0722-6691)*, **No. 114**, p. 20-24

Mayor, M. et al. 2007, From Hot Jupiters to Hot Super-Earths, In: it Extreme Solar Systems, Santorini, June 25-29, 2007

McArthur, B.E., et al. 2004, Detection of a Neptune-Mass Planet in the Cancri System Using the Hobby-Eberly Telescope, *ApJL*, **614**, 81

Melo, C. et al. 2007, A new Neptune-mass planet orbiting HD 219828, *A&A*, **467**, 721

Palmer, B.A., & Engleman, R. Jr. 1983, *Atlas of the Thorium Spectrum (Los Alamos Rep. LA-9615)*

Pepe, F. et al. 2000, HARPS: a new high-resolution spectrograph for the search of extrasolar planets, In: *Proc. SPIE*, **4008**, 582

Pepe, F. et al. 2002, HARPS: ESO's coming planet searcher. Chasing exoplanets with the La Silla 3.6-m telescope, *The Messenger, (ISSN 0722-6691)*, **No. 110**, 9

Pepe, F. et al. 2007, The HARPS search for southern extra-solar planets. VIII. Arae, a system with four planets, *A&A*, **470**, 721

Rivera, E.J., et al. 2005, A 7.5 Earth Mass Planet Orbiting the Nearby Star, GJ 876, it ApJ, **634**, 625

Rupprecht, G. et al. 2004, The exoplanet hunter HARPS: performance and first results, In: *Proc. SPIE*, **5492**, 148

Santos, N.C. et al. 2001, The metal-rich nature of stars with planets, *A&A*, **373**, 1019

Santos, N.C., Israelian, G., & Mayor, M. 2004, Spectroscopic [Fe/H] for 98 extra-solar planet-host stars. Exploring the probability of planet formation, *A&A*, **415**, 1153

Sato, B. et al. 2007a, A Planetary Companion to the Hyades Giant Tauri, *ApJ*, **661**, 527

Sato, B. et al. 2007b, Properties of Planets around G and K Giants, In: *Extreme Solar Systems, Santorini, June 25-29, 2007*

Sozzetti, A. 2005, Astrometric Methods and Instrumentation to Identify and Characterize Extrasolar Planets: A Review, *PASP*, **117**, 1021

Struve, O. 1952, Proposal for a project of high-precision stellar radial velocity work, *The Observatory*, **72**, 199

Udry, S., et al. 2006, The HARPS search for southern extra-solar planets. V. A 14 Earth-masses planet orbiting HD4308, *A&A*, **447**, 361

Udry, S., Fischer, D., & Queloz, D. 2007, A Decade of Radial-Velocity Discoveries in the Exoplanet Domain, In: *Protostars and Planets V*, B. Reipurth, D. Jewitt, & K. Keil, eds. (Tucson: University of Arizona Press)

Udry, S., et al. 2007, The HARPS search for southern extra-solar planets. XI. Super-Earths (5 and 8 Earth masses) in a 3-planet system, *A&A*, **469**, L43

van Eyken, J.C. et al. 2004, First Planet Confirmation with a Dispersed Fixed-Delay Interferometer, *ApJ*, **600**, L79

Vogt, S.S. et al. 2000, Six New Planets from the Keck Precision Velocity Survey, *ApJ*, **536**, 902

Vogt, S.S. et al. 2005, Five New Multicomponent Planetary Systems, *ApJ*, **632**, 638

Walker, G.A.H., et al. 1995, A search for Jupiter-mass companions to nearby stars, *Icarus*, **116**, 359

Wan, X.K., et al. 2006, A fiber feed system for a multiple object Doppler instrument at Sloan Telescope, In: *Proc. SPIE*, **6269**, 88

Wright, J. et al. 2007, Multiple-Planet Systems, In: *Extreme Solar Systems, Santorini, June 25-29, 2007*

Wu, Y. , Murray, N. W., & Ramsahi, J. M., 2007, Hot Jupiters in binary star systems, *ApJL*, submitted (astro-ph/07006.0732)

Zhao, B., & Ge, J. 2006, An optical spectrograph design for a new-generation multiple object Doppler instrument, In: *Proc. SPIE*, **6269**, 89

Zucker, S. & Mazeh, T. 2002, On the Mass-Period Correlation of the Extrasolar Planets, *ApJ*, **568**, L113

3 Detection of Extrasolar Planets by Gravitational Microlensing

David P. Bennett

Summary. Gravitational microlensing provides a unique window on the properties and prevalence of extrasolar planetary systems because of its ability to find low-mass planets at separations of a few AU. The early evidence from microlensing indicates that the most common types of exoplanet yet detected are the so-called super-Earth planets of 10 Earth-masses at a separation of a few AU from their host stars. The detection of two such planets indicates that roughly one third of stars have such planets in the separation range 1.5-4 AU, which is about an order of magnitude larger than the prevalence of gas-giant planets at these separations. We review the basic physics of the microlensing method, and show why this method allows the detection of Earth-mass planets at separations of 2-3 AU with ground-based observations. We explore the conditions that allow the detection of the planetary host stars and allow measurement of planetary orbital parameters. Finally, we show that a low-cost, space-based microlensing survey can provide a comprehensive statistical census of extrasolar planetary systems with sensitivity down to 0.1 Earth-masses at separations ranging from 0.5 AU to infinity.

3.1 Introduction

The gravitational microlensing method relies upon chance alignments between background source stars and foreground stars, which may host planet systems. These background source stars serve as sources of light that are used to probe the gravitational field of the foreground stars and any planets that they might host. The relative motion of the source star and lens system allows the light rays from the source to sample different paths through the gravitational field of the foreground system, and it is changing total gravitational lens magnification of the source star with time that provides the observable gravitational microlensing signal.

The microlensing method is unique among exoplanet detection methods in a number of respects:

1. The amplitude of planetary microlensing signals is large (typically $\gtrsim 10\%$) and is approximately independent of the planetary mass. Instead, the source-lens alignment necessary to give a detectable planetary signal depends on the planet-

star mass ratio, q, and so the probability of a detectable planetary signal scales as $\sim q$.
2. This scaling of the probability of planet detection with the mass ratio, q, is shallower than the sensitivity curves for other methods, so microlensing is more sensitive to low-mass planets than other methods that are sensitive to planetary mass. The sensitivity of the microlensing planet search method extends down to $0.1 M_{\oplus}$.
3. Microlensing is most sensitive to planets at orbital separations of 1.5-4 AU, which corresponds to the vicinity of the Einstein ring radius. This range of separations also corresponds to the "snow line" where planet formation is most efficient according to the leading core accretion model of planet formation. Thus, microlensing complements the Doppler radial velocity and transit methods, which are most sensitive to planets in very short period orbits.
4. Microlensing is the only planet detection method that is sensitive to old, free-floating planets, which have been ejected from the gravitational potential well of their parent stars through planet-planet scattering. Theory predicts that such planets may be quite common, and ground-based microlensing can detect free-floating gas giant planets, while a space-based survey is needed to detect free-floating terrestrial planets.
5. Since the microlensing method doesn't rely upon light from the host star in order to detect its planets, it can detect planets orbiting unseen stars. This can make it difficult to determine the properties of the host stars, but space-based follow-up observations can detect the host stars for most planets discovered by microlensing.
6. A space-based microlensing survey would provide a nearly complete statistical census of extrasolar planets with masses down to $0.1 M_{\oplus}$ at all separations ≥ 0.5 AU. This includes analogs of all the Solar System's planets, except for Mercury.

Gravitational microlensing differs from other extrasolar planet search techniques in a number of aspects. It is a purely gravitational method that doesn't rely upon on detecting photons from either the planet or its host star.

While most of the known exoplanets have been discovered with the Doppler radial velocity method, the early results from the microlensing method indicate that cool, super-Earth or sub-Neptune mass planets are more representative of typical extrasolar planets than any of the 200+ exoplanets discovered by radial velocities.

3.2 Gravitational Microlensing Theory

3.2.1 The Single Lens Case

The basic physics of gravitational lensing depends only a single input from General Relativity, the deflection angle, α, for a light ray passing a mass, M, with an impact

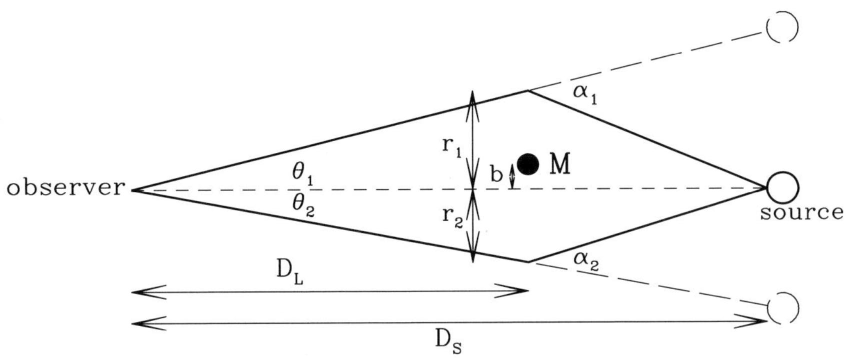

Fig. 3.1. The geometry of gravitational lens of mass, M, that is offset by a distance, b, from the line of sight to the source. The observer sees two images that are offset by angles, θ_1 and θ_2 from the line of sight to the source star.

parameter, r:

$$\alpha = \frac{4GM}{c^2 r} . \tag{3.1}$$

With the lens geometry shown in Fig. 3.1, we have

$$\alpha_i = \frac{4GM}{c^2(r_i - b)} = \frac{r_i D_S}{D_L(D_S - D_L)} , \tag{3.2}$$

in the small angle approximation. If the lens and source are perfectly aligned, the two images merge to form a ring of radius

$$R_E \equiv \theta_E D_L \equiv \sqrt{\frac{4GMD_L(D_S - D_L)}{c^2 \ D_S}} , \tag{3.3}$$

known as the Einstein ring radius. (θ_E is the angular Einstein radius.) We can now rewrite the single lens equation as

$$r_i = \frac{R_E^2}{r_i - b} , \tag{3.4}$$

and it has two solutions: $r_{+,-} = 0.5(b \pm \sqrt{b^2 + 4R_E^2})$. The lensed images are also magnified, and the magnification of a source of infinitesimal size can be computed using area elements obtained by differentiating eq. 3.4. This yields

$$A_{+,-} = \frac{1}{2}\left(\frac{u^2+2}{u\sqrt{u^2+4}} \pm 1\right) , \tag{3.5}$$

where $u \equiv b/R_E$ is the dimensionless lens-source separation. The total magnification of both images is given by

$$A = A_+ + A_- = \frac{u^2+2}{u\sqrt{u^2+4}} . \tag{3.6}$$

For a lens of $M = 1M_\odot$, that is half-way to a source in the Galactic center (at $D_S = 8\,\text{kpc}$), we find

$$R_E = 4.04\,\text{AU}\sqrt{\frac{M}{M_\odot}\frac{D_S}{8\,\text{kpc}}\,4x(1-x)}\ , \tag{3.7}$$

where $x = D_L/D_S$, so R_E is similar to the orbital radius of planets in our own Solar System. This also implies that $\theta_E \sim 1\,\text{mas}$. Since the image separation is of order $\sim \theta_E$, this implies that images will not generally be resolved with virtually all planned and future astronomical instruments (with a few exceptions (Delplancke et al., 2001)). On the other hand, if we assume a typical Galactic velocity of $v_\perp = 100\,\text{km/sec}$ for the relative velocity between the lens star and the line-of-sight to the source, then the typical Einstein radius crossing time for a lens in the Galactic disk and a bulge source is $t_E = R_E/v_\perp \sim 2\,\text{months}$. Thus, the main observational effect for lensing by stars within the Milky Way is the time varying magnification instead of the image separation, and this is why it is referred to as microlensing instead of lensing.

The microlensing light curve is generally described by eq. 3.6 with the lens-source separation given by

$$u = \sqrt{\left(\frac{t-t_0}{t_E}\right)^2 + u_0^2}\ , \tag{3.8}$$

assuming that the relative motion between the lens and the observer-source line-of-sight. Thus, a single-lens microlensing light curve is described by three parameters, the time of peak magnification, t_0, the Einstein radius crossing time or width, t_E, and the minimum separation, u_0, which determines the peak magnification. u_0 is the only parameter that affects the intrinsic light curve shape, as shown in Fig. 3.2, but only t_E constraints the physically interesting parameters of the event: the lens mass, M, the lens distance, D_L, and the relative velocity, $v_\perp$.

The first microlensing events were discovered in 1993 towards the Large Magellanic Cloud (LMC) by the MACHO Project (Alcock et al., 1993) and towards the Galactic bulge by the OGLE Collaboration. (Udalski et al., 1993). The early emphasis of microlensing surveys was the search for dark matter in the Milky Way's halo (Paczyński, 1986), but this issue has been largely resolved with the demonstration that the excess microlensing seen toward the LMC by the MACHO group (Alcock et al., 2000b; Bennett, 2005) requires at most 20% of the Milky Way's dark matter in the form of stellar mass objects, while the results of the EROS group (Tisserand et al., 2007) suggest that much of this microlensing excess may be caused by stars associated with the LMC itself (Sahu, 1994), perhaps in the LMC halo (Wu, 1994).

With the dark matter microlensing question mostly resolved, the prime focus of microlensing observations has shifted to the detection of extrasolar planets. Microlensing was first suggested as a method to find planets by Liebes (1964), but as Mao & Paczyński (1991) pointed out, this requires a consideration of multiple lens systems.

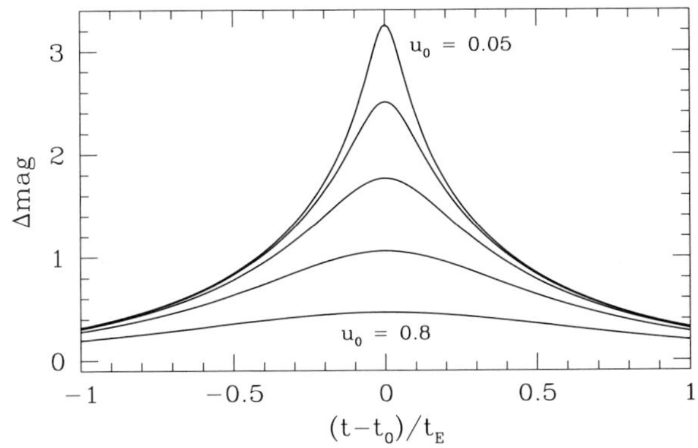

Fig. 3.2. Example microlensing light curves for a point source and a single lens that moves with a constant lens velocity with respect to the observer-source line-of-sight. Light curves with $u_0 = 0.05, 0.1, 0.2, 0.4$, and 0.8 are shown.

3.2.2 Multiple Lens Systems

The lens equation for a multiple lens system is a straight forward generalization of the single lens equation, eq. 3.4, but with more than one lens mass, we can no longer assume that the source, observer and lens system all lie in a single plane. However, as long as the distances to the lens and source (D_L and D_S) are much larger than the extent of the lens system, we can assume that the lens system resides a single distance, and define the Einstein radius of the total lens system mass using eq. 3.3. So, as before, we will rescale all the length variables with R_E.

Because we can no longer define a source-lens-observer plane, we must now define the lens and source positions in the 2-dimensional "lens-plane" perpendicular to the line-of-sight and projected to the distance of the lens (or equivalently, we can just use angular variables for the positions of the source and lenses on the sky). The double-lens system was first solved using two real coordinates for the lens plane (Schneider & Weiss, 1986), but the algebra is much simpler if we describe the lens plane with complex coordinates following Bourassa et al. (1973) and Rhie (1997). The generalization of eq. 3.4 is

$$w = z - \sum_i \frac{\epsilon_i}{\bar{z} - \bar{x}_i} , \tag{3.9}$$

where w and z are the complex positions of the source and image, respectively, and x_i are the complex positions of the lens masses. The individual lens masses are represented by ϵ_i, which is the mass fraction of the ith lens mass, so that $\sum_i \epsilon_i = 1$. The appearance of the complex conjugates in the denominator in the sum on the right side of eq. 3.9 is simply a reflection of the fact that the lens deflection is the in the direction from the source to the lens with a magnitude of the inverse of that distance. With real coordinates, we would express this as the vector difference of the positions divided by this vector squared, but with complex coordinates, we can divide through by this vector leaving only its complex conjugate in the denominator.

If we knew the position of the images, z, in eq. 3.9, then it would be trivial to solve for the position of the source. But this is the inverse of the problem that we will usually want to solve, which is to find the positions of the images based on a known position for the source. However, the solution of this "inverse" problem is the basis of the brute-force, ray-shooting method (Schneider & Weiss, 1987) for solving eq. 3.9. This method involves taking a large grid of points in the "image plane" and propagating them back to the source plane using eq. 3.9. This method has the advantage that it can handle very complicated lens mass distributions, but it is usually not the method of choice for the analysis of microlensing events.

The most successful method for calculating multi-lens microlensing light curves (Bennett & Rhie, 1996) involves solving eq. 3.9 for the positions of the point-source images and invoking the ray-shooting method only in the vicinity of images that are affected by finite-source size effects. For the majority of the light curve, the finite-source calculations are not needed, and we can use the point source magnification formula. This formula can be derived from the Jacobian determinant of the lens equation (and its complex conjugate):

$$J = \frac{\partial w}{\partial z}\frac{\partial \bar{w}}{\partial \bar{z}} - \frac{\partial w}{\partial \bar{z}}\frac{\partial \bar{w}}{\partial z} = 1 - \left|\frac{\partial w}{\partial \bar{z}}\right|^2 , \tag{3.10}$$

where

$$\frac{\partial w}{\partial \bar{z}} = \sum_i \frac{\epsilon_i}{(\bar{z} - \bar{x}_i)^2} . \tag{3.11}$$

Because eq. 3.10 gives the Jacobian determinant of the inverse mapping from the image plane to the source plane, the magnification of each image is given by

$$A = \frac{1}{|J|} , \tag{3.12}$$

evaluated at the position of each image.

The solution of the lens equation, 3.9, is non-trivial. For the case of two lens masses, this equation can be embedded into a fifth order polynomial equation in z, which can be solved numerically. This equation has either 3 or 5 solutions (Rhie, 1997) that correspond to solutions of eq. 3.9, which means that a double lens system must have either 3 or 5 images depending on the configuration of the lens system and the location of the source. For the triple lens case (which is relevant for at least one planetary microlensing event), the lens equation can be embedded in a rather complicated tenth order polynomial that has 4, 6, 8, or 10 solutions that correspond to physical images (Rhie, 2002). This tenth order polynomial equation can be solved numerically, although it may require extended precision numerical calculations in order to avoid serious round-off errors (Bennett et al., in preparation). The case of 4 lens masses, has also been investigated (Rhie, 2001), but the lens equation has not been converted to a polynomial.

The most important feature of lensing by multiple masses occurs at the locations where $J = 0$. From eq. 3.12, this implies infinite magnification for a point source. (The magnification is always finite for the realistic case of a source of finite angular size.) For a single lens, $J = 0$ only occurs at a single point in the source plane, the location of the lens mass, but for lens systems with more than one mass, there are a set of one or more closed curves with $J = 0$, known as critical curves. The source positions corresponding to the critical curves are obtained by applying the lens equation, 3.9, and they are referred to as caustic curves. When the source passes to the interior of a caustic curve, two new images are created, and it is these new images that have infinite magnification for the (unphysical) case of a point source. The shape of the light curve of a (point) source crossing a caustic has a characteristic form:

$$A = \frac{F_c \Theta(x - x_c)}{\sqrt{x - x_c}} + A_{nc} , \tag{3.13}$$

where F_c gives the amplitude of the caustic, A_{nc} gives the magnification of the images that are not associated with the caustic and $\Theta(x) = 1$ for $x \geq 0$ and $\Theta(x) = 0$ otherwise. x is the distance perpendicular to the direction of the caustic curve, and x_c is the location of the caustic curve. Eq. 3.13 is a good approximation to the magnification for a point source when the curvature of the caustic curve can be neglected. Note, that the singularity in eq. 3.13 is weak enough so that the integral of this formula will yield a finite magnification for a finite size source star.

Caustic crossings that follow the form of eq. 3.13 are often referred to as *fold* caustic crossings, and they have the feature that there is essentially no warning that the caustic crossing is imminent when the caustic curve is approached from the outside (*i.e.* $x < x_c$). This is because the magnification pattern for a fold caustic extends only to the interior of the caustic since it involves the magnification of images that only exist inside the caustic curve. However, each caustic curve also has at least three sharp pointy features, known as cusps, and the magnification pattern extends outward from the cusps on a caustic curve. The magnification scales as the inverse of the distance to the cusp, just as in the single lens case, eq. 3.6.

The path of the source with respect to the caustic curves provides the basic characteristics of a multiple lens microlensing light curve. Multiple lens light curves frequently have features which match the expected $A \sim \Theta(x)x^{-1/2}$ shape of a caustic crossing or the $A \sim r^{-1}$ shape of a cusp approach. But, there are additional complications, as the strength of a caustic crossing (F_c in eq. 3.13) can vary and the angular size of the source star can sometimes be larger than the entire caustic curve for a planetary microlensing event.

3.3 Planetary Microlensing Events

Planetary microlensing events are a subset of multiple lens events where the mass ratio is quite small. Planetary events have light curves that appear quite similar to the single lens light curves shown in Fig. 3.2, but for a brief period of time they deviate from the single lens form and display the characteristics of a binary lens light curve. We will define a requirement on the mass ratio $q \equiv \epsilon_2/\epsilon_1 < 0.03$ to separate planetary microlensing events from stellar binary events following Bond et al. (2004), because $q \approx 0.03$ is the approximate location of the "brown dwarf desert" that appears to separate stellar from planetary secondaries. We will also initially only consider events with only one detectable planet, as these represent the majority of planetary microlensing events and this will simplify the discussion.

The caustic structure of a binary lens system is determined by the mass ratio and the separation of the lenses (Schneider & Weiss, 1986). For a separation $d \ll 1$ (in units of R_E), there are three caustics, two triangular caustics with 3 cusps each and a caustic close to the center of mass which has 4 cusps. These merge into a single caustic with 5 cusps at $d \sim 1$, which splits into two caustic curves with 4 cusps each for $d \gg 1$. For small values of the mass ratio, q, the division between these regimes occurs near $d \approx 1$, so most events have multiple caustic curves. As shown in Fig. 3.3.

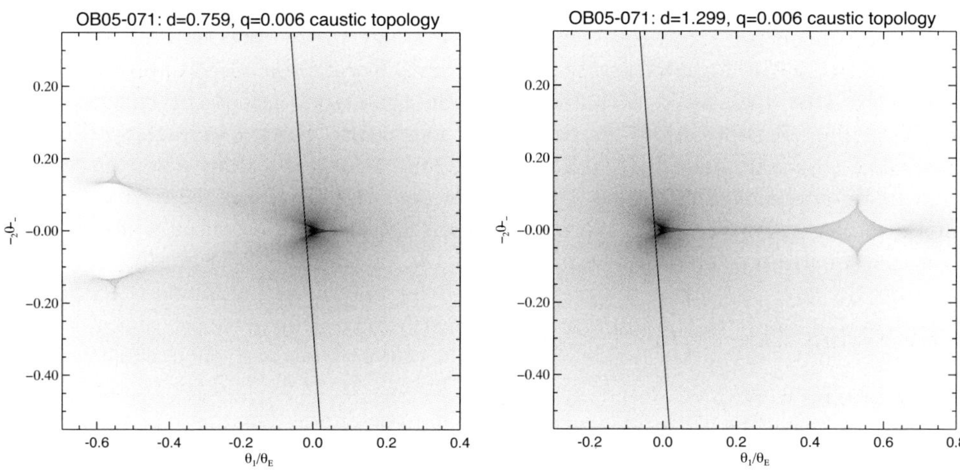

Fig. 3.3. Grey-scale magnification patterns for two models of the planetary microlensing event OGLE-2005-BLG-71. The darkness of the image is proportional to the logarithm of the magnification. The nearly vertical lines in each panel indicate the path of the source star for each model. The source passes close to the central, or stellar, caustic, and as discussed in Sect. 3.4, the magnification pattern in the vicinity of the central caustic is similar for planetary systems related by $d \leftrightarrow 1/d$, The magnification patterns for the planetary caustics (on the outside of each panel) are clearly very different and easy to distinguish. (These images are provide courtesy of Daniel Kubas.)

3.3.1 Planetary Caustic Perturbations

There are two classes of caustic curves in planetary microlensing events: planetary caustics and the stellar or central caustic. The planetary caustics result when one of the two light rays in Fig. 3.1 passes close to the planet and is deflected by the planetary gravitational field of the planet. The locations of the caustics are given by

$$s_c = d - 1/d \ . \tag{3.14}$$

If we see a planetary deviation at a point where the best fit single lens light curve predicts a magnification, A, we can find the corresponding u value by inverting eq. 3.6 to get $u_c = s_c$, and then solve for d by inverting eq. 3.14 to give

$$d = \frac{1}{2}\left(u_c \pm \sqrt{u_c^2 + 4}\right) . \tag{3.15}$$

The two solutions to eq. 3.15 for a given u_c value are referred to as major and minor image perturbations. They were first studied in detail by Gould & Loeb (1992) who showed that many features of the planetary caustics and their magnification pattern could be explained by the simpler Chang-Refsdal lens system (Chang & Refsdal, 1979, 1984). Several important features of the planetary caustic light curve perturbations can be seen in Fig. 3.3. The left side of the left panel of this figure shows the magnification pattern of the two roughly triangular minor image caustic, which are generated by a planet with $d < 1$. As with all caustics, there is excess magnification in the interior of the caustic curve, as well as extending outward from the cusps. But there is also a very pronounced magnification deficit in between the two minor image caustics, where the magnification is substantially below the single lens magnification. In contrast, the magnification pattern for the major image caustic (shown on the right side of the right panel in Fig. 3.3) is predominantly positive, with only small magnification deficits very close to the caustic curve, away from the 4 cusps.

3.3.2 Stellar Caustic Perturbations

It was originally suspected that the planetary caustic perturbations would be the best way to detect planetary signals in microlensing events, but Griest & Safizadeh (1998) argued that there were a number of advantages to searching for planetary light curve perturbations due to the stellar caustic. They showed that the planet detection efficiency for each high magnification event was substantially higher than for events of more modest magnification. While the higher planet detection efficiency for higher magnification events was seen in previous work (Bolatto & Falco, 1994; Bennett & Rhie, 1996), Griest & Safizadeh (1998) emphasized that this effect is quite dramatic and that this fact could be used to increase the observational planet detection efficiency. In the same year that the Griest & Safizadeh paper was published, the MPS and MOA Collaborations demonstrated this method with observations of the MACHO-98-BLG-35 event. The subsequent analysis showed (Rhie et al., 2000) that the lens star for this event did not have any Jupiter-mass planets with a projected separation of 0.6-8 AU.

The high planet detection efficiency for high magnification events is particularly useful when a large number of microlensing events are discovered by the microlensing survey groups. This is the current situation, as the OGLE-3 and MOA-2 surveys combine to detect > 700 microlensing events in progress toward the central regions of the Milky Way between February and October of each year. Relatively sparse monitoring of events (*i.e.* one or two observations per day) is required to predict most high magnification events in advance, and this allows observing resources to be focused on events with a high planet detection efficiency.

One important consequence of the high planet detection efficiency for high magnification events is that the chances of detecting multiple planets in such events are greatly enhanced (Gaudi et al., 1998). Indeed, the first multi-planet system discovered by microlensing is shown below in Sect. 3.5.2. There is, however, a potential downside to this higher sensitivity to multiple lens masses. The signals for all the detectable lens masses will be concentrated in the very high magnification part of the light curve, and this could make it difficult to work out the details of multiple planet systems that are detected in microlensing events. Thus, the development of efficient light curve modeling methods for lens systems with three or more masses is an important active area of current research.

A final advantage of high magnification events is that they allow planet detection with relatively faint source stars. This makes it much easier to detect the planetary host star with follow-up observations (Bennett et al., 2007a) as explained in Sect. 3.4.1.

3.3.3 Finite Source Effects

Microlensing is arguably the exoplanet search technique that is most sensitive to low-mass planets, and the lower limit in sensitivity of the microlensing method is set by the finite angular size of the source stars. Roughly speaking, when the angular radius of the planetary Einstein ring, $\sqrt{\epsilon_p}\theta_E$, is much smaller than the source star angular radius, θ_*, we expect that planetary signal to be washed out. But this is only a crude, order-of-magnitude estimate, and a full finite source solution to the lens equation, 3.9, is required to determine the precise limits on the microlensing planet detection method set by the finite angular size of the sources.

Full finite source planetary microlensing light curves were first calculated by Bennett & Rhie (1996) using the methods described in Sect. 3.2.2. Results of these calculations are reproduced in Figs. 3.4 and 3.5. Figure 3.4 shows a series of planetary light curves with planetary mass fractions of $\epsilon = 10^{-4}$ and 10^{-5}. For a typical lens star mass of $\sim 0.3M_\odot$, these correspond to $1M_\oplus$ and $10M_\oplus$, respectively. The finite source light curves are characterized by the source star radius in Einstein ring units: $\rho \equiv \theta_*/\theta_E$. The ρ values shown in Fig. 3.4 are 0.003, 0.006, 0.013, and 0.03, and these span the expected range of ρ for a low mass planetary host star in the Galactic bulge with a source star ranging in radius from $1R_\odot$ to $10R_\odot$, which is a typical radius for a "red clump" K-giant in the bulge. A number of the planet detections to date actually have ρ values in the 0.4-1$\times 10^{-3}$ range because the lens stars reside in the bulge and have a larger than average mass. (Both of these imply a larger θ_E.)

Several general trends are apparent from Fig. 3.4. First, the planetary deviations are easily detectable for $\rho = 0.003$, but the signals are much weaker for $\rho = 0.03$. This implies that $1M_\oplus$ planets are easily detected with main sequence source stars, but planets of $10M_\oplus$ are close to the lower limit of detectability for giant source stars. Another notable feature of these light curves is that the planetary signals with $d = 0.8$ are more easily washed out by the finite source effects than those with

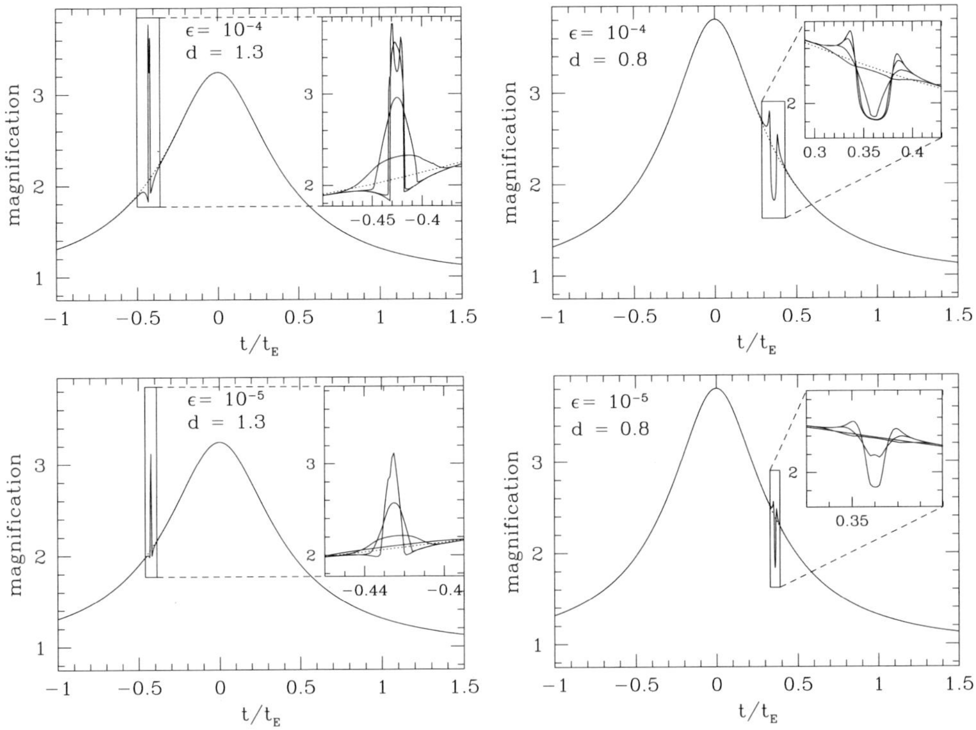

Fig. 3.4. Microlensing lightcurves which show planetary deviations are plotted for a mass ratio of $\epsilon = 10^{-4}$ & 10^{-5} and separations of $d = 1.3$ & 0.8. The main plots are for a stellar radius of $\rho = 0.003$ while the insets show light curves for radii of 0.006, 0.013, and 0.03 as well. The dashed curves are the unperturbed single lens lightcurves. For each of these lightcurves, the source trajectory is at an angle of $\sin^{-1} 0.6$ with respect to the star-planet axis. The impact parameter $u_0 = 0.27$ for the $d = 0.8$ plots and $u_0 = 0.32$ for the $d = 1.3$ plots.

$d = 1.3$. This is a consequence of the large magnification deficit seen between the two planetary caustics for a minor image perturbation, as shown in the left hand panel of Fig. 3.3. When a large finite source effectively averages over the vicinity of a minor image planetary caustic, the positive and negative deviations effectively cancel each other out (Bennett & Rhie, 1996; Gould & Gaucherel, 1997). In contrast, the major image planetary caustic magnification deviation pattern is mostly positive, so the finite source effect merely smoothes it out.

Figure 3.5 shows how the planet detection probability varies as a function of d for the same ϵ and ρ values used for Fig. 3.4. The greater tolerance of deviations with $d > 1$ to finite source effects is apparent for $\epsilon = 10^{-4}$, $\rho = 0.03$ and $\epsilon = 10^{-5}$, $\rho = 0.013$. The behavior of Fig. 3.5 near $d = 1$ is even more interesting. For $\epsilon = 10^{-4}$ and $\rho = 0.003$ or 0.006, the detection probability reaches a maximum at $d \approx 1$, but for $\rho = 0.013$, the probability has a local minimum at $d = 1$, and for $\rho = 0.03$ or any of the ρ values with $\epsilon = 10^{-5}$, the detection probability $= 0$ at $d = 1$. This

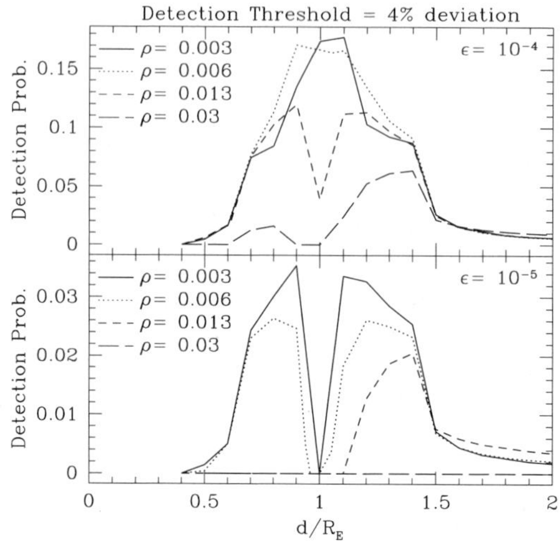

Fig. 3.5. The planetary deviation detection probability is plotted for different values of the planetary mass ratio, ϵ, and the stellar radii, ρ (in units of R_E. A planet is considered to be "detected" if the lightcurve deviates from the standard point lens lightcurve by more than 4% for a duration of more than $t_E/200$.

is due to the fact that for $d \approx 1$ the planetary and stellar caustics merge for form a relatively large single caustic that is extended along the lens axis. This caustic is large, but relatively weak, and it has associated positive and negative deviation regions that tend to cancel when averaged over by a moderately large finite source.

However, some features of Fig. 3.5 are dependent on the somewhat arbitrary choice of the event detection threshold, and the sensitivity of a real observing strategy can differ from this. In fact, the planetary deviation detected in event OGLE-2005-BLG-169 (Gould et al., 2006) would not have passed the selection criteria for Fig. 3.5, but the planet is nevertheless detected with a strong signal. The reason for this is that it was identified as a very high magnification event with a very high sensitivity to planets, and for this reason it was observed much more frequently than most events with potential planetary signals. The additional observations provided enough additional signal to allow the definitive detection of a relatively low-amplitude signal.

As a practical matter, finite source effects imply a lower planetary mass limit of $M_p \gtrsim 5M_\oplus$ for giant source stars in the bulge, and a limit of $M_p \gtrsim 0.05M_\oplus$ for bulge main sequence stars. Thus, searches for terrestrial exoplanets must focus on main sequence source stars.

3.4 Planetary Parameters from Microlensing Events

The determination of the properties of the lens systems that are detected in microlensing events is often a serious challenge. The simple form of the microlensing light curves shown in Fig. 3.2 is an advantage when trying to identify microlensing events, but as I mentioned in Sect. 3.2.1, in a single lens event it can also be a drawback when trying to interpret observed microlensing events. For most single lens events, it is only the t_E parameter that constrains the physically interesting parameters of the event: the lens mass, M, the lens distance, D_L, and the relative velocity, $v_\perp$. The single lens parameters u_0 and t_0 don't constrain lens system parameters that are of much interest.

In addition to the parameters needed to describe a single lens event, a planetary microlensing event must have three additional binary lens parameters: the planetary mass ratio, $q = \epsilon/(1-\epsilon)$, the star-planet separation, d, (which is in units of R_E), and the angle between the star-planet axis and the trajectory of the source with respect to the lens system, θ. So, two of these new parameters, q and d, directly constrain planetary parameters of interest, although d is normalized to R_E, which may not be known. Most planetary light curves, at least those for low-mass planets, also have caustic crossings or a close approach to a cusp that reveal light curve features due to the finite size of the source star. This enables the source radius crossing time, t_*, to be measured.

The determination of the star-planet separation and the planetary mass fraction is usually quite straightforward from the microlensing light curve. For events at moderate magnification, due to the planetary caustic, the separation can be determined by the magnification predicted by the single lens model that describes the event outside the region of the planetary deviation following eq. 3.15. This still leaves an ambiguity between the $d < 1$ and $d > 1$ solutions, but this is easily resolved by the drastically different magnification patterns in the vicinity of major image and minor image caustics, as shown in Fig. 3.3. The planetary mass fraction, q, can generally be determined by the duration of the planetary perturbation. In some cases, if the time scale of the deviation is similar to or smaller than t_*, both q and t_* determine the deviation time scale, but good light curve coverage with moderately precise photometry allow both q and t_* to be determined (Gaudi & Gould, 1997).

The situation is somewhat different for high magnification, stellar caustic deviation events. Dominik (1999) pointed out an approximate degeneracy in the properties of the stellar caustic under the transformation $d \to 1/d$, which means that there may be a $d \leftrightarrow 1/d$ ambiguity in the modeling of stellar caustic planetary events. This is apparent from the magnification patterns shown in Fig. 3.3. For a source trajectory nearly parallel to the lens axis or for $d \sim 1$, this degeneracy breaks down, so the ambiguity disappears. With precise photometry it is usually possible to distinguish between the $d < 1$ and $d > 1$ solutions, and this has been the case for all events observed to date.

3.4.1 Angular Einstein Radius

A large fraction of planetary light curve deviations exhibit finite source effects that allow the source radius crossing time, t_*, to be measured. This is the case for most detectable events with a planetary mass, $M_p \lesssim 10M_\oplus$, but for gas giant planets of $M_p \gtrsim 300 M_\oplus$, it is possible to detect a planetary deviation without the source crossing a caustic or coming close enough to a cusp to display finite source effects. So, t_* is measurable for most, but not all, planetary microlensing events.

When t_* is measured, it is possible to place an additional constraint, as long as the angular radius of the source star, θ_*, can be estimated because the angular Einstein radius is given by

$$\theta_E = \frac{\theta_* t_E}{t_*} . \tag{3.16}$$

The angular radius of the source star can be measured if the brightness and color of the source are known with the use of empirical color-angular radius relations (van Belle, 1999; Kervella et al., 2004). In the crowded fields where microlensing events are observed, the most reliable measure of the source star brightness and color comes from the light curve models, which include the source brightness as a model parameter. So, although it is sufficient to measure the detailed light curve shape in a single passband, it is important to obtain a few measurements during the microlensing event in at least one additional passband so that the light curve fit will also reveal the color of the source. It is also important to estimate the extinction towards the source. With measurements in only two colors, such as V and I, the extinction can be estimated by comparison to the red clump giant stars within an arc minute or two of the target star (Yoo et al., 2004). While this does not yield a precise measure of the extinction to the source, note that an error in the extinction to the source will affect both the estimated intrinsic brightness and color of the source. Fortunately, the extinction-induced brightness and color errors have the opposite effect on the estimated source star radius. This partial cancelation implies that the estimated θ_* value is not very sensitive to to the uncertainty in the extinction.

A more precise estimate of θ_* can be obtained with observations during the microlensing event in more than two passbands, particularly if one of the passbands is in the infrared because the optical-IR color-radius relations are much more precise than the optical ones (Kervella et al., 2004) and because extinction is much lower in the IR than in the optical. Observations in 3 or more colors also allow an estimate of extinction that doesn't depend on the nearby clump giants, with the use of empirical color-color relations (Bessell & Brett, 1998).

When the angular Einstein radius is measured, we have the following relation

$$M_L = \frac{c^2}{4G} \theta_E^2 \frac{D_S D_L}{D_S - D_L} , \tag{3.17}$$

which can be considered to be a mass-distance relation because D_S is generally known (approximately) from the brightness and color of the source. (The high density of stars in the Galactic bulge means that the source is almost always a bulge star.) Eq. 3.17 provides a one-parameter family of solutions to the microlensing event, and this can be converted to a measurement of the planetary host star properties with one additional piece of information. Since the brightness of the source star can be determined by the light curve fit, the brightness of the lens star can be

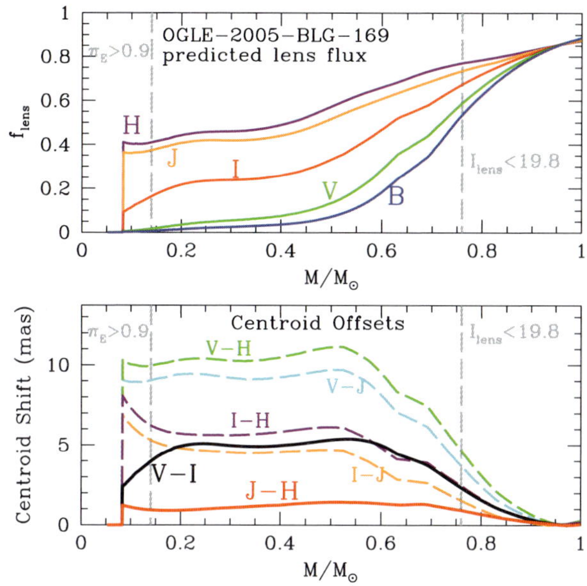

Fig. 3.6. The predicted fractional brightness, $f_{\rm lens} = F_L/(F_S + F_L)$, of the OGLE-2003-BLG-169 lens is plotted in the top panel as a function of mass in the $BVIJH$ passbands. The predicted offsets of the centroids of the blended source+lens images in different passbands are shown in the bottom panel, assuming that the images are taken 2.4 years after peak magnification.

determined with an image that has sufficient angular resolution to resolve the source and lens stars from the unrelated stars in the field. This generally requires space-based imaging with the Hubble Space Telescope (HST), or possibly ground-based adaptive optics imaging because of the extreme crowding in the Galactic bulge fields where microlensing events are most easily found. (The lens-source relative proper motion has typical value $\mu_{\rm rel} \sim 5\,{\rm mas/yr}$, so the lens and source are not typically resolved from each other until a decade or more after the event.) If the combined lens-plus-source image is significantly brighter than the brightness of the source from the microlensing fit, then the difference determines the brightness of the lens. This then allows the mass of the planetary host (lens) star to be determined using a main sequence star mass-luminosity relation (Bennett et al., 2007a).

The top panel of Fig. 3.6 shows the predicted brightness of the lens for the OGLE-2005-BLG-169 event in the $BVIJH$ passbands. This indicates that the lens star will easily be detected if it is a main sequence star, since even a $0.08M_\odot$ lens star will contribute $\gtrsim 40\%$ of the H-band flux and $\gtrsim 10\%$ of the I-band flux. This case is more favorable than most because of a relatively large θ_E value, but in most cases, the lens star will be detectable in the H-band unless it is a late M-dwarf located in the bulge. However, for Galactic disk lenses at a certain range of distances (corresponding to $0.2M_\odot \lesssim M \lesssim 0.4$ for OGLE-2005-BLG-169 in the IJH-bands)

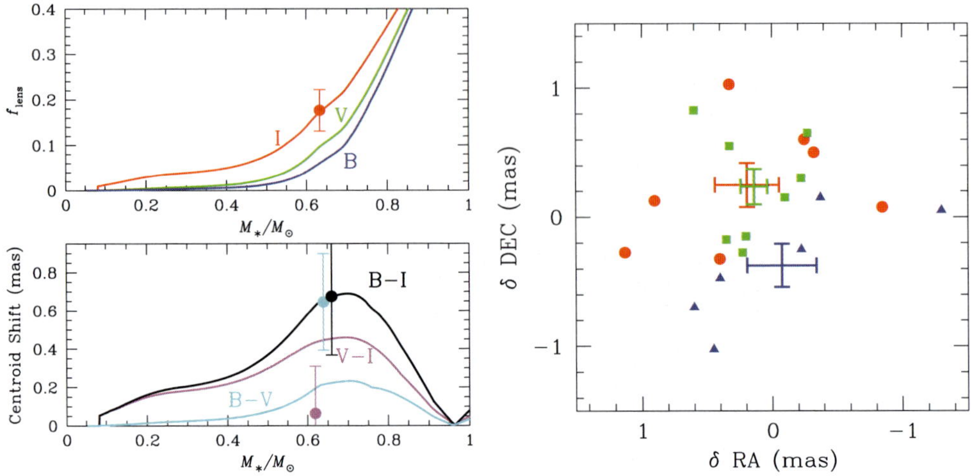

Fig. 3.7. The top-left panel shows the fraction of the source+lens flux for event OGLE-2003-BLG-235/MOA-2003-BLG-53 that is predicted to come from the lens in the HST-I, V, and B passbands as a function of lens mass. The bottom-left panel shows the predicted color-dependent centroid shifts as a function of mass for 1.78 years of relative proper motion at $\mu_{\rm rel} = 3.3\,{\rm mas/yr}$. The measured values of $f_{\rm lens}$ in the I-band and the color dependent centroid shifts and error bars are indicated with their error bars. These are plotted at an arbitrary value for the stellar mass (M_*). The centroids of the source+lens star blended images in the individual HST/ACS/HRC images are shown in the right panel as red circles (I), green squares (V), and blue triangles (B). The crossed error bars are the average centroid in each passband.

the mass-distance relation, eq. 3.17, combines with the mass-luminosity relation to yield a nearly flat mass-brightness relation for the planetary host star. In these cases, it is useful to have images in shorter wavelength bands, such as V and B because this cancelation does generally not occur in the optical and infrared passbands for the same range of lens star masses.

High resolution images in multiple colors also allow an independent method for estimating the lens star brightness, as shown in the bottom panel of Fig. 3.6 and the bottom-left panel of Fig. 3.7. Because the lens and source stars usually have different colors, the centroid of the blended source+lens image will usually be color dependent. So, an additional constraint on the lens star is obtained by measuring the centroid offset between the centroids of the blended source+lens in different passbands. As indicated in Fig. 3.7, this effect was marginally detected for the first planet detected by microlensing (Bennett et al., 2006) with HST images taken only 1.8 years after peak magnification. Also, because this color dependent centroid shift depends on the relative lens-source proper motion, $\mu_{\rm rel}$, it can be used to help determine θ_E for planetary events with no finite source effects, and hence, no measurement of t_*.

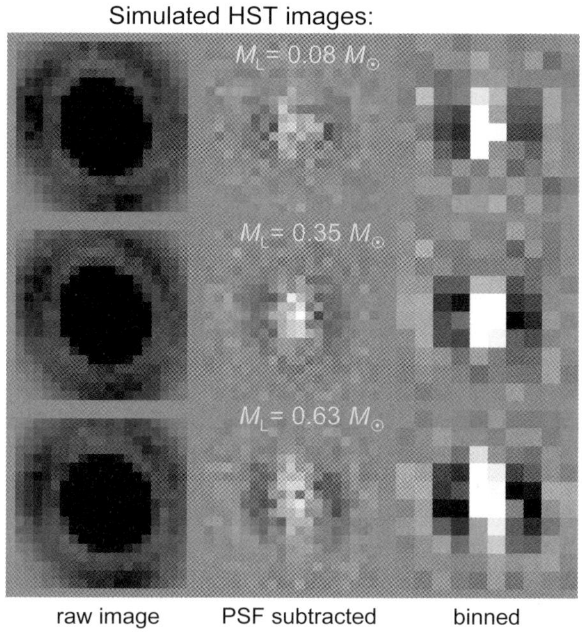

Fig. 3.8. Simulated image stacks of multiple dithered exposures of the OGLE-2005-BLG-169 source and lens star 2.4 years after peak magnification using the HST/ACS High Resolution Camera (HRC) in the F814W filter band. The top row of images assumed a host star mass of $M_* = 0.08 M_\odot$, the middle row assumes $M_* = 0.35 M_\odot$, and the bottom row assumes $M_* = 0.63 M_\odot$. In each row, the image on the left shows the raw image stack sampled at one half the native HRC (28 mas) pixel size. The central column shows the residuals after subtraction of the best fit PSF model, showing the blended image elongation along the x-axis due to the lens-source separation. The right hand column shows these residuals rebinned to the 28 mas native pixel scale.

The stable point-spread function (PSF) of space-based telescopes, such as HST, allows the measurement of the image elongation due to the growing separation of the lens and source stars after the microlensing event. Simulations of this effect for the OGLE-2005-BLG-169 event are shown in Fig. 3.8 for three different cases: $M_L = 0.08 M_\odot$, $M_L = 0.35 M_\odot$, and $M_L = 0.63 M_\odot$. This event has a higher relative proper motion than most events, but this simulation assumes images taken only 2.4 years after peak magnification, so for other events, it may be necessary to obtain the follow-up space-based images ~ 4 years after peak magnification.

When $\mu_{\rm rel}$ can be measured from image elongation and/or the color dependent centroid shift, then the angular Einstein radius can be determined via

$$\theta_E = \mu_{\rm rel} t_E \ , \tag{3.18}$$

so that mass-distance relation can be determined even when t_* cannot be measured.

3.4.2 Microlensing Parallax

Another method to "solve" a microlensing event involves the microlensing parallax effect. This refers to measurements of $\tilde{r}_e = R_E D_S/(D_S - D_L)$, the Einstein radius projected to the position of the Solar System. $\tilde{r}_e$ can be measured with the help of observations of microlensing events by observers at different locations. Because the Einstein radius is typically of order $R_E \sim 1\,\text{AU}$, the observers must generally be separated by a distance of $\sim 1\,\text{AU}$. The conceptually simplest way to do this is to observe an event simultaneously with a satellite in a heliocentric orbit, (Refsdal, 1966; Gould, 1992), as has recently been done with Spitzer (Dong et al., 2007). However, it is much more common to use the orbital motion of the Earth to measure the microlensing parallax effect (Alcock et al., 1995; Mao, 1999; Smith et al., 2002; Bennett et al., 2002; Mao et al., 2002), but for events of very high magnification it is possible to measure this effect with observations from different observatories on Earth (Gould, 1997), as has recently been done by Gould et al.(2007, in preparation).

A potential complication with this method is that the orbital motion of the source star can mimic the effect of the orbital motion of the Earth, but if the signal is strong, it is generally possible to detect the characteristic features of the Earth's orbit (Poindexter et al., 2005). Another potential complication is that for events with $t_E \ll 1\,\text{yr}$, it is often possible to measure only a single component of the two-dimensional $\tilde{\mathbf{r}}_\mathbf{e}$ vector (Smith et al., 2003). But, for events with detectable lens stars, the two-dimensional relative proper motion, $\boldsymbol{\mu}_{\text{rel}}$, can be measured and it is possible to determine the full $\tilde{\mathbf{r}}_\mathbf{e}$ vector because $\tilde{\mathbf{r}}_\mathbf{e} \parallel \boldsymbol{\mu}_{\text{rel}}$.

When $\tilde{r}_e$ and θ_E are both measured, the lens system mass is given by

$$M_L = \frac{c^2}{4G}\tilde{r}_e\theta_E \; . \tag{3.19}$$

This method has been used to determine the lens mass for a binary star lens system towards the Galactic bulge (An et al., 2002) and a low-mass M-dwarf lens towards the Large Magellanic Cloud (Gould, Bennett, & Alves, 2004). The first use of this method in a planetary microlensing event is the case of the double planet event OGLE-2006-BLG-109, to be published later this year (Gaudi et al. 2007, in preparation; Bennett et al. 2007, in preparation).

3.4.3 Planetary Orbits

The final property of a planetary system that can be measured is the orbital motion of the planet with respect to the star. This is a lower order effect than microlensing parallax because we see the effects of both the planet and the star in the light curve. So, we are sensitive to the relative velocity between the star and planet, whereas the velocity of the Earth around the Sun cannot be separated from the lens-source relative velocity. However, the time scale of the planetary deviation is generally only a small fraction of the microlensing light curve, and this limits the amount of time over which we can detect the orbital motion effects. Also, the typical orbital period of a planet detected by microlensing is $\sim 10\,$yrs, so the orbital velocities are generally lower than that of the Earth around the Sun.

For a planetary deviation of duration Δt and an orbital period, P, the orbital motion during the planetary deviation causes a shift in the planetary lens position with respect to the source of order

$$\Delta u \approx \Delta t \frac{2\pi}{P} \approx 0.002 - 0.02 \ , \tag{3.20}$$

assuming a planetary deviation duration of $1 - 10\,$days. In order to determine whether eq. 3.20 indicates that the effect of orbital motion is detectable, we need to know what value of Δu is measurable. One thing that limits our resolution in Δu is the finite angular size of the source star. The typical angular size for a bulge main sequence source is $\theta_* \sim 0.5\,\mu$as, and a typical angular Einstein radius for a bulge event is $\theta_E \sim 0.5\,$mas, so the source radius is typically of order $\rho = \theta_*/\theta_E \sim 0.001$. So, if we can detect Δu as small as 0.1ρ, then we could be sensitive to $\Delta u \sim 10^{-4}$.

In practice, it can be difficult to do this well in the measurement of orbital effects because changes in other model parameters can often compensate for the change in Δu due to orbital motion. In order to retain a constraint on the orbital motion, it is generally necessary to have a relatively complicated planetary deviation with more than a single caustic crossing or cusp passage that is well sampled by the data. Finally, for events with relatively long planetary signals, the orbital acceleration can be as large as $\Delta u \approx (\Delta t 2\pi/P)^2 \approx 4 \times 10^{-4}$. So, with a very well sampled planetary deviation it is also possible to measure the orbital acceleration, as well as the velocity.

3.5 Observational Programs

There are a variety of different observing programs that contribute to the detection of planets via gravitational microlensing. The most basic requirement is to be able to identify microlensing events, as was first done by the MACHO Collaboration towards the LMC (Alcock et al., 1993) and OGLE group toward the Galactic bulge (Udalski et al., 1993). Because microlensing observing programs do not yet have the resources to observe $\gtrsim 10\,$square degrees of the Galactic bulge several times per hour, it has been necessary to follow a strategy first suggested by Gould & Loeb (1992). Stellar microlensing events must be identified in progress, and then followed with a global network of telescopes on an ~hourly time scale. The MACHO (Alcock et al., 1994, 1996) and OGLE (Udalski et al., 1994) groups developed real-time microlensing detection systems within a year after the first microlensing events were discovered, and this led to the first spectroscopic confirmation of a microlensing event (Benetti et al., 1995). The MOA group began real time detections in 2000 (Bond et al., 2001) and was the first group to employ real time event detection with the more advanced difference imaging photometry method (Bond et al., 2002a).

The first microlensing follow-up projects were the Probing Lensing Anomalies NETwork or PLANET group (Albrow et al., 1998) and the Global Microlensing

Alert Network, or GMAN, (Pratt et al., 1995), which both began taking data in 1995. The PLANET team followed the Gould & Loeb (1992) strategy, but the GMAN group focused more on non-planetary microlensing. A second follow-up group focused on exoplanets, the Microlensing Planet Search (MPS) collaboration began in 1997 (Rhie et al., 1999), but MPS merged with PLANET in 2004. The final microlensing follow-up group is the Microlensing Follow-up Network or MicroFUN (Yoo et al., 2004), which began observations in 2003. MicroFUN does not follow the Gould & Loeb (1992) strategy, but instead focuses on high magnification microlensing events as suggested by Griest & Safizadeh (1998).

3.5.1 Early Observational Results

The most definitive of the early planetary microlensing observational results involved limits on the presence of planets based on the lack of detection of planetary signals. The MPS and MOA groups reported the first planetary limits from a high magnification event (Rhie et al., 2000). This was the first demonstration of sensitivity to Earth-mass planets by any method, except for pulsar timing (Wolszczan & Frail, 1992). The PLANET group followed with limits from a lower magnification event (Albrow et al., 2000b), and then a systematic analysis of five years worth of null detections (Albrow et al., 2001; Gaudi et al., 2002). They found that less than 33% of the lens stars in the inner Galactic disk and bulge have companions of a Jupiter mass or greater between 1.5 and 4 AU. These papers claim that their limits apply to Galactic bulge M-dwarfs, but this summary of the PLANET result neglects an important bias in the events that have been searched for planets. The microlensing teams are more efficient at finding long time scale microlensing events (Alcock et al., 2000a; Sumi et al., 2003; Popowski et al., 2005; Sumi et al., 2006; Hamadache et al., 2006). The long events are also more likely to be discovered prior to peak magnification, so they can be more efficiently searched for planetary signals. As a result, the median time scale of the events search for planets in Gaudi et al. (2002) is $\langle t_E \rangle = 37\,$days, while the actual efficiency corrected median time scale is $\langle t_E \rangle = 16\,$days. This implies that the events that have been searched for planets have more massive lens stars and are more likely to reside in the disk than the typical Galactic bulge microlensing event. Thus, it is probably the case that most of the events searched by Gaudi et al. (2002) have lens stars that are either more massive than an M-dwarf, or reside in the Galactic disk, or both.

In addition to these upper limits on the planetary frequency, there were also a number of less-than-certain planet detections. Bennett & Rhie (1996) showed that the very first microlensing event discovered showed a light curve feature that could be explained by a planet, but there was a near equal mass binary lens fit that could also explain the data. The MACHO group (Bennett et al., 1997) pointed out that there is a good chance that event MACHO-95-BLG-3 was caused by a free-floating Jupiter-mass planet. The MPS group found that their data for MACHO-97-BLG-41 was best explained by a Jupiter-mass planet orbiting a binary star system (Bennett et al., 1999), but the PLANET data for this event favored an orbiting binary star interpretation (Albrow et al., 2000a). (Some of the MPS data are now known to be contaminated by moonlight reflecting off the telescope optics.) An analysis by the MOA group (Bond et al., 2002b) showed that the combined MACHO, MOA, MPS, and PLANET data for MACHO-98-BLG-35 was consistent with the low S/N detection of a terrestrial planet. Finally, Jaroszynski & Paczyński (2002) showed that the event OGLE-2002-BLG-55 had a signal consistent with a planet detection, but Gaudi & Han (2004) pointed out that there were other possible explanations.

Table 3.1. Exoplanets Discovered by Microlensing

Event Name	Star Mass	Planet Mass	Semi-Major Axis	Lead Group
OGLE-2003-BLG-235Lb/ MOA-2003-BLG-53Lb	$0.63^{+0.07}_{-0.09} M_\odot$	$830^{+250}_{-190} M_\oplus$	$4.3^{+2.5}_{-0.8}$ AU	MOA
OGLE-2005-BLG-71Lb	$0.5 \pm ? M_\odot$	$1000^{+700}_{-500} M_\oplus$	3.0 AU	OGLE
OGLE-2005-BLG-390Lb	$0.22^{+0.21}_{-0.11} M_\odot$	$5.5^{+5.5}_{-2.7} M_\oplus$	$2.6^{+1.5}_{-0.6}$ AU	PLANET
OGLE-2005-BLG-169Lb	$0.49^{+0.14}_{0.18} M_\odot$	$13^{+4}_{-5} M_\oplus$	$3.2^{+1.5}_{-1.0}$ AU	MicroFUN
OGLE-2006-BLG-109Lb	$0.6 \pm ? M_\odot$	$102 \pm ? M_\oplus$	$5.2 \pm ?$AU	MicroFUN
OGLE-2006-BLG-109Lc	$0.6 \pm ? M_\odot$	$278 \pm ? M_\oplus$	$2.5 \pm ?$AU	MicroFUN

3.5.2 Microlensing Planet Detections

Table 3.1 summarizes the properties of the planets discovered by microlensing to date, including four published microlensing exoplanet discoveries (Bond et al., 2004; Udalski et al., 2005; Beaulieu et al., 2006; Gould et al., 2006) plus a 2-planet system that will soon be published (Gaudi et al. 2007, Bennett et al. 2007, both in preparation). The microlensing discoveries are compared to other known exoplanets in Fig. 3.9.

The first planet discovered by microlensing is shown in Fig. 3.10. The light curve is plotted in units of the source star flux, which is determined by the best microlensing model to the event, because the star field is too crowded to determine the unmagnified stellar flux directly. This event was first discovered by the OGLE group and announced via their "early warning system" as event OGLE-2003-BLG-235 on 2003 June 22. On 2003 July 21, the alert system of the MOA-I microlensing survey detected this event and reported it as MOA-2003-BLG-53. The MOA detection came later because the MOA-I telescope had only a 0.61 m aperture and has worse seeing conditions than are typical at the 1.3 m OGLE telescope in Chile. However, the MOA telescope had a larger field-of-view (FOV), and this enabled them to image each of their survey fields ~ 5 times per clear night. As a result, MOA was able to detect the second caustic crossing for this event, and arrange for the additional observations that caught the caustic crossing endpoint (thanks to first author, Ian Bond, who was monitoring the photometry in real time).

The naming convention for planets discovered is that the name from the first team to find the microlensing event is used for the event, so in this case OGLE-

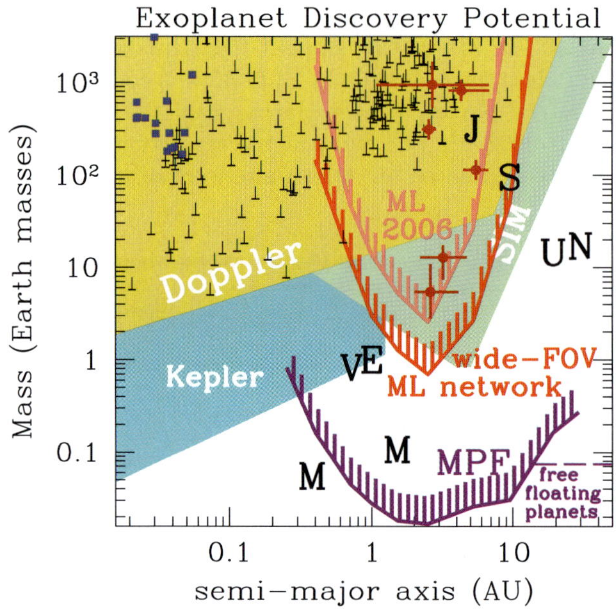

Fig. 3.9. The sensitivity of various exoplanet detection methods is plotted in the mass vs. semi-major axis plane. Doppler radial velocity detections are shown in black, with 1-sided error bars for the $m \sin i$ uncertainty. Planets first detected by transits are shown in blue, and the microlensing planet discoveries are shown in red. The gold, cyan and green shaded regions show the sensitivity of the radial velocity method and NASA's Kepler and SIM missions, respectively. The light red and red curves show the sensitivity of current and future microlensing planet search programs, and the purple curve gives the sensitivity of the proposed Microlensing Planet Finder (MPF) mission.

2003-BLG-235 takes precedence over MOA-2003-BLG-53. When referring to the lens system, we add a suffix "L", and when referring to the source, we add an "S". For a lens or source system that is multiple, we add an additional capital letter suffix for a stellar mass object or a lower case letter for a planetary mass companion. So, OGLE-2006-BLG-109LA, OGLE-2006-BLG-109Lb, and OGLE-2006-BLG-109Lc, refer to the star and two known planets of the OGLE-2006-BLG-109 lens system. This convention provides names for multiple components of the source star system. For example, OGLE-2022-BLG-876Sb would refer to a planetary companion to the source star which would be difficult, but not impossible (Graff & Gaudi, 2000; Lewis, 2001) to detect.

It is interesting to note that this event was discovered by a procedure that differs from both the alert-plus-follow up strategy suggested by Gould & Loeb (1992) and the high magnification strategy suggested by Griest & Safizadeh (1998). Instead, the planetary deviation was detected in the observations of one of the survey teams, and identified in time to obtain additional data to confirm the planetary nature of

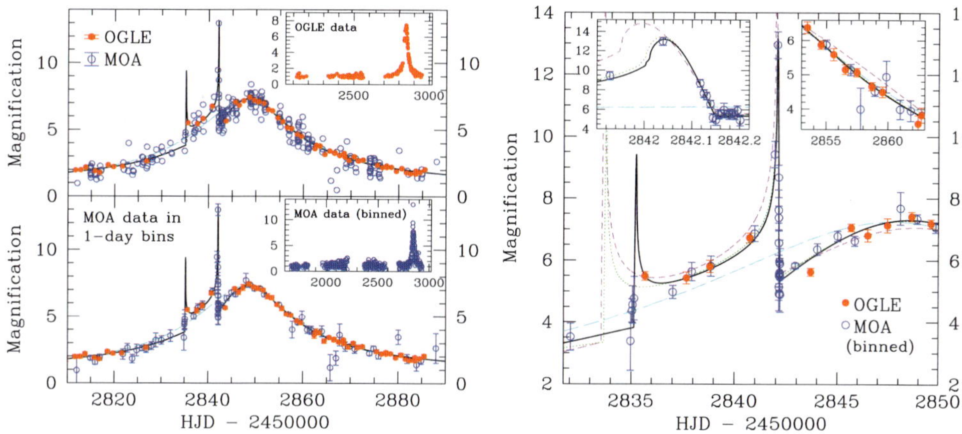

Fig. 3.10. The OGLE-2003-BLG-235/MOA-2003-BLG-53 light curve with OGLE data in red and MOA data in blue. The top-left panel presents the complete data set during 2003 (main panel) and the 2001–2003 OGLE data (inset). The median errors in the OGLE and MOA points are indicated in the legend. The bottom panel is the same as the top panel, but with the MOA data grouped in 1 day bins, except for the caustic crossing nights, and with the inset showing MOA photometry during 2000–2003. The binary- and single-lens fits are indicated by the solid black and cyan dashed curves, respectively. The right panel shows the light curve and models during caustic traverse. These models are the single-lens case (*cyan, long-dashed curve*), the best binary lens with $q \gtrsim 0.03$ (*magenta, short-dashed line*), the planetary lens with caustic entry before day 2835 (*green, dotted line*), and the best overall fit with $q = 0.0039$(*black, solid line*). The insets show the second caustic crossing and a region of the declining part of the light curve where the best-fit nonplanetary binary-lens model fails to fit the data. MOA data on days other than the caustic entry and exit (days 2835 ± 0.5 and 2842 ± 0.5) are placed in 1 day bins.

the light curve deviation. We will return to this strategy later in the discussion of future microlensing projects given in Sect. 3.6.1.

Another notable feature of this event is that the lens star has been identified in HST images (Bennett et al., 2006). As indicated in Fig. 3.7, there is an extra source of light superimposed at the location of the source star. This is very likely to be the lens star, and if so, the HST photometry implies that a fraction, $f_{\rm lens} = 0.18 \pm 0.05$, of the total source plus lens flux comes from the lens. During the microlensing event, the lens and source were separated by < 0.1 mas, but by the time of the HST images, $\Delta t = 1.78$ years after peak magnification, the lens-source separation should have grown to $\Delta t \mu_{\rm rel} = 5.9 \pm 0.7$ mas. ($\mu_{\rm rel} = 3.3 \pm 0.4$ mas/yr was determined from eq. 3.16 with input parameters from the light curve model.) This separation, plus the mass-distance relation, eq. 3.17, enable to derivation of the curves shown in the bottom left panel of Fig. 3.7. These show the amplitude for the offset of the centroids of the blended lens plus source images in different color bands. The HST data indicate a marginal detection of this color-dependent centroid shift at a level consistent with the assumption that the excess flux is due to the lens.

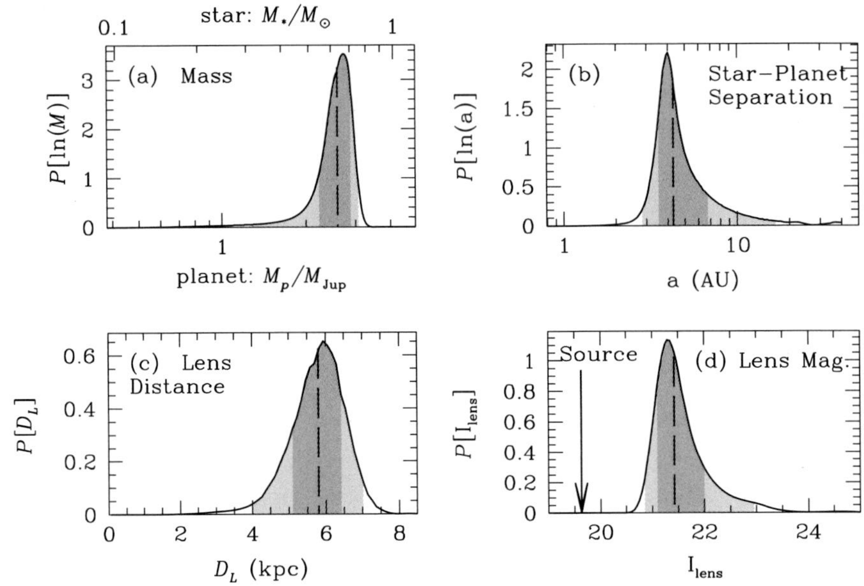

Fig. 3.11. Bayesian probability densities for the properties of the planet, OGLE-2003-BLG-235Lb, and its host star if it is a main sequence star. (a) The masses of the lens star and its planet (M_* and M_p respectively); (b) the separation; (c) their distance from the observer (D_L); and (d) the I-band brightness of the host star. The dashed vertical lines indicate the medians, and the shading indicates the central 68.3% and 95.4% confidence intervals. All estimates follow from a Bayesian analysis assuming a standard model for the disk and bulge population of the Milky Way, the stellar mass function of Bennett & Rhie (2002).

With this marginal detection of the color-dependent centroid shift, we can't be absolutely sure that the lens star has been detected because it is possible that the excess flux could be due to a companion to the source star. It is straight forward to deal with this uncertainty with a Bayesian analysis (Bennett et al., 2006), and the results of such an analysis are shown in Fig. 3.11. The resulting most likely parameter values for the event parameters are a host star mass of $M_* = 0.63^{+0.07}_{-0.09} M_\odot$, a planet mass of $M_\mathrm{p} = 2.6^{+0.8}_{-0.6} M_\mathrm{Jup}$, and an orbital semi-major axis of $a = 4.3^{+2.5}_{-0.8}$ AU. The distance to the lens system is $D_L = 5.8^{+0.6}_{-0.7}$ kpc, and the lens star magnitude is $I_L = 21.4^{+0.6}_{-0.3}$.

The light curve of the second planet discovered by microlensing, OGLE-2005-BLG-71Lb, is shown in Fig. 3.12 (Udalski et al., 2005). This was a moderately high magnification event that would have reached a maximum magnification of $A_\mathrm{max} \approx 42$ if the lens star had no planets. Because of the $d \leftrightarrow 1/d$ ambiguity discussed in Sect. 3.4, this event has two models that explain the major features of the light curve quite well. Fig. 3.3 shows the magnification patterns for these models, and for the trajectory of the lens, which is nearly perpendicular to the lens axis, the

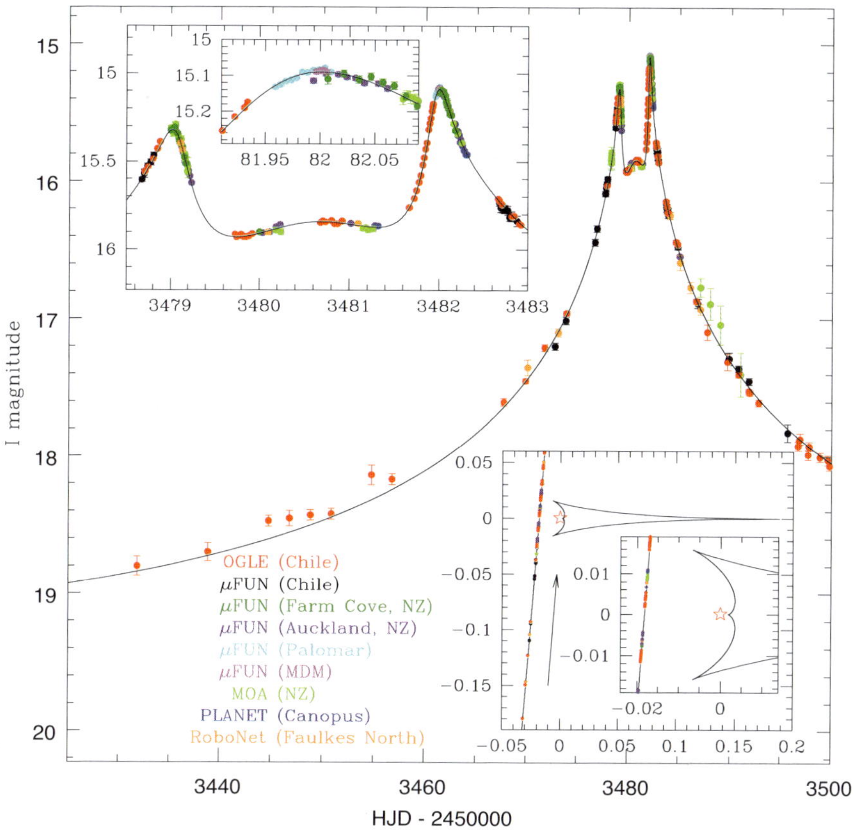

Fig. 3.12. The OGLE-2005-BLG-71 light curve showing the planetary anomaly near the peak. The triple peak (two large symmetric peaks surrounding a small peak) indicates that the source passed three cusps of a caustic, the middle one being weak (*insets*). The interval between peaks (and so cusps) is $\Delta t = 3$ days, implying that the companion mass must be small.

light curves for these different models are quite similar. From Udalski et al. (2005) the physically interesting parameters of the best fit models are $t_E = 70.9 \pm 3.3$, $q = 7.1 \pm 0.3 \times 10^{-3}$, and $d = 1.294 \pm 0.002$ for the "wide" model and $t_E = 73.9 \pm 3.5$, $q = 6.7 \pm 0.3 \times 10^{-3}$, and $d = 0.758 \pm 0.002$ for the "close" model. However, the χ^2 difference between these two models is $\Delta\chi^2 = \chi^2_{\rm close} - \chi^2_{\rm wide} = 22.0$, so the "wide" model is strongly preferred.

OGLE-2005-BLG-71Lb was the first planet discovery with significant contributions from amateur astronomers, with critical observations near the two strong cusp approach peaks by Grant Christie of the Auckland Observatory and Jennie McCormick of the Farm Cove Observatory.

With a mass ratio of $q = 7.1 \pm 0.3 \times 10^{-3}$, OGLE-2005-BLG-71Lb must certainly be a gas giant planet, but without further information such as measurement of finite

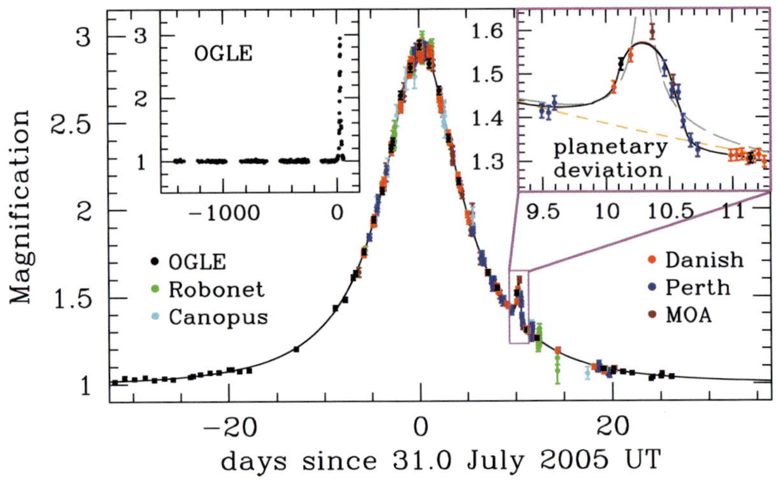

Fig. 3.13. The observed light curve of the OGLE-2005-BLG-390 microlensing event and best-fit model plotted as a function of time. The data set consists of 650 data points from PLANET Danish (*red points*), PLANET Perth (*blue*), PLANET Canopus (Hobart, *cyan*), RoboNet Faulkes North (*green*), OGLE (*black*), and MOA (*brown*). The top left inset shows the OGLE light curve extending over the previous 4 years, whereas the top right one shows a zoom of the planetary deviation, covering a time interval of 1.5 days. The solid curve is the best binary lens model described in the text with $q = 7.6 \pm 0.7 \times 10^{-5}$, and a projected separation of $d = 1.610 \pm 0.008 R_E$. The dashed grey curve is the best binary source model that is rejected by the data, and the dashed orange line is the best single lens model.

source effects, the detection of the lens star or a measurement of the microlensing parallax effect, we cannot determine the properties of the host star or the planetary mass with much precision. Fortunately, we are able to detect the lens star in a set of HST images, and the light curve yields weak detections of both a finite source size and the microlensing parallax effect. So, we expect to determine the host star and planet masses and to convert their separation into physical units, but this analysis is not yet complete (Dong et al 2007, in preparation).

The first low-mass planet discovered by microlensing was OGLE-2005-BLG-390Lb (Beaulieu et al., 2006), led by the PLANET Collaboration. This planet is currently tied with Gl 581c (Udry et al., 2007) as the lowest mass exoplanet orbiting a normal star yet to be discovered[1]. This event was detected through a planetary caustic deviation, and the amplitude of the deviation was significantly reduced by the finite angular size of the clump giant source star. If the planet were smaller by a factor of ~ 2, it would not have been detected in this event. As originally pointed

[1]The minimum mass of $M_p \geq 5.03 M_\oplus$ is often quoted for Gl 581c, but the $M_p \sin i$ ambiguity of the radial velocity method implies that the median predicted mass is $M_p = 5.5 M_\oplus$. This is the appropriate number to compare to other detection methods.

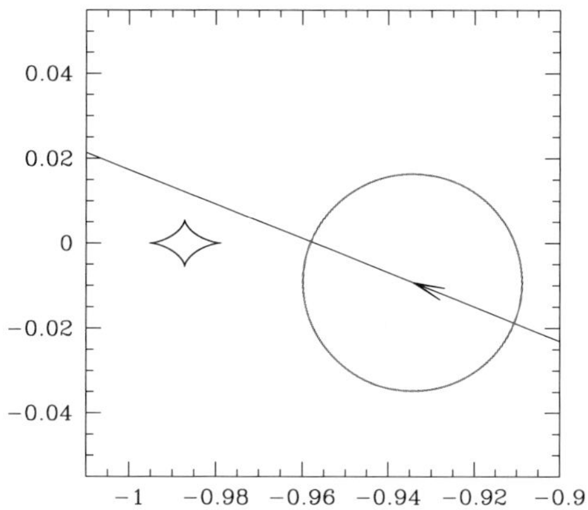

Fig. 3.14. Comparison of the OGLE-2005-BLG-390Lb planetary caustic (*the black diamond shaped curve*) with the source star size (*black circle*). The black line with the arrow show the motion of the source star.

out in Bennett & Rhie (1996) and discussed in Sect. 3.3.3, a microlensing search for Earth-mass planets should focus on events with main sequence source stars.

The OGLE-2005-BLG-390Lb light curve deviation does not show the characteristic features of a fold caustic crossing, like OGLE-2003-BLG-235, or of a cusp approach, like OGLE-2005-BLG-71. This is because the planetary caustic is smaller than the source star's angular radius of $\theta_* = 5.3 \pm 0.7\mu$as, as shown in Fig. 3.14. Because the light curve does not show these characteristic binary-microlensing features, we must consider a non-planetary explanation for the light curve involving the lensing of a binary source star by a single star lens. Gaudi (1998). However, as the Fig. 3.13 shows, a binary source model is a poor fit to the data, as it fails to account for the Perth and Danish data near the end of the perturbation. Formally, the binary source model increases the fit χ^2 by $\Delta\chi^2 = 46.25$ with one fewer degree of freedom. These data are also sufficient to avoid a possible degeneracy in the planetary parameters for such events that occurs when the wings of the deviation are poorly sampled (Gaudi & Gould, 1997).

The microlensing model for this event directly determines the planet-star mass ratio, $q = 7.6 \pm 0.7 \times 10^{-5}$, the projected planet-star separation, $d = 1.610 \pm 0.008$, the Einstein radius crossing time, $t_E = 11.03 \pm 0.11\,$days, and the source radius crossing time, $t_* = 0.282 \times 0.010\,$days. With the value for θ_* mentioned above, this yields the angular Einstein radius, $\theta_E = 0.21 \pm 0.03\,$mas, from eq. 3.16 and the mass-distance relation from eq. 3.17. This mass-distance relation can be combined with a standard Galactic model in a Bayesian analysis to estimate the probability distribution of the lens system parameters (Alcock et al., 1995, 1996; Poindexter et al., 2005; Dominik, 2006). The results of such an analysis are shown in Fig. 3.15 following the method of Dominik (2006), and nearly identical results are obtained using the Galactic model and mass function parameters of Bennett & Rhie (2002). This analysis gives a 95% probability that the planetary host star is a main-sequence

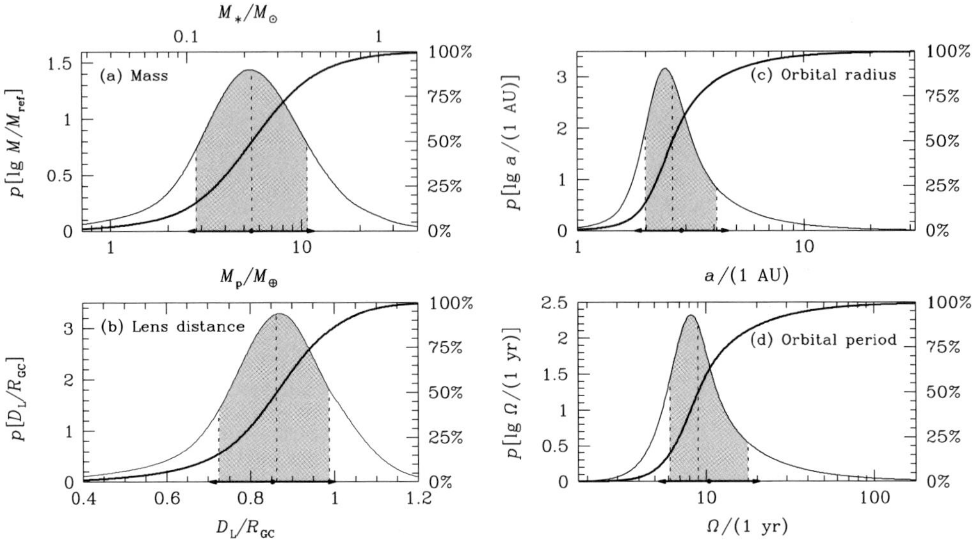

Fig. 3.15. Bayesian probability densities for the properties of the planet and its host star: (a), the masses of the lens star and its planet (M_* and M_p respectively), (b), their distance from the observer D_L, (c), the three dimensional separation or semi-major axis a of an assumed circular planetary orbit; and (d), the orbital period Ω of the planet. The bold, curved line in each panel is the cumulative distribution, with the percentiles listed on the right. The dashed vertical lines indicate the medians, and the shading indicates the central 68.3% confidence intervals, while dots and arrows on the abscissa mark the expectation value and standard deviation. The medians of these distributions yield a $M_p = 5.5^{+5.5}_{-2.7} M_\oplus$ planetary companion at a separation of $d = 2.6^{+1.5}_{-0.6}$ AU from a $M_* = 0.22^{+0.21}_{-0.11} M_\odot$ Galactic Bulge M-dwarf at a distance of $D_L = 6.6 \pm 1.0$ kpc from the Sun. The median planetary period is $\Omega = 9^{+9}_{-3}$years.

star, a 4% probability that it is a white dwarf, and a probability of,1% that it is a neutron star or black hole. The median parameters shown in Fig. 3.15 imply that the planet receives radiation from its host star that is only 0.1% of the radiation that the Earth receives from the Sun, so the probable surface temperature of the planet is 50 K, similar to the temperature of Neptune.

As discussed in Sect. 3.4.1, the lens star mass can be determined directly if the lens star is detected. However, this will be quite difficult for OGLE-2005-BLG-390L, because the source is a giant star. For the median mass and distance to the lens system, the lens star would be fainter than the source by a factor of 2000 in the K-band. So, the detection of the lens star may require many years for the relative proper motion of $\mu_{\rm rel} = 6.8$ mas/yr, and the development of new instruments for large ground-based or space telescopes.

OGLE-2005-BLG-169 was the third published event from the 2005 season and the second low-mass planet found by microlensing (Gould et al., 2006). This was a very high magnification event, with a peak magnification of $A_{\rm max} \simeq 800$, and

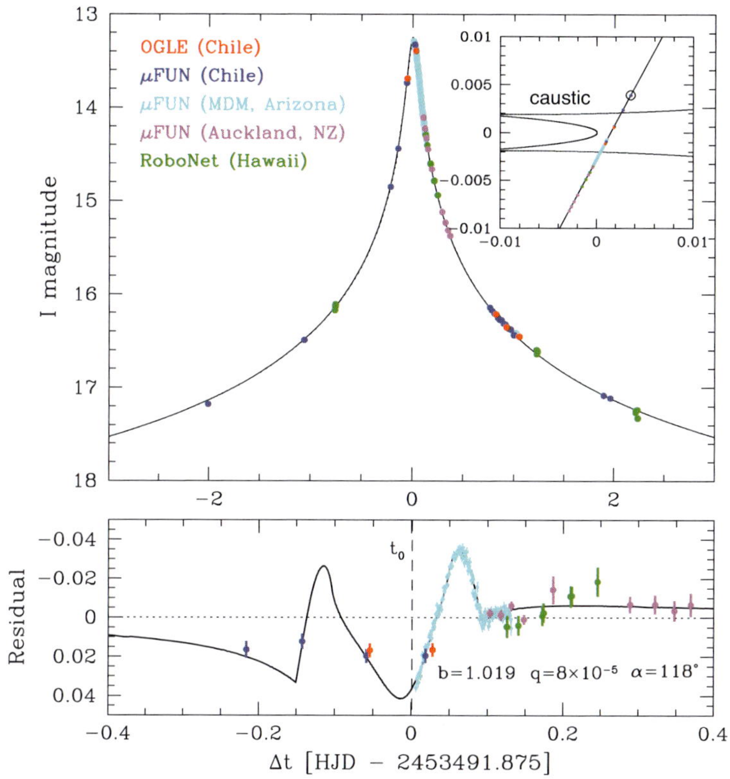

Fig. 3.16. *Top*: Data and best-fit model for OGLE-2005-BLG-169. *Bottom*: Difference between this model and a single-lens model with the same single lens parameters (t_0, u_0, t_E, and ρ). It displays the classical form of a caustic entrance/exit that is often seen in binary microlensing events, where the amplitudes and timescales are several orders of magnitude larger than seen here. MDM data trace the characteristic slope change at the caustic exit ($\Delta t = 0.092$) very well, while the entrance is tracked by a single point at $\Delta t = -0.1427$. The dashed line indicates the time t_0. *Inset*: Source path through the caustic geometry. The source size, ρ, is indicated.

it's light curve is shown in Fig. 3.16. The bottom panel of Fig. 3.16 indicates that the planetary deviation has a maximum amplitude of about 4% compared to the light curve of the same event without a planet. Such low amplitude deviations are characteristic of the very weak caustics due to low-mass planets near the Einstein ring. However, it is only part of the caustic curve that is so weak. If the source would have passed on the other side of the host star and crossed the backwards "C" shaped part of the caustic in the inset of Fig. 3.16, the planetary signal would have been very much stronger. But, it order to detect the low amplitude signal due to the caustics actually crossed by the source star, it was quite helpful to have continuous observations over the course of three hours from the 2.4 m MDM telescope.

The analysis of Gould et al. (2006) indicated a super-Earth mass planet with $M_p = 13^{+6}_{-8} M_\oplus$ orbiting a star of $M_* \simeq 0.49 M_\odot$. Such a planet, like OGLE-2005-BLG-390Lb, would be invisible to other planet detection methods. High magnification events also place tight constraints on the presence of additional planets, and in the case of OGLE-2005-BLG-169, Jupiter-mass planets can be exclude from the separation range 0.6-18 AU and Saturn-mass planets can be excluded from the range 1-11 AU.

The precise masses of the host star and planet have not yet been determined because the host star has not been detected. Thus, the lens system properties can only be determined by a Bayesian analysis, as was done for OGLE-2005-BLG-390Lb in Fig. 3.15. This analysis uses the parameters from the microlensing light curve, including the Einstein radius crossing time of $t_E = 43 \pm 1.3$ days, the source radius crossing time of $t_* = 0.019 \pm 0.01$ days, and the lens-source relative proper motion of $\mu_{\rm rel} = 8.4 \pm 0.6$ mas/yr. The results of this analysis are presented in Fig. 3.17. These have assumed a Han-Gould model for the Galactic bar (Han & Gould, 1995), a double-exponential disk with a scale height of 325 pc, and a scale length of 3.5 kpc, as well as other Galactic model parameters as described in Bennett & Rhie (2002).

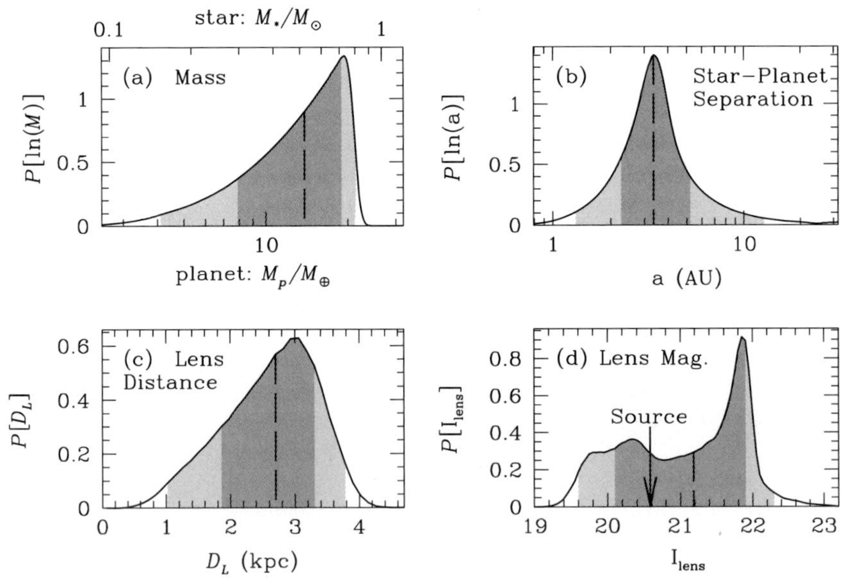

Fig. 3.17. OGLE-2005-BLG-169 lens property figure. Bayesian probability densities for the properties of the planet and its host star if it is a main sequence star. (a) The masses of the lens star and its planet (M_* and M_p respectively). (b) the separation, (c) their distance from the observer (D_L); and (d) the I-band brightness of the host star. The dashed vertical lines indicate the medians, and the shading indicates the central 68.3% and 95.4% confidence intervals. All estimates follow from a Bayesian analysis assuming a standard model for the disk and bulge population of the Milky Way, the stellar mass function of Bennett & Rhie (2002).

Because this model is different from the Galactic model used by Gould et al. (2006), the resulting parameters differ slightly from their results. We find a lens system distance of $D_L = 2.7^{+0.6}_{-0.9}$ kpc, a three dimensional star-planet separation of $a = 3.3^{+1.9}_{-0.9}$ AU and main sequence stellar and planetary masses of $M_* = 0.52^{+0.19}_{-0.22} M_\odot$ and $M_p = 14^{+5}_{-6} M_\oplus$. If we assume that white dwarfs have an *a priori* probability to host planets that is equal to that of main sequence stars (at the separations probed by microlensing), then there is a 35% probability that the host star is a white dwarf. The possibility of a brown dwarf host star is excluded by the light curve limits on the microlensing parallax effect (Gould et al., 2006).

Fig. 3.17(d) shows the probability distribution of the I-band magnitude of the planetary host star compared to the source star at $I = 20.58 \pm 0.10$. The implied planetary host star brightness distribution has a median and 1-σ range of $I_{\rm lens} = 21.9^{+0.7}_{-1.1}$, but the most interesting feature of this figure is that the probability of a main sequence lens fainter than $I = 23$ vanishes. This is because the mass-distance relation, eq. 3.17 ensures that the lens star will be nearby and at least at bright as $I = 23$, even if it is at the bottom of the main sequence at $M_* = 0.08 M_\odot$. In fact, the microlensing parallax constraint from the light curve yields a lower limit for the lens star mass of $M_* \gtrsim 0.14 M_\odot$. Thus, the planetary host star must be at least 16% of the brightness of the combined lens plus source star blended image, and this implies that it will be detectable if it is not a stellar remnant. Plus, the relatively rapid relative proper motion, $\mu_{\rm rel} = 8.4 \pm 0.6$ mas/yr, of OGLE-2005-BLG-169L, implies that the lens-source separation is already detectable with HST (Bennett et al., 2007a), as discussed in Sect. 3.4.1.

One of the most interesting consequences of the discoveries of OGLE-2005-BLG-390Lb and 169Lb is that super-Earth planets of $\sim$ 5-15$M_\oplus$ are likely to be quite common. Gould et al. (2006) combined these detections with null results from very high magnification events (Abe et al., 2004; Dong et al., 2006) plus samples of lower magnification events (Albrow et al., 2001; Gaudi et al., 2002) to solve for the fraction, $f_{\rm se}$, of stars with planets of mass ratio $\sim 8 \times 10^{-5}$ at the separations of 1.5-4 AU, where microlensing is most sensitive. They found that the median and 90% confidence level upper and lower limits are $f_{\rm se} = 0.38^{+0.31}_{-0.22}$, based on the two planets discovered and the accumulated null results. The 90% c.l. lower limit is $f_{\rm se} \geq 16\%$. This is significantly higher than the fraction of F, G, and K stars with Jupiter-mass planets in this 1.5-4 AU separation range. This fraction of stars with Jupiters at this separation can be estimated from Butler et al. (2006) to be $f_{\rm J} \simeq 3\%$. Thus, these cool, super-Earth planets appear to represent the most common type of exoplanet yet discovered. This would seem to confirm the prediction of the core-accretion theory that $\sim 10 M_\oplus$ planets form much more frequently than gas giants, like Jupiter (Ida & Lin, 2004; Laughlin, Bodenheimer & Adams, 2004), although this may not be incompatible with the disk instability theory (Boss, 2006).

The final event that we will present is OGLE-2006-BLG-109, which is much more complicated than the other events (Gaudi et al. 2007; Bennett et al. 2007, both in preparation). The light curve for this event is shown in Fig. 3.18, while the central caustic configuration is shown in Fig. 3.19. This is the first microlensing event with

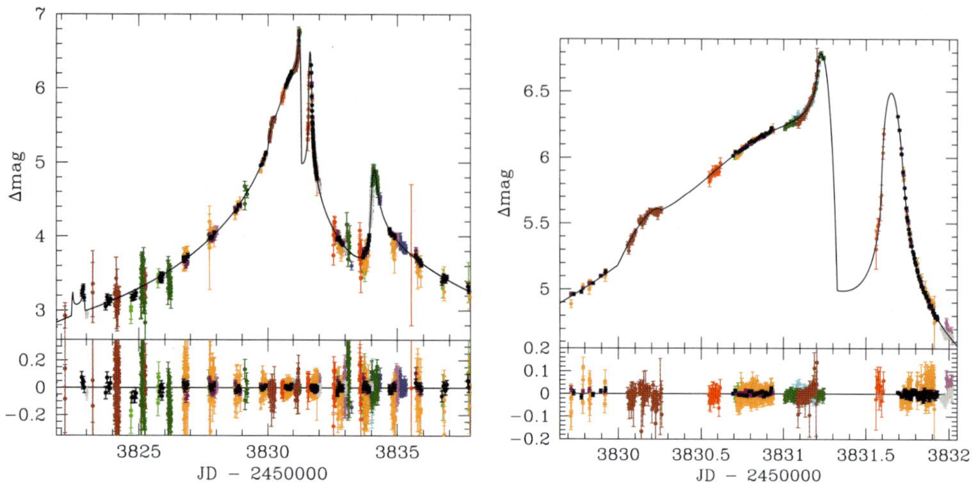

Fig. 3.18. Two views of the OGLE-2006-BLG-109 light curve, which is the first multi-planet microlensing event with a planet of slightly less than a Jupiter mass ($q = 1.35 \times 10^{-3}$ at $\sim 2.7\,\mathrm{AU}$ and a planet of slightly more than a Saturn mass ($q = 4.9 \times 10^{-4}$) at $\sim 5.4\,\mathrm{AU}$. The signal is dominated by the Saturn-mass planet because it is close to the Einstein ring at $d = 1.4$, and there are two pairs of caustic crossing features (at $t = 3822.5$, 3822.9 and $t = 3830.2$, 3831.2) and a cusp approach (at $t = 3834.1$) due to the Saturn-mass planet. The Jupiter-mass planet planet is further from the Einstein ring at $d = 0.63$, so its signal is limited to the highest magnification part of the light curve and is responsible for the cusp approach feature at $t = 3831.65$. Both planetary orbital motion and microlensing parallax must be included to obtain an acceptable model for this event.

two detected planets, and it also shows clear signals of planetary orbital motion and microlensing parallax. These effects are detectable because the Saturn-mass planet has a projected separation that is close to the Einstein ring, which causes its caustic to become quite extended. Its effects are visible for 11 days.

Another notable feature of OGLE-2006-BLG-109 is that the lens is > 5 times brighter than the source. It is detectable (although not completely resolved) in the best seeing (0.7") OGLE images and is clearly visible in K and H-band adaptive optics images from the Keck telescope. As a result, there are two methods to determine the lens star mass: the combination of the θ_E determination from the finite source effects and the microlensing parallax effect yields the lens mass via eq. 3.19, while the lens star detection give the lens mass with the help of mass-luminosity relations, as discussed in Sect. 3.4.1. However, one complication is that there is some degeneracy in the effects of microlensing parallax and the planetary orbital motion on the microlensing light curve. On the other hand, the planetary orbital motion parameters yield information about the orbits that haven't been detected before in a microlensing event. So, this event will yield much more information about the OGLE-2006-BLG-109L planetary system than was anticipated for any planetary microlensing event.

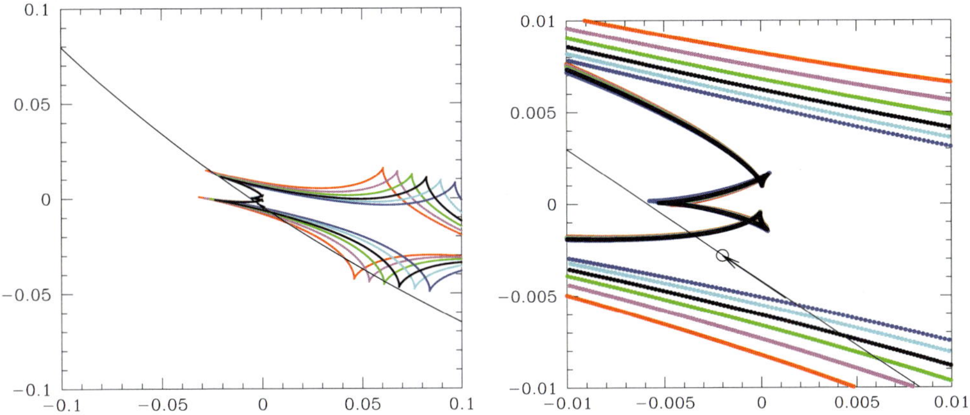

Fig. 3.19. The central caustic OGLE-2006-BLG-109 configuration for is shown at 3-day intervals from $t = 3820$ (shortly before the first caustic crossing) through $t = 3835$ (a day after the final cusp approach). The time-order of the different color caustic curves is *red, magenta, green, black, cyan, blue.* The grey curve is the source trajectory, which is curved due to the microlensing parallax effect (*i.e.* the orbital motion of the Earth) and the small circle that the source trajectory in the left, close-up panel shows the source star radius.

3.6 Future Programs

Our experience with the existing microlensing planet search programs provides indications of how the sensitivity of future microlensing surveys can be improved. At present, the OGLE and MOA groups are each able to independently discover more than 500 microlensing events per year. There is a great overlap between the discoveries of these two groups, but the total number of events discovered every year is probably about 700. This is at least an order of magnitude larger than the global follow-up groups can hope to follow. The follow-up groups optimize their observations by focusing on high magnification events. However, many of the shorter time scale high magnification events are not recognized as such in time, and so a large fraction of the high magnification events are not searched for planets.

The solution to this problem is to observe many microlensing events in each image with a global network of very wide FOV telescopes that can observe 10 square degrees or more of the Galactic bulge at 15-20 intervals. The new 1.8m MOA-II telescope (Hearnshaw et al., 2005) with a 2.2 square degree FOV CCD camera (Yanagisawa et al., 2000) that began operation in 2006 is the first telescope that meets this requirement, and the OGLE group plans to upgrade to a 1.4 square degree OGLE-4 camera in time for the 2009 Galactic bulge observing season. WIth MOA-II in New Zealand, and OGLE-IV in Chile, all that is needed is a very wide-FOV microlensing survey telescope in Southern Africa. A number of groups are pursuing funding for such a telescope.

Simulations of such a system have been performed by Bennett (2004) and Gaudi (2007, private communication), and estimates of the sensitivity of the global network

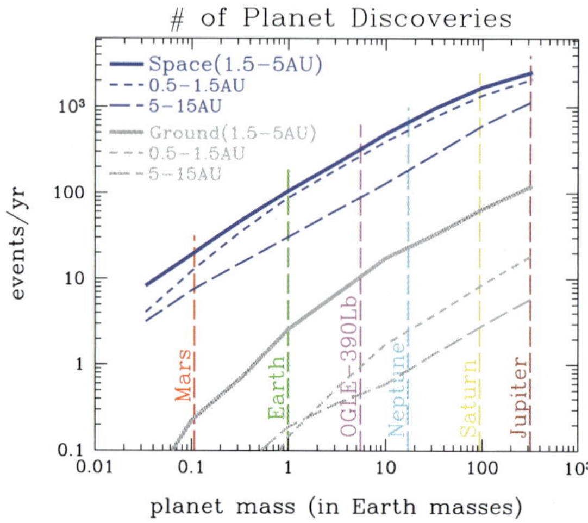

Fig. 3.20. The number of planet detections expected per year as a function of planet mass is shown for proposed future space and ground-based microlensing surveys under the assumption of one planet per star in the indicated separation ranges. The space-based survey has its most significant advantage over the ground-based survey at separations smaller (0.5-1.5 AU) and larger (5-15 AU) than the Einstein radius, because a space-based survey is able to resolve bulge main sequence stars and detect moderate amplitude planetary signals when the magnification due to the stellar lens is small.

consisting of MOA-II, OGLE-IV and an OGLE-IV-like system in South Africa are presented in Fig. 3.9 and Fig. 3.20. The improvement in sensitivity with such a network in the mass vs. semi-major axis plane is shown in Fig. 3.9 with the light and dark red curves showing the sensitivity of the current surveys and the future very wide-FOV network, respectively. This network will extend the sensitivity of the microlensing method down to an Earth mass at planet-star separations close to the Einstein ring radius ($\sim$ 2-3 AU).

The separation range where ground-based microlensing is most sensitive, 1-5 AU corresponds to the vicinity of the so-called "snow-line," which is the region of the proto-planetary disk where it is cold enough for water-ice to condense. The density of solids in the proto-planetary disk increases by a factor of ~ 4 across the "snow-line," and as a result, the core accretion theory predicts that this is where the most massive planets will form (Ida & Lin, 2004; Laughlin, Bodenheimer & Adams, 2004; Kennedy et al., 2006). According to this theory, giant planets form just outside the "snow line" where they can accrete $\sim 10 M_\oplus$ of rock and ice to form a core that grows into a gas giant like Jupiter or Saturn via the run-away accretion of hydrogen and helium onto this core. However, this theory also predicts that the hydrogen and helium gas can easily be removed from the proto-planetary disk during the millions of years that it takes to build the rock-ice core of a gas-giant. Thus, if the core accretion theory is correct, rock-ice planets of $\sim 10 M_\oplus$ that failed to grow into gas giants should be quite common, although it is possible to form such planets in the competing gravitational instability theory (Boss, 2006).

The number of planetary microlensing event detections expected per year is shown in Fig. 3.20 assuming an average of one such planet per star, with conservative assumptions regarding photometric precision. The assumption of an average of one such planet per star is certainly too optimistic for Jupiter mass planets (Gaudi et al., 2002; Butler et al., 2006), but it is closer to reality for super-Earths, like OGLE-2005-BLG-390Lb and OGLE-2005-BLG-169Lb (Beaulieu et al., 2006; Gould et al., 2006). It could very well be accurate for Earth-mass planets where the weaker two-body gravitational interactions allow two planets to orbit in the separation range corresponding to the bins in Fig. 3.20. (Our own Solar System is an example of this.)

Another future development that is already funded is a global network of robotic telescopes dedicated to monitoring transient events like planetary microlensing events, known as the Las Cumbres Global Telescope Network (Brown et al., 2007). Ideally, this network would routinely observe high magnification microlensing events and planetary deviations discovered in progress with a very high cadence, such as that provided by the MDM telescope for OGLE-2005-BLG-169 (see Fig. 3.16). This would enable the very wide-FOV survey telescopes to maintain their normal sampling strategy so that other planetary microlensing events would not be missed. This might add to the planet detection efficiency substantially, but such a system is more difficult to model.

3.6.1 The Ultimate Exoplanet Census: Space-Based Microlensing

The ultimate census for virtually all types of exoplanets would be a space-based microlensing survey (Bennett & Rhie, 2002; Bennett et al., 2007b). Such a survey could provide a statistical census of exoplanets with masses $\geq 0.1 M_\oplus$ and orbital separations ranging from 0.5 AU to ∞. This includes analogs to all the Solar System's planets except for Mercury, as well as most types of planets predicted by planet formation theories. This survey would determine the frequency of planets around all types of stars except those with short lifetimes. Close-in planets with separations < 0.5 AU are invisible to a space-based microlensing survey, but these can be found by Kepler (Basri et al., 2005). Other methods, including ground-based microlensing, cannot approach the comprehensive statistics on the mass and semi-major axis distribution of extrasolar planets that a space-based microlensing survey will provide. Detailed simulations of a space-based microlensing survey (Bennett & Rhie, 2002) have been used to determine the sensitivity of such a mission, and Figs. 3.9 and 3.20 show the sensitivity of the proposed Microlensing Planet Finder (MPF) mission (Bennett et al., 2004). These figures also show that the sensitivity of a ground-based microlensing survey to terrestrial planets is limited to the vicinity of the Einstein radius at 2-3 AU. This is because ground-based survey generally requires moderately high magnification $A \gtrsim 10$ in order to resolve the source star well enough to get the moderately precise photometry that is needed to detect planets with the microlensing method. A space-based microlensing survey would generally resolve the source stars, so planets further from the Einstein radius can be detected

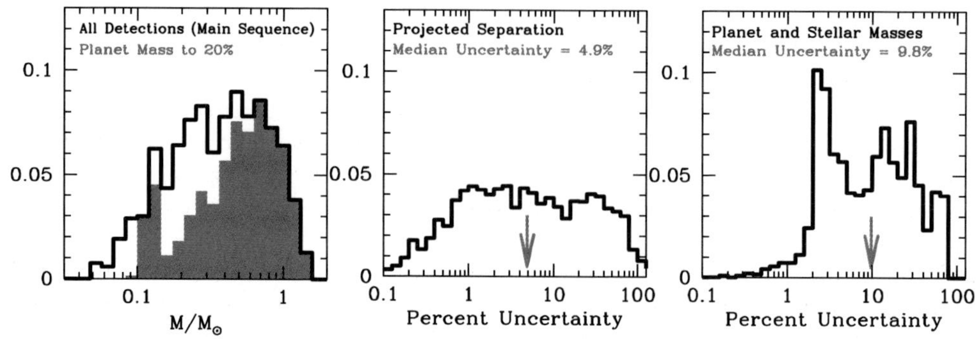

Fig. 3.21. (a) The simulated distribution of stellar masses for stars with detected terrestrial planets. The grey histogram indicates the subset of this distribution for which the masses can be determined to better than 20%. (b) The distribution of uncertainties in the projected star-planet separation. (c) The distribution of uncertainties in the star and planet masses. Note that it is the two-dimensional projected separation that is presented here, and we have not included the uncertainty in the separation along the line-of-sight as was done in Fig. 3.17.

via their light curve perturbations at relatively low magnification from the lensing effect of the planetary host star.

A space-based microlensing survey is also able to detect most of the planetary host stars for most planetary microlensing events. Using the methods described in Sect. 3.4.1 and in more detail in Bennett et al. (2007a), this allows the determination of the star and planet masses and separation in physical units. This can be accomplished with HST observations for a small number of planetary microlensing events (Bennett et al. 2006), but only a space-based survey can do this for hundreds or thousands of planetary microlensing events that future surveys would expect to discover. Fig. 3.17 shows the distribution of planetary host star masses and the predicted uncertainties in the masses and separation of the planets and their host stars (Bennett et al., 2007a) from simulations of the MPF mission. The host stars with masses determined to better than 20% are indicated by the red histogram in Fig. 3.17(a), and these are primarily the host stars that can be detected in MPF images. Ground-based microlensing surveys also suffer significant losses in data coverage and quality due to poor weather and seeing. As a result, a significant fraction of the planetary deviations seen in a ground-based microlensing survey will have poorly constrained planet parameters due to poor light curve coverage (Peale, 2003). (These poorly characterized detections are not included in Fig. 3.20, however.)

Proposed improvements to ground-based microlensing surveys can detect Earth-mass planets in the vicinity of the "snow-line," which is critical for the understanding of planet formation theories (Gould et al., 2007). But such a survey would have its sensitivity to Earth-like planets limited to a narrow range of semi-major axes, so it would not provide the complete picture of the frequency of exoplanets down to $0.1M_\oplus$ that a space-based microlensing survey would provide. Such a survey would probably not detect the planetary host stars for most of the events, and so it cannot provide the systematic data on the variation of exoplanet properties as a function of host star type that a space-based survey will provide.

A space-based microlensing survey, such as MPF, will provide a census of extrasolar planets that is complete (in a statistical sense) down to $0.1 M_{\oplus}$ at orbital separations ≥ 0.5 AU, and when combined with the results of the Kepler mission a space-based microlensing survey will give a comprehensive picture of all types of extrasolar planets with masses down to well below an Earth mass. This complete coverage of planets at all separations can be used to calibrate the poorly understood theory of planetary migration. This fundamental exoplanet census data is needed to gain a comprehensive understanding of processes of planet formation and migration, and this understanding of planet formation is an important ingredient for the understanding of the requirements for habitable planets and the development of life on extrasolar planets (Bennett et al., 2007b).

The basic requirements for a space-based microlensing survey are a 1-m class wide field-of-view space telescope that can image the central Galactic bulge in the near-IR or optical almost continuously for periods of at least several months at a time. This can be accomplished as a NASA Discovery mission, as the example of the MPF mission shows, but it could also be combined with other programs that require an IR-optimized wide-FOV space telescope, as long as a large fraction of the observing time is devoted to Galactic bulge observations. As Fig. 3.9 shows, there is no other planned mission that can duplicate the science return of a space-based microlensing survey, and our knowledge of exoplanets and their formation will remain incomplete until such a mission is flown.

Thus, a space-based microlensing survey is likely to be the only way to gain a comprehensive understanding of the nature of planetary systems, which is needed to understand planet formation and habitability. The proposed Microlensing Planet Finder (MPF) mission is an example of a space-based microlensing survey that can accomplish these objectives with proven technology and a cost that fits comfortably under the NASA Discovery Program cost cap.

References

Abe, F., et al. 2004, Search for Low-Mass Exoplanets by Gravitational Microlensing at High Magnification, *Science*, **305**, 1264

Albrow, M., et al. 1998, The 1995 Pilot Campaign of PLANET: Searching for Microlensing Anomalies through Precise, Rapid, Round-the-Clock Monitoring, *Astrophys. J.*, **509**, 687

Albrow, M. D., et al. 2000a, Detection of Rotation in a Binary Microlens: PLANET Photometry of MACHO 97-BLG-41, *Astrophys. J.*, **534**, 894

Albrow, M. D., et al. 2000b, Limits on Stellar and Planetary Companions in Microlensing Event OGLE-1998-BUL-14, *Astrophys. J.*, **535**, 176

Albrow, M. D., et al. 2001, Limits on the Abundance of Galactic Planets From 5 Years of PLANET Observations, *Astrophys. J. Lett.*, **556**, L113

Alcock, C., et al. 1993, Possible Gravitational Microlensing of a Star in the Large Magellanic Cloud, *Nature*, **365**, 621

Alcock, A., et al. 1994, Possible Gravitational Microlensing Event, *I.A.U.C.*, **6068**, 1

Alcock, C., et al. 1995, First Observation of Parallax in a Gravitational Microlensing Event, *Astrophys. J.*, **454**, L125

Alcock, C., et al. 1996, Real-Time Detection and Multisite Observations of Gravitational Microlensing, *Astrophys. J. Lett.*, **463**, L67

Alcock, C., et al. 2000a, The MACHO Project: Microlensing Optical Depth toward the Galactic Bulge from Difference Image Analysis, *Astrophys. J.*, **541**, 734; (E) **557**, 1035

Alcock, C., et al. 2000b, The MACHO Project: Microlensing Results from 5.7 Years of Large Magellanic Cloud Observations, *Astrophys. J.*, **542**, 281

An, J., et al. 2002, First Microlens Mass Measurement: PLANET Photometry of EROS BLG-2000-5, *Astrophys. J.*, **572**, 521

Basri, G., Borucki, W. J., & Koch, D. 2005, The Kepler Mission: A wide-field transit search for terrestrial planets, *New Astronomy Rev.*, **49**, 478

Beaulieu, J.-P., et al. 2006, Discovery of a Cool Planet of 5.5 Earth Masses Through Gravitational Microlensing, *Nature*, **439**, 437

Benetti, S., Pasquini, L., & West, R. M. 1995, First spectroscopic confirmation of a microlensing event towards the galactic bulge, *Astron. & Astrophys.*, **294**, L37

Bennett, D. P. 2004, The Detection of Terrestrial Planets via Gravitational Microlensing: Space vs. Ground-based Surveys, *Extrasolar Planets: Today and Tomorrow*, **321**, 59

Bennett, D. P. 2005, Large Magellanic Cloud Microlensing Optical Depth with Imperfect Event Selection, *Astrophys. J.*, **633**, 906

Bennett, D.P. & Rhie, S.H. 1996, Detecting Earth-Mass Planets with Gravitational Microlensing, *Astrophys. J.*, **472**, 660

Bennett, D. P. & Rhie, S. H., 2002, Simulation of a Space-based Microlensing Survey for Terrestrial Extrasolar Planets, *Astrophys. J.*, **574**, 985

Bennett, D. P., et al. 1997, Planetary Microlensing from the MACHO Project, *Planets Beyond the Solar System and the Next Generation of Space Missions*, **119**, 95

Bennett, D. P., et al. 1999, Evidence for a Planet Oribiting a Binary Star System from Gravitational Microlensing, *Nature*, **402**, 57

Bennett, D. P., et al. 2002, Gravitational Microlensing Events Due to Stellar-Mass Black Holes, *Astrophys. J.*, **579**, 639

Bennett, D. P., et al. 2004, The Microlensing Planet Finder: Completing the Census of Extrasolar Planets in the Milky Way *SPIE*, **5487**, 1453, (astro-ph/0409218)

Bennett, D. P., Anderson, J., Bond, I.A., Udalski, A., Gould, A. 2006, Identification of the OGLE-2003-BLG-235/MOA-2003-BLG-53 Planetary Host Star *Astrophys. J. Lett.*, **647**, L171

Bennett, D. P., Anderson, J., & Gaudi, B. S. 2007a, Characterization of Gravitational Microlensing Planetary Host Stars *Astrophys. J.*, **660**, 781

Bennett, D. P., et al. 2007b, An Extrasolar Planet Census with a Space-based Microlensing Survey *ArXiv e-prints*, **704**, arXiv:0704.0454

Bessell, M. S. & Brett, J. M. 1988, JHKLM photometry - Standard systems, passbands, and intrinsic colors*P.A.S.P.*, **100**, 1134

Bolatto, A. D., & Falco, E. E. 1994, The detectability of planetary companions of compact Galactic objects from their effects on microlensed light curves of distant stars, *Astrophys. J.*, **436**, 112

Bond, I. A., et al. 2001, Real-time difference imaging analysis of MOA Galactic bulge observations during 2000, *M.N.R.A.S.*, **327**, 868

Bond, I. A., et al. 2002a, Improving the Prospects for Detecting Extrasolar Planets in Gravitational Microlensing in 2002, *M.N.R.A.S.*, **331**, L19

Bond, I.A., et al. 2002b, Study by MOA of extra-solar planets in gravitational microlensing events of high magnification, *M.N.R.A.S.*, **333**, 71

Bond, I.A., et al. 2004, OGLE 2003-BLG-235/MOA 2003-BLG-53: A Planetary Microlensing Event, *Astrophys. J. Lett.*, **606**, L155

Bourassa, R. R., Kantowski, R., & Norton, T. D. 1973, The Spheroidal Gravitational Lens, *Astrophys. J.*, **185**, 747

Boss, A. P. 2006, Rapid Formation of Super-Earths around M Dwarf Stars, *Astrophys. J. Lett.*, **644**, L79

Brown, T., Rosing, W. E., Baliber, N., Hidas, M., & Street, R. 2007, Surveys, Temporal Variability, and the Las Cumbres Observatory Global Telescope, *American Astronomical Society Meeting Abstracts*, 210, #66.01

Butler, R. P., et al. 2006, Catalog of Nearby Exoplanets, *Astrophys. J.*, **646**, 505

Chang, K., & Refsdal, S. 1979, Flux variations of QSO 0957+561 A, B and image splitting by stars near the light path, *Nature*, **282**, 561

Chang, K., & Refsdal, S. 1984, Star disturbances in gravitational lens galaxies, *Astron. & Astrophys.*, **132**, 168

Delplancke, F., Górski, K. M., & Richichi, A. 2001, Resolving gravitational microlensing events with long-baseline optical interferometry. Prospects for the ESO Very Large Telescope Interferometer, *Astron. & Astrophys.*, **375**, 701

Dominik, M. 1999, The binary gravitational lens and its extreme cases, *Astron. & Astrophys.*, **349**, 108

Dominik, M. 2006, *M.N.R.A.S.*, Stochastic distributions of lens and source properties for observed galactic microlensing events, **367**, 669

Dong, S., et al. 2006, Planetary Detection Efficiency of the Magnification 3000 Microlensing Event OGLE-2004-BLG-343, *Astrophys. J.*, **642**, 842

Dong, S., et al. 2007, First Space-Based Microlens Parallax Measurement: Spitzer Observations of OGLE-2005-SMC-001, *Astrophys. J.*, **664**, 842

Gaudi, B. S. 1998, Distinguishing Between Binary-Source and Planetary Microlensing Perturbations, *Astrophys. J.*, **506**, 533

Gaudi, B. S. & Gould, A. 1997, Planet Parameters in Microlensing Events, *Astrophys. J.*, **486**, 85

Gaudi, B. S., & Han, C. 2004, The Many Possible Interpretations of Microlensing Event OGLE 2002-BLG-055, *Astrophys. J.*, **611**, 528

Gaudi, B. S., et al. 2002, Microlensing Constraints on the Frequency of Jupiter-Mass Companions: Analysis of 5 Years of PLANET Photometry, *Astrophys. J.*, **566**, 463

Gaudi, B. S., Naber, R. M., & Sackett, P. D. 1998, Microlensing by Multiple Planets in High-Magnification Events, *Astrophys. J. Lett.*, **502**, L33

Gould, A. 1992, Extending the MACHO search to about 10^6 solar masses, *Astrophys. J.*, **392**, 442

Gould, A. 1997, Extreme Microlensing toward the Galactic Bulge, *Astrophys. J.*, **480**, 188

Gould, A., Bennett, D. P., & Alves, D. R. 2004, The Mass of the MACHO-LMC-5 Lens Star, *Astrophys. J.*, **614**, 404

Gould, A., & Gaucherel, C. 1997, Stokes's Theorem Applied to Microlensing of Finite Sources *Astrophys. J.*, **477**, 580

Gould, A., Gaudi, B. S., & Bennett, D. P. 2007, Ground-based Microlensing Surveys, *ArXiv e-prints*, **704**, arXiv:0704.0767

Gould, A. & Loeb, A. 1992, Discovering planetary systems through gravitational microlenses, *Astrophys. J.*, **396**, 104

Gould, A., et al. 2006, Microlens OGLE-2005-BLG-169 Implies That Cool Neptune-like Planets Are Common, *Astrophys. J. Lett.*, **644**, L37

Graff, D. S., & Gaudi, B. S. 2000, Direct Detection of Large Close-in Planets around the Source Stars of Caustic-crossing Microlensing Events *Astrophys. J. Lett.*, **538**, L133

Griest, K., & Safizadeh, N. 1998, The Use of High-Magnification Microlensing Events in Discovering Extrasolar Planets, *Astrophys. J.*, **500**, 37

Hamadache, C., et al. 2006, Galactic Bulge Microlensing Optical Depth from EROS-2, *Astron. & Astrophys.*, **454**, 185

Han, C. & Gould, A. 1995, The Mass Spectrum of MACHOs from Parallax Measurements, *Astrophys. J.*, **447**, 53

Hearnshaw, J. B., et al. 2005, The MOA 1.8-metre alt-az wide-field survey telescope and the MOA project, *ArXiv Astrophysics e-prints*, arXiv:astro-ph/0509420

Ida, S., & Lin, D.N.C. 2004, Toward a Deterministic Model of Planetary Formation. II. The Formation and Retention of Gas Giant Planets around Stars with a Range of Metallicities, *Astrophys. J.*, **616**, 567

Jaroszynski, M., & Paczyński , B. 2002, A Possible Planetary Event OGLE-2002-BLG-055, *Acta Astronomica*, **52**, 361

Kennedy, G. M., Kenyon, S. J., & Bromley, B. C. 2006, Planet Formation around Low-Mass Stars: The Moving Snow Line and Super-Earths, *Astrophys. J. Lett.*, **650**, L139

Kervella, P., Thévenin, F., Di Folco, E., & Ségransan, D. 2004, The angular sizes of dwarf stars and subgiants. Surface brightness relations calibrated by interferometry, *Astron. & Astrophys.*, **426**, 297

Lauer, T. R. 1999, The Photometry of Undersampled Point-Spread Functions, *P.A.S.P.*, **111**, 1434

Laughlin, G., Bodenheimer, P., & Adams, F.C. 2004, The Core Accretion Model Predicts Few Jovian-Mass Planets Orbiting Red Dwarfs, *Astrophys. J. Lett.*, **612**, L73

Lewis, G. F. 2001, Gravitational microlensing of stars with transiting planets, *Astron. & Astrophys.*, **380**, 292

Liebes, S. 1964, Gravitational Lenses, *Physical Review*, **133**, 835

Mao, S. & Paczyński , B. 1991, Gravitational microlensing by double stars and planetary systems, *Astrophys. J. Lett.*, **374**, L37

Mao, S. 1999, An Ongoing OGLE Parallax Microlensing Event Toward Carina, *Astron. & Astrophys.*, **350**, L19

Mao, S. et al. 2002, Optical Gravitational Lensing Experiment. OGLE-1999-BUL-32: the Longest Ever Microlensing Event – Evidence for a Stellar Mass Black Hole?, *M.N.R.A.S.*, **329**, 349

Paczyński, B. 1986, Gravitational microlensing by the galactic halo, *Astrophys. J.*, **304**, 1

Peale, S. J. 2003, Comparison of a Ground-based Microlensing Search for Planets with a Search from Space, *Astron. J.*, **126**, 1595

Poindexter, S., Afonso, C., Bennett, D. P., Glicenstein, J.-F., Gould, A., Szymański, M. K., & Udalski, A. 2005, Systematic Analysis of 22 Microlensing Parallax Candidates, *Astrophys. J.*, **633**, 914

Popowski, P., et al. 2005, Microlensing Optical Depth toward the Galactic Bulge Using Clump Giants from the MACHO Survey, *Astrophys. J.*, **631**, 879

Pratt, M. R., et al. 1995, Real-time Detection of Gravitational Microlensing, *ArXiv Astrophysics e-prints*, arXiv:astro-ph/9508039

Refsdal, S. 1964, The gravitational lens effect, *M.N.R.A.S.*, **128**, 295

Refsdal, S. 1966, On the possibility of determining the distances and masses of stars from the gravitational lens effect, *M.N.R.A.S.*, **134**, 315

Rhie, S. H. 1997, Infimum Microlensing Amplification of the Maximum Number of Images of n-Point Lens Systems, *Astrophys. J.*, **484**, 63

Rhie, S. H. 2001, Can A Gravitational Quadruple Lens Produce 17 images?, *ArXiv Astrophysics e-prints*, arXiv:astro-ph/0103463

Rhie, S. H. 2002, How Cumbersome is a Tenth Order Polynomial?: The Case of Gravitational Triple Lens Equation, *ArXiv Astrophysics e-prints*, arXiv:astro-ph/0202294

Rhie, S. H., & Bennett, D. P. 1996, Search for Earth Mass Planets and Dark Matter Too, *Nuclear Physics B Proc. Suppl.*, **51**, 86

Rhie, S. H., Becker, A. C., Bennett, D. P., Fragile, P. C., Johnson, B. R., King, L. J., Peterson, B. A., & Quinn, J. 1999, Observations of the Binary Microlens Event MACHO 98-SMC-1 by the Microlensing Planet Search Collaboration, *Astrophys. J.*, **522**, 1037

Rhie, S. H. et al. 2000, On Planetary Companions to the MACHO 98-BLG-35 Microlens Star, *Astrophys. J.*, **533**, 378

Sahu, K. C. 1994, *Nature*, **370**, 275

Schneider, P., & Weiss, A. 1986, The two-point-mass lens - Detailed investigation of a special asymmetric gravitational lens, *Astron. & Astrophys.*, **164**, 237

Schneider, P., & Weiss, A. 1987, A gravitational lens origin for AGN-variability? Consequences of micro-lensing, *Astron. & Astrophys.*, **171**, 49

Smith, M. C., Mao, S., & Woźniak, P. 2002, Parallax Microlensing Events in the OGLE II Database Toward the Galactic Bulge, *M.N.R.A.S.*, **332**, 962

Smith, M. C., Mao, S., & Paczyński, B. 2003, Acceleration and Parallax Effects in Gravitational Microlensing, *M.N.R.A.S.*, **339**, 925

Sumi, T. et al. 2003, Microlensing Optical Depth toward the Galactic Bulge from Microlensing Observations in Astrophysics Group Observations during 2000 with Difference Image Analysis, *Astrophys. J.*, **591**, 204

Sumi, T., et al. 2006, Microlensing Optical Depth toward the Galactic Bulge Using Bright Sources from OGLE-II, *Astrophys. J.*, **636**, 240

Tisserand, P., et al. 2007, Limits on the Macho content of the Galactic Halo from the EROS-2 Survey of the Magellanic Clouds, *Astron. & Astrophys.*, **469**, 387

Udalski, A., Szymański, M., Kałużny, J., Kubiak, M., Krzmiński, W., Mateo, M., Preston, G. W., & Paczyński , B. 1993, The optical gravitational lensing experiment. Discovery of the first candidate microlensing event in the direction of the Galactic Bulge, *Acta Astronomica*, **43**, 289

Udalski, A., Szymański, M., Kałużny, J., Kubiak, M., Mateo, M., Krzmiński, W., & Paczyński , B. 1994, The Optical Gravitational Lensing Experiment. The Early Warning System: Real Time Microlensing, *Acta Astronomica*, **44**, 227

Udalski, A. et al. 2005, A Jovian-Mass Planet in Microlensing Event OGLE-2005-BLG-071, *Astrophys. J. Lett.*, **628**, L109

Udry, S., et al. 2007, The HARPS search for southern extra-solar planets. XI. Super-Earths (5 and 8 $M_\oplus$) in a 3-planet system, *Astron. & Astrophys.*, **469**, L43

van Belle, G. T. 1999, Predicting Stellar Angular Sizes, *P.A.S.P.*, **111**, 1515

Wolszczan, A., & Frail, D. A. 1992, A planetary system around the millisecond pulsar PSR1257 + 12, *Nature*, **355**, 145

Wu, X. 1994, Gravitational microlensing by the MACHOs of the Large Magellanic Cloud, *Astrophys. J.*, **435**, 66

Yanagisawa, T., et al. 2000, Wide-Field Camera for Gravitational Microlensing Survey: MOA-Cam2, *Experimental Astronomy*, **10**, 519

Yoo, J. et al. 2004, OGLE-2003-BLG-262: Finite-Source Effects from a Point-Mass Lens, *Astrophys. J.*, **603**, 139

4 Formation and Evolution of Terrestrial Planets in Protoplanetary and Debris Disks

George H. Rieke

Summary. This review discusses the properties of protoplanetary disks in the context of terrestrial planet formation, emphasising the general pattern of planet formation and evidence for similar patterns of evolution, and also for the diversity of starting points and hence probable conclusions. In addition, the process of terrestrial planet formation extends well beyond the protoplanetary stage, and produces disks of debris from the planetesimal collisions. The observed behaviour of these debris disks can test many hypotheses regarding the evolution of the Solar System. Debris disks also let us probe many different examples of how planetary systems generally like ours evolve (there are nearly 150 examples known within 50pc).

4.1 Overview

Current technology only allows us to study planets in significant numbers under special circumstances, such as Doppler-shift determinations of the velocity recoil, or when a planet orbit is aligned so we witness a transit. In comparison, planets swim in tenuous disks of dust (and early-on, gas) that are relatively easy to detect. The surface area of finely-divided material in these disks vastly exceeds that of any planets – an analogy is that it is easier to see paint when it is spread on a wall than when it is in the can. The dust intercepts energy from the star more efficiently than a planet can, and thus scatters and reradiates energy in far larger amounts than a planet could. In the process, it imposes its own signatures on this output. We know of hundreds of planetary systems through observation of circumstellar disks of dust. We can learn indirectly about them if we can read these signatures.

The infrared output of circumstellar dust plays a prominent role throughout the life of a young star. Star formation begins with a dark cloud with a cold ($\sim$ 10K) core supported against gravitational collapse by some combination of internal pressure, turbulence, and a magnetic field. If this stability is upset by an externally induced perturbation, or the magnetic field leaks out of the cloud, collapse toward a star proceeds quickly. These "Class 0" objects are very cold (a few tens of degrees K) and dusty. They are optically thick in the visible through mid-infrared, but are bright in the far-infrared and submm. They are rare, which implies they persist for only of order 10^4 years.

Accretion feeds the central object from the surrounding cloud, causing it to grow and warm up as the cloud becomes thinner and more transparent. The resulting "Class I" object consists of a core that will become a star surrounded by a disk where most of the angular momentum of the system resides. Energy released by the accretion powers strong emission from the combination of remaining cloud and optically thick protoplanetary disk, causing the spectral energy distribution (SED) to rise into the mid-infrared. Conventionally the SED is described in λF_λ units[1]; the resulting power law slope for a Class I object is > 0 (where, for comparison, a purely stellar SED would fall roughly as the Rayleigh-Jeans law, $\lambda F_\lambda \sim \lambda^{-3}$). This stage lasts of order 10^5 years.

The protoplanetary disk provides raw material for the continuing stellar accretion and for outflows. Dust and gas are processed in the disk, leading to the growth of planetesimals and gas giant planets. As accretion decreases and the other loss mechanisms continue to consume the disk, the infrared emission falls. When this emission has a power law slope (in λF_λ) between 0 and -1.5, the object is termed to be of "Class II," a phase that lasts a few million years. At the end of this phase, the disk material has either been lost or has gathered into rocks, planetesimals, and planets. As a result, the SED of the system becomes dominated by the star itself, with Rayleigh-Jeans behaviour. The star is then described as being of "Class III," with a SED that falls at least as fast as $\lambda^{-1.5}$ into the mid-infrared. The SED classes correlate well with more traditional spectral signatures (Padgett et al., 2006); thus, Class II sources are usually associated with Classical T Tauri stars (CTTS) defined by strong Hα emission lines indicative of accretion, while Class III properties generally accompany Weak Line T Tauri stars (WTTS). All of these steps occur well before the star settles on the main sequence which, depending on mass, requires $\sim$ 10–100 Myr for stars within a factor of two of 1 $M_\odot$.

If planet formation has been successful through its initial stages in the protoplanetary disk, the Class III object will be accompanied by planets and planetesimals. However, they will be essentially invisible, lost in the glare of the star and with negligible effect on its SED. The formation and evolution of gas giant planets will have run its course and nearly all the gas will have left the system. Growth of terrestrial planets, asteroids, and analogues to Kuiper Belt objects continues for a few tens of millions of years. The planetesimals collide vigorously during this period and continue to do so at a lower rate through the following very long settling-down period. Cascades of collisions grind the planetesimals down, and the resulting large numbers of small particles form a debris disk. These particles are warmed by the star and produce a measurable excess in the mid- and far-infrared (by factors of $\sim$ 1.2 to $\sim$ 10 or more relative to the stellar photosphere). Because the small particles are lost relatively quickly, continued collisional activity is necessary to sustain a debris disk. Nonetheless, debris disks can persist for many Gyr and provide evidence for planetary systems around a broad variety of main sequence stars.

---

[1]A source with equal output per logarithmic wavelength interval has constant λF_λ.

4.2 Protoplanetary Disks

4.2.1 Disk Behaviour

A typical cloud core has a velocity gradient across its (nominally ~ 0.1 pc) diameter of about 1 km s^{-1} pc^{-1}, although accurate determinations are difficult because of turbulence. If this gradient is attributed to rotation, the angular momentum exceeds that of the Sun by more than a factor of 100,000 (Armitage, 2007). To allow collapse, the angular momentum is predicted to be deposited into a massive circumstellar disk (e.g., Terebey et al. (1984); Hogerheijde (2001)).

Stars emerge from this phase surrounded by disks with sizes of 70 to 1000 AU (e.g., Kitamura et al. (2002); Qi et al. (2003); Semenov et al. (2005); Andrews & Williams (2007)). Hillenbrand (2002) reviews the evolution of these disks. Initially, their inner regions ($\leq$ 1AU) are heated by continuing accretion, while the heating in the outer zones is dominated by absorption of stellar radiation. The energy deposited by accretion can be estimated from the release of potential energy, showing that it is dominant at accretion rates larger than about 3×10^{-8} $M_{\odot}$/yr. Measured rates range from well above this value to two orders of magnitude below (Gullbring et al., 1998). Thus, as the star ages and accretion dies out, the entire disk begins to emit passively, powered by radiative energy absorbed from the star. During this later phase, the disks absorb and reradiate in the infrared about 1-10% of the stellar output[2]. This ratio of infrared excess to total luminosity is termed the fractional luminosity of the disk.

Initially, the inner zones of these disks out to a few stellar radii are cleared by the effects of the magnetic field. Another fixed boundary is set by the radius within which the dust is destroyed by the radiation field of the star. Because the inner rim of the region where the dust can survive is directly exposed to the stellar radiation, it is much hotter than the rest of the disk and may puff up. This effect is proposed to be dramatic for luminous Herbig Ae stars (where the rim is near 0.5 AU) (Dullemond et al. (2001), but see also Muzerolle et al. (2004)). It is more subdued for lower mass T Tauri stars (where the rim is within 0.1 AU: Muzerolle et al. (2003)). Terrestrial planets cannot form in these inner cleared regions, and the temperatures are too high for gaseous planets.

Outside this inner rim, the stellar energy is absorbed in the outer layers of the optically thick disk; half is promptly radiated away into space and the other half is radiated inward to heat the interior of the disk. Models combining both radiative transfer and hydrostatic equilibrium in the disk interior show that the disk flares at increasing distances from the star, such that it absorbs more energy than would be the case for a flat disk (Kenyon & Hartmann, 1987; Chiang & Goldreich, 1997). Scattered light images of young disks confirm that they have significant depth and, in favourable cases, show the flaring (e.g., Burrows et al. (1996); Stapelfeldt et al. (1998); Padgett et al. (1999)), see Fig. 4.1. Analysis of the scattered light images

[2]Although the role of protoplanetary disks in young stellar infrared excesses now seems obvious, it took more than a decade after the first infrared detections for this explanation to be proposed (Grasdalen et al., 1984; Beall, 1987)

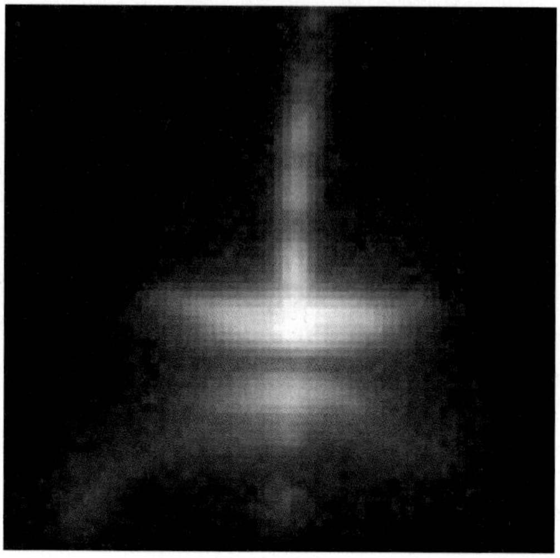

Fig. 4.1. Image of protoplanetary disk HH30 (Burrows et al., 1996). The edge-on disk occults the star to act as a natural coronagraph and allow detailed structures to be seen in both the disk and the jets.

can provide constraints on the disk inclination, the wavelength-dependent opacity (and hence can search for evidence of grain growth), and the scale height of the disk. Flaring can be measured directly by analysis of the dark, obscured lane. Further discussion can be found in the reviews by Dullemond et al. (2007) and Armitage (2007).

From this state, protoplanetary disks clear progressively to larger radii as the system evolves. There are a variety of dispersal mechanisms, such as photoevaporation, grain growth, accretion onto planetesimals, and ejection from the system. Clearing times for the inner disks (order of 1 AU) in these systems are typically 3 Myr (Haisch et al., 2001). *Spitzer* observations provide good statistics to track this behaviour out to its final stages. Lada et al. (2006) find 30 $\pm$ 4% optically thick disks (indicated by excesses at 5.8 and 8 μm) in the 2–3 Myr old IC 348 cluster. Hernandez et al. (2007) report that about 35% of the roughly solar-mass members of the 3 Myr old σ Ori cluster have excesses at 5.8 and 8 μm (see also Oliveira et al. (2006)). Dahm & Hillenbrand (2007) find 7 $\pm$ 2% optically thick disks left in NGC 2362 at 5 Myr. Currie et al. (2007a) find in h and χ Per at 13 Myr that no more than half this many stars still have excesses at both 5.8 and 8 μme. Gorlova et al. (2007) find an upper limit of about 1% for disks emitting at 5.8 and 8 μm in NGC 2547 at 25 Myr. For additional examples, see Sicilia-Aguilar et al. (2006) and Bonatto et al. (2006), but note Megeath et al. (2005) and Haisch et al. (2005); Hillenbrand (2005) reviews observations of disk dispersal.

Snapshots of the evolution of young disks are provided by images in scattered light, both with HST and from the ground. Because such observations are limited

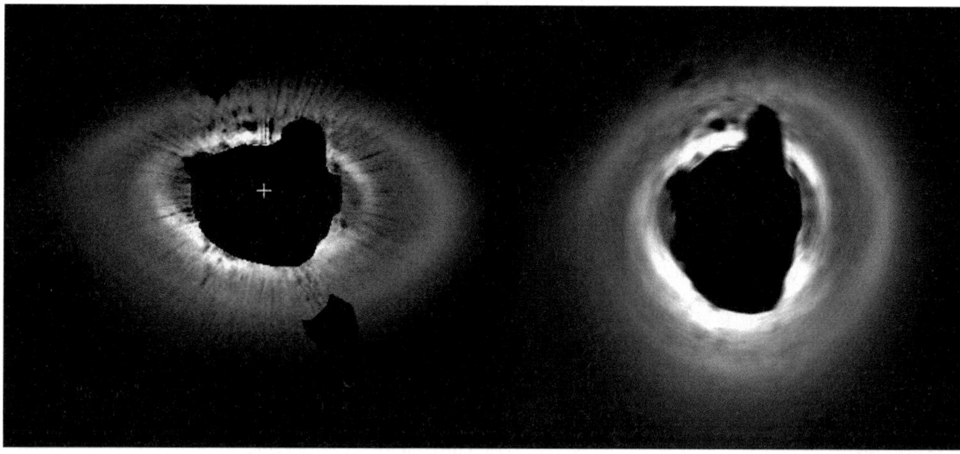

Fig. 4.2. Scattered light image of the disk around the 5 Myr old star HD 141569A (Clampin et al., 2003). The left panel is the HST/ACS image; the star has been occulted by the instrument coronagraph. In the right image, the disk is viewed as if face-on to show the structure in more detail (the image has also been filtered to remove diffraction artifacts). The structures in the disk may arise from planets, or from interactions with passing stars, or a combination.

in surface brightness sensitivity, they have been successful primarily on the relatively dense disks that are emerging from the protoplanetary stage. Among other examples, images have been obtained for T Tauri stars such as GG Tau and UY Aur (Krist et al., 2005; Close et al., 1998), pre-main-sequence stars such as TW Hya (Roberge et al. (2005) and references therein), and very young more massive stars such as HD 141569A (see Fig. 4.2) and AB Aur (Ardila et al. (2005); Mariñas et al. (2006)). The number of imaged systems is now adequate to support a vigorous field of comparative anatomy and physical analysis of the dynamics of young disks. It is often found that they have marked asymmetries and complex structures possibly associated with gravitational instabilities, either within the disk or due to perturbations by passing stars. This subject is reviewed by Watson et al. (2007).

Despite the trend of clearing from the inner to the outer regions, the dense outer disk zones also disperse rapidly: Andrews & Williams (2005) report that only 10% of young stars have evidence for a cold and dense outer disk in submm emission yet do not have near-infrared excesses indicative of material in the $\sim$ 1 AU zone. They also tentatively conclude that the dense outer disks dissipate on a $\sim$ 6 Myr time scale, as predicted by some models of disk photoevaporation (Alexander et al., 2006a,b).These trends also depend on stellar mass, with disks disappearing more rapidly around stars significantly more massive than the Sun (Carpenter et al., 2006; Hernandez et al., 2007; Dahm & Hillenbrand, 2007; Currie et al., 2007a).

The amount of raw material in protoplanetary disks can be determined by measurements in the submillimetre. There, the disks are generally optically thin and the emission is in the Rayleigh-Jeans regime, so the flux density and disk mass

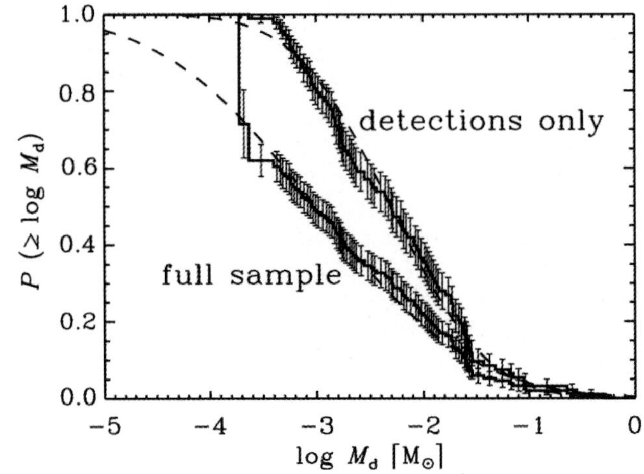

Fig. 4.3. Cumulative distribution of disk masses, from Andrews & Williams (2005). The best fitting log-normal distributions are shown as dashed lines. The best estimate of the intrinsic distribution is the fit for the full sample, including upper limits in the fit.

are directly proportional (Hildebrand, 1983). Andrews & Williams (2005) report submillimetre measurements of 153 young stellar objects. The data are fitted by a lognormal distribution with a mean value of 3×10^{-3} $M_\odot$ and a variance of 1.31 dex, i.e. there is a very broad range of masses (see Fig. 4.3). This result is reinforced by a study of 336 stars in the Orion Trapezium region (Eisner & Carpenter, 2006). Although less than 10% of the systems were detected directly, stacking the data yielded an estimate of 5×10^{-3} $M_\odot$ for the average disk mass. The wide range of disk mass for a narrow range of stellar mass indicates that there must be important variables in disk formation such as the residual angular momentum of the collapsing cold cloud core. As a result, there is little reason to expect planetary system formation to have a strong dependence on stellar type.

The conditions that might lead to a planetary system similar to ours make an interesting benchmark. The minimum mass of the solar protoplanetary disk can be derived by taking each planet and adding to its mass sufficient material to represent the solar composition (Weidenschilling, 1977; Hayashi, 1981). Thus, a substantial mass must be added for each terrestrial planet to account for the missing hydrogen and helium, while for Jupiter and Saturn the correction is small. The resulting mass surface density is roughly

$$\Sigma = 1000 \left(\frac{r}{AU}\right)^{-1.5} g \ \ cm^{-2} \tag{4.1}$$

from about 0.7 to 30 AU. From integrating this profile, the minimum mass required to form the planets is about 0.01 $M_\odot$. Based on the submm emission, Andrews & Williams (2005) found that only 37% of the disks exceed the minimum mass required for the planets in the Solar System; in agreement, Eisner & Carpenter (2006) found an average disk mass slightly below this minimum mass. These estimates are subject to significant systematic errors (largely because of uncertainties in the grain optical constants due to grain growth in the disk environment (Andrews & Williams, 2007)). However, due to the inefficiencies in planet formation, it is also likely that the Solar System started with a substantially more massive disk than the minimum possible. It appears likely that a substantial fraction of stars do not have disks massive enough to replicate our Solar System.

The evolution of the innermost parts of disks is generally correlated with the accretion rates onto their stars. About 80% of binary pairs are either both CTTS or both WTTS (e.g., Hartigan et al. (1994); Prato & Simon (1997); Duchêne et al. (1999); Hartigan & Kenyon (2003)), showing that the inner disks of both stars have evolved at similar rates. However, disk structures are seen to vary at a given age due to variations in the clearing time of the $\sim$ 1 AU disk zones prominent in the mid-infrared (Hernandez et al. (2007) and references therein). Variations in the infrared characteristics of very young binary pairs are also common (Haisch et al., 2006). Much of the variety may result from the range of initial disk properties, as is apparent from the range of masses. Another contributor may be that the transition in disk properties occurs rapidly and therefore may not synchronise precisely with emission line properties. In addition, a small number of systems appear to retain warm, primordial dust for perhaps 10 Myr, significantly longer than is typical (Silverstone et al., 2006).

In conclusion, there is a well-defined overall pattern of protoplanetary disk characteristics. However, equally striking is the wide range of starting conditions, e.g. disk masses, along with some variation in evolutionary time scales. These differences presumably translate into a wide range of properties for the planetary systems that develop within these disks.

4.2.2 Terrestrial Planet Formation

The steps that lead to the formation of planets in protoplanetary disks are poorly accessed by observation and hence are largely the province of theory. They are reviewed by Chambers (2004) and Nagasawa et al. (2007). Gas-rich giant planets must form very early on, well before the 3Myr time scale for disk dissipation and the accompanying escape of most of the gas from the system. These steps are described in the chapter by Hugh Jones et al. elsewhere in this volume. However, it is believed that terrestrial planets have a much longer incubation period, which we describe here.

We start with a dense, gas-rich disk. The pressure of the warm gas inflates and supports the disk in the vertical direction. The gas pressure decreases with distance from the star, producing a net outward force that causes the gas to orbit the star slightly more slowly than the Keplerian velocity appropriate for individual solid objects. The dust grains are also slowed through interactions with the gas. However, the dust grains are not well coupled to the gas pressure, and they migrate toward the mid-plane of the disk by the effects of the vertical component of the gravity of the star and viscous dissipation in the disk (for detailed modelling, see

Garaud & Lin (2004)). As a result, the dust-to-gas ratio increases along the mid-plane. Consequently, near the mid-plane the dust grains dominate the rotational dynamics and the gas is pulled along with them close to the Keplerian velocity. The result of the differing gas velocities within the disk is turbulence that supports and thickens the dust-rich central layer.

A possible path to planets is for dust grains to stick together and gradually aggregate into larger bodies. Numerical simulations indicate that colliding micron-sized grains have a high probability of sticking (Dominik & Tielens, 1997), a conclusion that is confirmed by experiment (Poppe et al., 2000a). This tendency can be further enhanced by electrostatic attraction, enabled by charge exchange between grains and the production of electrical dipoles (Poppe et al., 2000b; Marshall et al., 2005). The overall process of grain agglomeration can be quite complex, depending on the relative masses and energies of collision of the two input grains. Dominik et al. (2007) provide a review.

The next steps are poorly understood. Meter-sized boulders are large enough such that their motions will be very close to Keplerian orbits, so they will move through the gas and experience a headwind. Small particles in the gas will be brought to the surfaces of the boulders continuously. If the collisional velocities are too high, an erosion of the surfaces analogous to sand blasting will occur, but if the velocities are low there is a high probability the grains will stick and cause the boulders to grow. In addition, the boulders will lose orbital energy to the headwind and spiral in toward the star, eventually evaporating. The smaller boulders may have to grow rapidly to survive, since the gas headwind becomes more effective with the decreasing volume to surface area ratio of smaller objects.

The difficulties in gradual growth while avoiding spiralling inward too close to the star can be circumvented by another growth mode. The dust-rich layer may become sufficiently dense that local gravitational instabilities cause rapid collapse into solid bodies up to a few kilometres in size (e.g., Haghighipour & Boss (2003); Rice et al. (2004); Tanga et al. (2004); Boss (2005)). However, full treatment of this possibility is quite complex and therefore the conclusions are uncertain (Boley et al., 2006).

By either route, eventually, objects are built up to a few kilometres in diameter. Models indicate that most collisions between such bodies will lead to further accretion (e.g., Benz & Asphaug (1999); Leinhardt et al. (2000)). The collisional rates are enhanced by the tendency of gravitational interactions to reduce the relative velocities of large bodies. At the same time, gas drag tends to make the orbits more circular, decreasing the relative velocities and increasing the probability for accretion in encounters. The growth process favours the emergence of a small number of relatively large objects, termed planetary embryos. Eventually, these bodies grow to two orders of magnitude more mass than a typical planetesimal, and their gravitational effects rather than collective ones start to dominate the behaviour. The result is a period of oligarchic growth (Leinhardt et al., 2000). Planet embryos continue to grow, but at a rate regulated by feedback from gravitational effects: as an embryo grows overly massive, it perturbs nearby planetesimals in ways that increase the violence of any collisions and thus slow the growth. As a result, each embryo has a

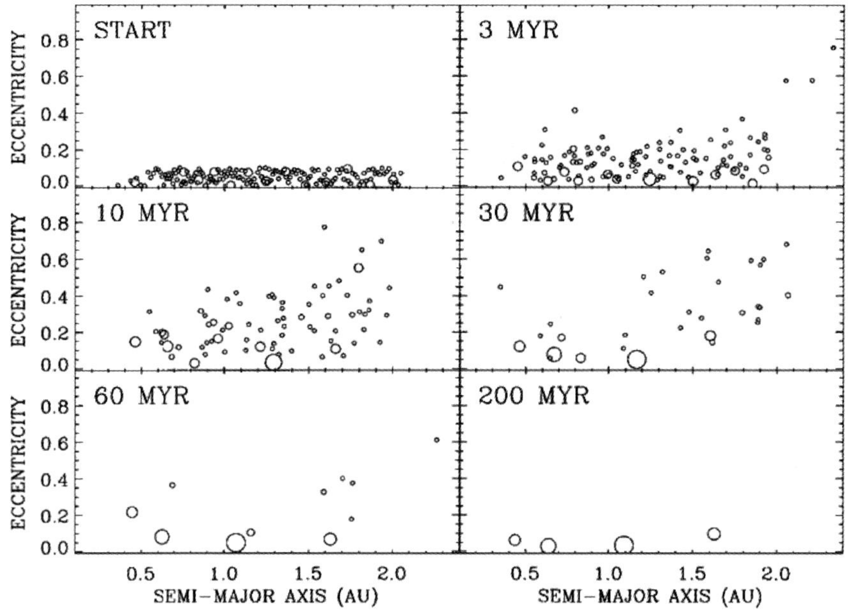

Fig. 4.4. Outcome of an N-body simulation of planet formation, from Chambers (2001). In the first panel, the figure shows the orbital eccentricities and semi-major axes for 153 equal-mass embryo planets, and it then follows their merging and growth with time. The size of the symbols is proportional to the size of the objects.

zone of influence, called a feeding zone, of width of order 0.01 AU (but proportional to the embryo mass), and within which it consumes most planetesimals. Over a period of about 1 Myr, the number of planetesimals dwindles to the point where the feedback is no longer effective and the growth stalls (e.g., Weidenschilling et al. (1997)). However, with the feedback turned off, gravitational interactions among the embryos (Kenyon & Bromley, 2006) or larger-scale perturbations due to the effects of residual gas in the system (Nagasawa et al., 2005) cause the embryos to stray from their feeding zones and interact. They grow into a small number of truly planet-sized objects (but not without setbacks due to inopportune collisions: Agnor & Asphaug (2004)). The final stages have been modelled by Chambers (2001), showing end points that are reasonably similar in overall characteristics to the Solar System. Figure 4.4 shows an example.

4.3 Debris Disks

Even after the protoplanetary disk has faded away, terrestrial planet formation continues to be marked by collisions, which produce debris particles. These particles are heated by the star and can be detected as an excess of infrared output above the level of the stellar photosphere. However, these particles are lost quickly due to non-gravitational forces, resulting either in their being ejected from the system or spiralling into the star. As a result, maintaining debris disks requires that there be continuing collisions among larger bodies to replenish the particles. Such collisions continue for a long time after a planetary system has formed and stabilised. For example, within the Solar System, collisional debris continues to be produced in the asteroid and the Kuiper belts. The characteristics of the resulting debris systems provide a measure of the evolution of planetary systems into the Gyr age range. In fact, debris disks are at present virtually the only way to probe events in large numbers of planetary systems outside the zones (within a few AU) that we can study through Doppler recoil and planetary transit measurements.

4.3.1 Debris in the Solar System

The behaviour of debris in the Solar System gives us a foundation to compare with other systems. From the crater record on the Moon and other arguments, we know that giant impacts occurred frequently over the first 100Myr (Canup, 2004). A catastrophic collision of this type led to the birth of the Moon. It must also have produced a huge cloud of debris that escaped from the Earth-Moon system, to orbit the Sun. Other collisions continued over the entire period when the terrestrial planets were forming and evolving toward their current form, as described by Chambers (2004). This process was punctuated by the Late Heavy Bombardment about 700 Myr after the formation of the Sun (Gomes et al., 2005). Throughout, there would have been a continuous production of debris, probably with large spikes in grain production whenever a particularly large collision occurred.

Although – thank goodness – the violent era in the Solar System is largely behind us, it has not stopped altogether. Gravitational stirring by Jupiter keeps the asteroid belt in a mild state of turmoil, and major collisions occurred there 5 and 8 million years ago. Despite the passage of time, these events account for up to 30% of the dust currently orbiting in the inner Solar System (Nesvorný et al., 2002, 2003). Detailed modelling of the surface area of dust in the system (Grogan et al., 2001) shows that such events are unexceptional. The inner edge of the Kuiper Belt is maintained by the gravitational action of the giant planets, particularly Neptune (Liou & Zook, 1999). Gravity also stirs the Kuiper Belt to produce collisions that replenish the small-sized debris there (Brown et al., 2007).

By now, the level of debris generation in the Solar System has declined to very low levels. The excess emission associated with the asteroid belt is no more than 10^{-7} of the solar luminosity (Backman et al., 1995). One of the biggest discoveries with the IRAS mission was that a number of nearby stars have excess emission at fractional luminosities of $\sim 10^{-4}$ (Aumann et al., 1984), that is, three orders of magnitude greater than the estimates for the Solar System. Most of these systems have radii of roughly 100 AU and therefore corresponded to the Kuiper Belt, which had not yet been discovered at the time of IRAS.

4.3.2 Theoretical Background

The general behaviour of debris disks is discussed by Backman & Paresce (1993) and Dominik & Decin (2003). The motions of grains larger than 100 μm (for an A star) are dominated by gravity and follow nearly Keplerian orbits. When they collide, the fragmentation products act as projectiles to shatter additional particles and (for the large ones) planetesimals and create more debris. The products of the resulting collisional cascades follow a size distribution proportional to $a^{-3.5}$, where a represents the object radii (Dohnanyi, 1969). Debris is removed when it has been ground down to a fine level. Blowout occurs when the ratio of photon pressure to gravitational force on a grain,

$$\beta^a = 0.57\, Q^a_{pr}\, \frac{L/L_\odot}{M/M_\odot} \left(\frac{a}{\mu\ m}\right)^{-1} \left(\frac{\delta}{g\ cm^{-3}}\right)^{-1}, \tag{4.2}$$

exceeds 0.5 (Burns et al., 1979). In this equation, L and M are the stellar luminosity and mass, δ is the density of the grain material, and Q^a_{pr} is the radiation pressure efficiency averaged over the stellar spectrum. Very small grains are blown away in this way very quickly, on a time scale of thousands of years. In dilute disks, collisions may not occur rapidly enough to grind the grains down very quickly to the photon-pressure-loss size range. In that case, Poynting-Robertson drag – the effect of photons being absorbed preferentially from the forward direction in the frame of an orbiting particle – can cause loss of orbital energy, leading to spiralling into the star. For a grain at distance r from a star, the timescale for this process is

$$\tau_{PR} = \frac{800\ \ yrs}{\beta^a} \left(\frac{r}{1AU}\right)^2 \left(\frac{M_\odot}{M}\right) \tag{4.3}$$

and can be thousands (e.g., for a 1 μm particle 1 AU from the Sun) to millions of years. However, in the dense disk zones that are bright in the infrared, the grains are reduced in size by collisions sufficiently quickly that they are ultimately expelled by photon pressure. If the particle loss mechanism depends on mutual particle interactions such as collisional cascades, the number of grains and fractional excess luminosity will decay as t^{-1} where t is time, while if the loss is independent of mutual interactions such as P-R drag, the fractional luminosity will decay as t^{-2} (Dominik & Decin, 2003).

It is implicit in analytic treatments that the debris evolve in a smooth and continuous fashion. Numerical simulations can introduce other processes. Large planets can carve out broad gaps in a disk of debris, and in doing so may sling many asteroid-sized objects as well as debris out of the plane of the system and possibly into escape orbits (e.g., Moro-Martín & Malhotra (2005) and references therein). Also, collisions of large bodies can generate enhanced grain populations for a period of time (e.g., Kenyon & Bromley (2004); Grigorieva et al. (2007)).

4.3.3 Evolution

IRAS and ISO were used to show that, even at the relatively high level of $\sim 10^{-4}$ fractional luminosity, debris disks occur frequently (a general summary of debris disk studies after the completion of these two missions can be found in Caroff et al. (2004)). Initial efforts were made with these data to trace the time evolution of debris production (e.g., Habing et al. (2001); Spangler et al. (2001)). A major advance with *Spitzer* is that debris disks can be detected in sufficient numbers and in complete samples, so the evolution of debris generation can be traced accurately. Strong excesses, suggesting active terrestrial planet building (and destruction) are common around young stars, less than 100 Myr in age (Rieke et al., 2005; Chen et al., 2005). A-type stars are attractive targets to track disk evolution beyond this initial stage both because of their high luminosity and because their main sequence lifetimes span the key period of disk evolution (discovered after the fact, of course). They have been studied most extensively at 24 μm, which tracks the roughly terrestrial planet zone. At this wavelength, they show an envelope to the infrared excess that decays roughly as t^{-1} (Rieke et al., 2005; Su et al., 2006; Rhee et al., 2007), with a characteristic time of about 150 Myr: see Fig. 4.5. However, as many as half of the sample have no detectable excess, even at the youngest age range (5 to 20 Myr) (Rieke et al., 2005). The behaviour at 70 μm appears to be similar in both regards (Su et al., 2006), except that the decay of the excesses is much slower (but still consistent with t^{-1}).

The broad range of excesses around young stars can be explained consistently as arising from the broad range of protoplanetary disk mass apparent from submm observations (Wyatt et al., 2007a) (see Fig. 4.5). That is, to first order, the fate of a planetary system as measured by its debris content is probably determined by the mass of its protoplanetary disk. Given this conclusion, the upper envelope of the excesses should be a true indication of the time dependence of their decay (since it traces the most massive systems at each age). Thus, the inverse time dependence is a confirmation of the theoretical prediction for collisional cascade generation of debris. The difference in decay rates at 24 and 70 μm shows that planetary systems evolve from the inside outward, that is, collisional activity dies down in the terrestrial planet zone more quickly than in the Kuiper Belt one.

About 17% of the stars more than 1 Gyr in age have significant 70 μm excesses (Trilling et al., 2007a). The persistence of debris systems indicates that the t^{-1} decay must slow substantially beyond 1 Gyr. Kim et al. (2005) and Bryden et al. (2006) make two different types of comparison of external debris systems with the Solar System, and each concludes that the Kuiper Belt is likely to have a far infrared output within the distribution of that from similar stars. COBE data have been used to place an upper limit of 10^{-6} for the fractional luminosity of the Kuiper Belt in the far infrared (Backman et al., 1995), so our system would fall slightly below the current detection limits for exo-solar systems at $\sim 10^{-5}$ fractional luminosity. That is, the detected systems represent the bright end of a distribution that includes the behaviour of the Solar System.

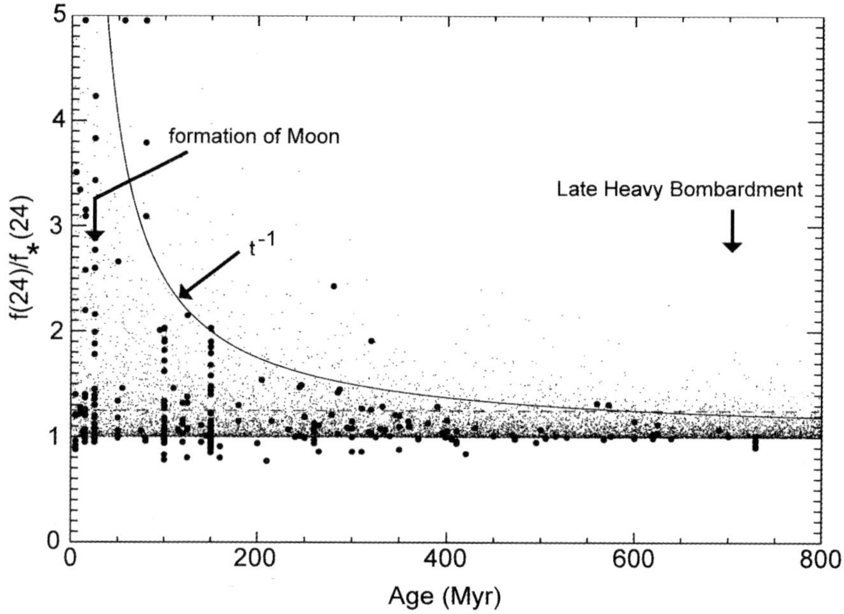

Fig. 4.5. Evolution of 24 μm excesses of A stars with time, after Wyatt et al. (2007a). The figure shows the ratio of measured 24 μm flux density to the predicted level for the stellar photosphere alone; therefore, a ratio of 1 indicates no excess. The large dots are measured values, from Rieke et al. (2005). The small dots are from a model that assumes the variation in excess goes as the disk masses deduced from submm measurements (Andrews & Williams, 2005; Wyatt et al., 2007a). Two milestones in the formation of the Solar System are indicated to put the evolution into perspective.

4.3.4 Spectral Energy Distributions

The SEDs of the great majority of debris disks are remarkably similar, with shapes indicating that the material is at a temperature of about 70K. The SEDs encode the disk structure, since the equilibrium temperature of a grain is given by

$$\frac{R_*^2}{r^2}\int Q_{ab}^a\left(\lambda\right)B\left(T_*,\lambda\right)d\lambda = 4\int Q_{ab}^a\left(\lambda\right)B\left(T_g,\lambda\right)d\lambda. \tag{4.4}$$

Here, $B(T,\lambda)$ is the blackbody function for temperature T and at wavelength λ, $Q_{ab}^a(\lambda)$ is the grain absorption coefficient at λ, T_* and T_g are the temperatures of the star and grain, respectively, R_* is the stellar radius, and r is the distance of the grain from the star. The left side of this equation represents the energy absorbed by the grain from the star, and the right side that emitted by the grain. Applying this equation, the SEDs of the vast majority of debris disks indicate that the grain population is at tens to a couple of hundred AU from the star (depending on type). In fact, for stars of roughly solar type (late F to early K), *Spitzer* SEDs generally indicate little detectable debris emission at wavelengths short of 25 μm (Kim et al., 2005; Beichman et al., 2005), showing that these rings of debris are usually terminated at their inner edges with little material inside. The limits on the mass of interior material are very strong because at the higher equilibrium temperature for such grains they would be easily detectable. The most likely explanation for these inner clearings is the presence of massive planets (e.g., Liou & Zook (1999); Moro-Martín & Malhotra (2005)).

However, it is difficult to draw more specific conclusions from photometric-resolution SEDs, since a variety of models can be made consistent with the sparsely spaced measurements (Moro-Martín et al. (2005); see also Su et al. (2006)). Typically, a power law disk density behaviour similar to eq. 4.1 is assumed (but with alternative choices for the spectral index). This degeneracy of SED models is a significant obstacle to making further progress in understanding debris disk structure on a broad basis (Moro-Martín et al., 2005).

In general, the spectra of debris disks are featureless (Beichman et al., 2005; Chen et al., 2006), indicating that the particles are large enough to be optically thick (i.e., of order 10 μm or more). However, there are exceptions. The strong crystalline features in the 10–35 μm spectrum of HD 69830 show its excess is due almost entirely to a large population of extremely tiny, crystalline grains with very short lifetimes against loss or destruction (Beichman et al., 2006). Their presence in such numbers requires that they have been generated recently as part of a transient phenomenon such as a super-comet being deflected into an orbit approaching the star, or a collision in an asteroid belt more than an order of magnitude more densely populated than ours (Beichman et al., 2006). The discovery of a complex planetary system around this star, with three Neptune-mass members (Lovis et al., 2006), may help account for the peculiarities of its debris system. Song et al. (2005) have found similar properties in the infrared excess of BD +20.307, and they conclude that this star must also have its debris system dominated by a recent large collision.

A minority of systems also have very different SEDs, with strong excesses at 24 μm. The infrared outputs of many of these systems are likely to be dominated by recent massive collisions or other transient events. The best-studied example, ζ Lep, has its debris concentrated within 3AU of the star (Chen & Jura, 2001; Moerchen et al., 2007). Wyatt et al. (2007a) conclude that its peculiar characteristics may well stem from a recent planetesimal collision. Wyatt et al. (2007b) show that, in addition to ζ Lep, HD 69830 and BD +20.307, the hot dust around η Corvi and HD 72905 indicates the presence of planetesimals recently scattered inward from an outer belt of such bodies.

Objects with extreme excesses are also found in young clusters: up to eight in h and χ Per at 13 Myr (Currie et al., 2007b); one or two in NGC 2547 at 25 Myr (Gorlova et al., 2007); and one in M47 at 100 Myr (Gorlova et al., 2004). A plausible explanation is that they represent collisions of very large bodies, perhaps analogous to the collision that formed our Moon.

4.3.5 Imaging

Sample images of debris disks in the submillimetre can be found in Holland et al. (1998, 2003): Vega, Fomalhaut, β Pic, Wilner et al. (2002): Vega, Greaves et al. (2004): τ Ceti, Greaves et al. (2005): ϵ Eri, and Marsh et al. (2006): Vega. The images uniformly show prominent rings of material about 100 AU from the stars. These rings, analogous to the Kuiper Belt, are presumably where most of the mass in the debris system lies. In some cases, structures in the submm rings have been attributed to interactions with massive planets (e.g., Wilner et al. (2002); Marsh et al. (2006)). In comparison, infrared images can show a variety of structures, some of which represent smaller particles derived from those in the rings and that move to different zones under nongravitational forces.

Four nearby stars of ages only about 10 Myr have spectacular debris disks. β Pic is the best studied and, at $\sim$ 19pc, the closest (Telesco et al. (2005); Golimowski et al. (2006) and references therein). HR 4796A and 49 Cet resemble β Pic but are about four times more distant (Wahhaj et al. (2005) and Wahhaj et al. (2007) and references therein). All three are characterised by huge excesses already at 24 μm, clearly the products of intense terrestrial planet building (and destruction). AU Mic ((Metchev et al., 2005) and references therein) is at similar distance as β Pic, but has a low mass central star (M1V as opposed to A-type stars for the other three). These systems let us probe additional aspects of planet building such as the detection of probable comets falling into β Pic (Karmann et al. (2003) and references therein), and the large number of grains below the blowout size that are being lost from the system (Krivov et al., 2000). There is substantial structure in the β Pic disk (Okamoto et al., 2004; Telesco et al., 2005), as well as ring-like structures in other young disks (see, e.g., Fig. 4.2), all of which are most logically explained as arising in part from the influence of massive planets through orbital resonant interactions with the disk particles (Ardila et al., 2005; Freistetter et al., 2007; Krivov et al., 2007).

The best images of mature debris disks (older than 30 Myr) obtained with *Spitzer* are of Fomalhaut (Fig. 4.6), Vega (Fig. 4.7), and ϵ Eridani. The variety of behaviour is a warning against relying on general application of the SED models described in the preceding section. Fomalhaut does indeed behave as expected, with a circumstellar ring inclined to the line of sight and of radius $\sim$ 140 AU, prominent at 70 μm and through the submm. The ring is filled in at 24 μm by dilute, warm dust (Staplefeldt et al., 2004). ϵ Eridani shows a 65 AU ring in the submm (Greaves et al., 2005), but the ring is already filled at 70 μm (D. Backman, private communication). However, the Vega image is astounding (Fig. 4.7). Debris in the nearly perfectly face-on disk can be traced to a radius of nearly 1000AU at 70 μm. Su et al. (2005) fit the characteristics of the system by appealing to small grains ($\sim$ 10 microns in size) that are being ejected by photon pressure. As with HD 69830 and the other systems bright at 24 μm, the loss rate for these grains is so high that it is implausible that the Vega system has always had its current appearance. Su et al.'s fits to the images indicate that the grains originate in a Keplerian ring of objects detected in the submm at a radius of about 90 AU, where they suggest a large collision has

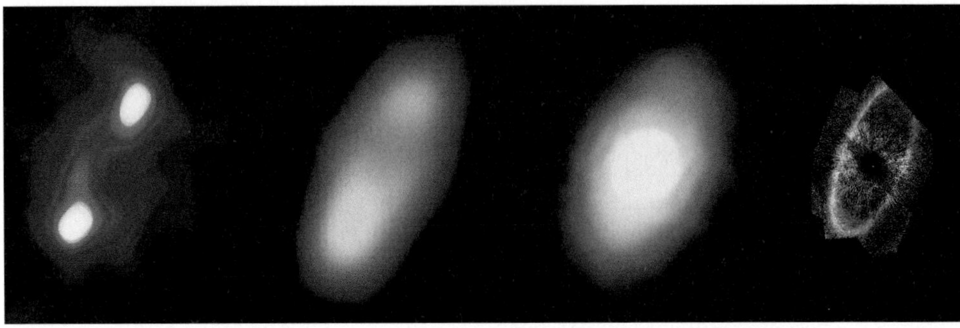

Fig. 4.6. Images of the Fomalhaut disk at different wavelengths. Left, 450 μm, beam 7.5", from Holland et al. (2003). Center left, 70 μm, beam 18"; center right 24 μm, beam 6", from Staplefeldt et al. (2004). Right, 0.8 μm, beam 0.07", from Kalas et al. (2005). The figure is approximately 70" high.

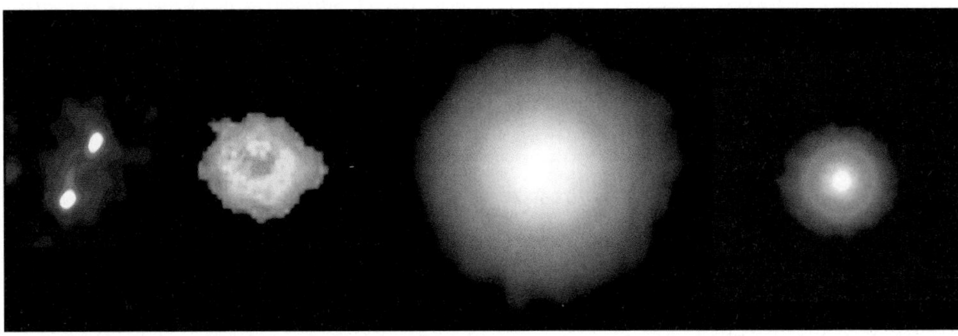

Fig. 4.7. Images of the Vega disk at different wavelengths. Left, 450 μm image of Fomalhaut to the same physical scale. Center left, 350 μm, beam 11", from Marsh et al. (2006). Center right 70 μm, beam 18"; right, 24 μm, beam 6", from Su et al. (2005). The figure is approximately $3'$ in height.

taken place on the order of a million years ago, setting up the collisional cascade that is responsible for the small grains.

Mature debris disks are difficult to image in scattered light because their surface brightness is generally too low. Fomalhaut is a dramatic exception (Kalas et al., 2005). The HST image reveals a ring that is about 25 AU wide and has a sharp inner edge at a radius from the star of 133 AU (Fig. 4.6). The ring center is offset by 15 AU from the position of the star, a configuration that probably can only be maintained by interaction with a massive planet. The planet presumably also is responsible for the sharp inner edge, either through gravitational resonances or through ejection of material (Kalas et al., 2005). Comparison of the scattered light image with the thermally radiated one from *Spitzer* (Staplefeldt et al., 2004) shows two interesting effects. First, the asymmetry in the *Spitzer* images of the ring at 24 and 70 μm is likely due to the greater heating closer to the star due to the 15 AU ring offset. Second, the grains responsible for the central peak in the 24 μm image, the asteroidal/zodiacal component of the system, are invisible in scattered light, presumably because the grain population there is very tenuous.

4.3.6 Dependence on Stellar Mass, Metallicity, and Presence of Companions

The detection rate for excesses around A stars is higher than that around solar-like stars. However, much of the difference appears to result from the lower average age of the A stars (with main sequence lifetimes of $\sim$ 800 Myr) than for solar-like ones (lifetimes $\sim$ 10 Gyr). Gorlova et al. (2006) and Siegler et al. (2007) show that the incidence of debris disks around young solar-like stars is remarkably similar to that around A stars of similar age. Within the Gyr age range, the incidence of debris disks also appears to be independent of stellar type and mass to first order, except for the lower incidence around M type stars (Trilling et al., 2007a). There have been suggestions that debris might be absent in the latter stellar type due to the effects of winds in expelling particles (Plavchan et al., 2005). To date, no mature M stars have been found to have excesses (Gautier et al., 2007), but the sample is not yet large enough to prove the wind hypothesis definitively. Thus, with the possible exception of the lowest mass range, there appears to be no strong dependence of debris disk behaviour on stellar mass.

Bryden et al. (2006) find no correlation of debris disk incidence with stellar metallicity, a surprising result given the strong correlation of metallicity with the presence of Doppler-recoil-detected planets (Fischer & Valenti, 2005). Moro-Martín et al. (2007) find no correlation of excess incidence with the presence of such planets, but Bryden et al. (2007) find some evidence for such a correlation after allowance for observational selection effects. Trilling et al. (2007b) find tentative evidence that the frequency of infrared excesses increases in close binary systems compared with single stars, a result that is perhaps related to the behaviour with massive planets. Thus, planetesimal formation is not inhibited in multiple systems, and collisional activity may be enhanced when the system components are both of substantial mass.

4.4 Conclusion

We have a general understanding of the initial collapse of cold cloud cores into young stars with protoplanetary disks, and how these disks evolve, largely constrained by observation. How planets form in these disks is the realm of theory. The timescales have to be short to reach a system with giant gas planets already in place. Terrestrial planets are thought to form through a series of stages, some of which are poorly understood. However, after a few tens of millions of years, there is general agreement between numerical simulations and observations that fully formed planetary systems are in place and the gas-rich disks have been completely dispersed.

The final stages of planet formation and evolution produce second-generation disks of dusty debris from collisions. Observations of them show behaviour that is generally analogous to our understanding of the evolution of the Solar System over its first few hundred million years. These steps include the initial decay from a very high level of collisional activity, a significant level of activity out to the time of the Late Heavy Bombardment, and eventual separation of asteroidal and Kuiper Belt zones. However, since we can study hundreds of debris disk systems, we are also beginning to explore the large variations on this overall theme, such as large collisional events at various times that dominate the dust production for a number of millions of years.

We have only been aware of protoplanetary and debris disks for about 30 years. They are very active areas of research that will provide unique insights to the overall process of planet formation and evolution, which are recognised as major themes both of research and broader human inquiry. Fortunately, a number of powerful new capabilities will continue the rapid advance in these areas over the next decade. Herschel will provide new observations of the cold outer regions of these disks and will take spectra that should reveal the chemistry and how it changes from the stages of cold cloud core through the final escape of gas from protoplanetary disks. Large groundbased optical/infrared telescopes will provide diffraction-limited images of protoplanetary disks, at resolutions on the order of 1 AU. Submillimetre arrays, especially ALMA, will yield high resolution images of the cold disk components and will also study gas features with sufficient spectral (and thus velocity) resolution to place them within the disk structure. JWST will let us image the terrestrial planet zones in many debris disks, as well as exploring gas chemistry in the mid-infrared where there are strong transitions of many molecules that play essential roles in the possible formation of life. Additional powerful observatories are under discussion that would keep this area advancing dramatically for many more years.

Acknowledgements

I thank Andras Gáspár, James Muzerolle, John Stansberry, Kate Su, and David Trilling for helpful discussions. This work was supported in part by NASA contracts 1255094 and NAG-12318.

References

Agnor, C. & E. Asphaug 2004, Accretion efficiency during planetary collisions, *ApJL*, **613**, L157

Alexander, R. D., C. J. Clarke, & J. E. Pringle 2006a, Photoevaporation of protoplanetary discs – I Hydrodynamic models, *MNRAS*, **369**, 216

Alexander, R. D., C. J. Clarke, & J. E. Pringle 2006b, Photoevaporation of protoplanetary disks – II. Evolutionary models and observable properties, *MNRAS*, **369**, 229

Andrews, S. M. & J. P. Williams 2005, Circumstellar dust disks in Taurus-Auriga: The submillimeter perspective, *ApJ*, **631**, 1134

Andrews, S. M. & J. P. Williams 2007, High-resolution submillimeter constraints on circumstellar disk structure, *ApJ*, **659**, 705

Ardila, D. R. et al 2005, A dynamical simulation of the debris disk around HD 141569A, *ApJ*, **627**, 986

Armitage, P. J. 2007, Lecture notes on the formation and early evolution of planetary systems, astro-ph/0701485, **recommended review**

Aumann, H. H. et al 1984, Discovery of a shell around Alpha Lyrae, *ApJL*, **278**, L23

Backman, D. E. & F. Paresce 1993, Main-sequence stars with circumstellar solid material – The VEGA phenomenon, in Protostars and Planets III, (University of Arizona Press, Tucson, AZ) pp. 1253–1304, **dated but classic review**

Backman, D. E., A. Dasgupta, & R. E. Stencel 1995, Model of a Kuiper Belt small grain population and resulting far-infrared emission, *ApJL*, **450**, L35

Beall, J. H. 1987, The observational appearance of protostellar accretion disks, *ApJ*, **316**, 227

Beichman, C. A. et al 2005, An excess due to small grains around the nearby K0 V star HD 69830: Asteroid or cometary debris?, *ApJ*, **626**, 1061

Beichman, C. A. et al 2006, IRS spectra of solar-type stars: A search for asteroid belt analogs, *ApJ*, **639**, 1166

Benz, W. & E. Asphaug 1999, Catastrophic disruptions revisited, *Icarus*, **142**, 5

Boley, A. C. et al 2006, The thermal regulation of gravitational instabilities in protoplanetary disks. III. Simulations with radiative cooling and realistic opacities, *ApJ*, **651**, 517

Bonatto, C., E. Bica, S. Ortolani, & B. Barbuy 2006, Detection of K_S-excess stars in the 14 Myr open cluster NGC 4755, *A&A*, **453**, 121

Boss, A. P. 2005, Evolution of the solar nebula. VII. Formation and survival of protoplanets formed by disk instability, *ApJ*, **629**, 535

Brown, M. E., K. M. Barkume, D. Ragozzine, & E. L. Schaller 2007, A collisional family of icy objects in the Kuiper Belt, *Nature*, **446**, 294

Bryden, G. et al 2006, Frequency of debris disks around solar-type stars: First results from a *Spitzer*/MIPS survey, *ApJ*, **636**, 1098

Bryden, G. et al 2007, Planets and debris disks: Results from a *Spitzer*/MIPS search for IR excess, submitted to *ApJ*

Burns, J. A., P. L. Lamy, & S. Soter 1979, Radiation forces on small particles in the solar system, Icarus, **40**, 1, **recommended review**

Burrows, C. J. et al 1996, Hubble Space Telescope observations of the disk and jet of HH 30, *ApJ*, **473**, 437

Canup, R. M. 2004, Dynamics of lunar formation, ARAA, **42**, 441, **recommended review**

Caroff, L., L. Juleen Moon, D. Backman, & E. Praton 2004, Debris disks and the formation of planets: A symposium in memory of Fred Gillett, ASP Conf. Series, No. 324

Carpenter, J. M., E. E. Mamajek, L. A. Hillenbrand, & M. R. Meyer 2006, Evidence for mass-dependent circumstellar disk evolution in the 5 Myr old Upper Scorpius OB Association, *ApJL*, **651**, L49

Chambers, J. E. 2001, Making more terrestrial planets, *Icarus*, **152**, 205

Chambers, J. E. 2004, Planetary accretion in the inner Solar System, Earth and Planetary Science Letters, **223**, 241, **recommended review**

Chen, C. H. & M. Jura 2001, A possible massive asteroid belt around ζ Leporis, *ApJL*, **560**, L171

Chen, C. H., M. Jura, K. D. Gordon, & M. Blaylock 2005, A *Spitzer* study of dusty disks in the Scorpius-Centaurus OB Association, *ApJ*, **623**, 493

Chen, C. H. et al. 2006, *Spitzer* IRS spectroscopy of IRAS-discovered debris disks, *ApJS*, **166**, 351

Chiang, E. I. & P. Goldreich 1997, Spectral energy distributions of T Tauri Stars with passive circumstellar disks, *ApJ*, **490**, 368

Clampin, M. et al. 2003, Hubble Space Telescope ACS coronagraph imaging of the circumstellar disk around HD 141569A, *AJ*, **126**, 385

Close, L. M. et al 1998, Adaptive optics imaging of the circumbinary disk around the T Tauri binary UY Aurigae: Estimates of the binary mass and circumbinary dust grain size distribution, *ApJ*, **499**, 883

Currie, T. et al 2007a, *Spitzer* IRAC and JHK$_S$ observations of h and χ Per: Constraints on protoplanetary disk and massive cluster evolution at $\sim 10^7$ years, *ApJ*, **659**, 599

Currie, T. et al 2007b, Terrestrial zone debris disk candidates in h and χ Persei, *ApJL*, **663**, L105

Dahm, S. E. & L. A. Hillenbrand 2007, *Spitzer* observations of NGC 2362: Primordial disks at 5 Myr, *AJ*, **133**, 2072

Dohnanyi, J. W. 1969, Collisional models of asteroids and their debris, *J. Geophys. Res.*, **74**, 2531

Dominik, C. & G. Decin 2003, Age dependence of the Vega Phenomenon: Theory, *ApJ*, **598**, 626, **recommended for an overview of debris disk evolution**

Dominik, C. & A. G. G. M. Tielens 1997, The physics of dust coagulation and the structure of dust aggregates in space *ApJ*, **480**, 647

Dominik, C., J. Blum, J. N. Cuzzi, & G. Wurm 2007, Growth of dust as the initial step toward planet formation, in *Protostars & Planets V,* pp 783–800, **recommended review**

Duchêne, G., J.-L. Monin, J. Bouvier, & F. Ménard 1999, Accretion in Taurus PMS binaries: a spectroscopic study, *A&A*, **351**, 954

Dullemond, C. P., C. Dominik, & A. Natta 2001, Passive irradiated circumstellar disks with an inner hole, *ApJ*, **560**, 957

Dullemond, C. P., D. Hollenbach, I. Kamp, & P. D'Alessio 2007, Models of the structure and evolution of protoplanetary disks, in *Protostars and Planets V,* pp 555–572, **recommended review**

Eisner, J. A. & J. M. Carpenter 2006, Massive protoplanetary disks in the Trapezium region, *ApJ*, **641**, 1162

Fischer, D. A. & J. Valenti 2005, The planet-metallicity connection, *ApJ*, **622**, 1102

Freistetter, F., A. V. Krivov, & T. Löhne 2007, Planets of β Pictoris revisited, *A&A*, **466**, 389

Garaud, P. & D. N. C. Lin 2004, On the evolution and stability of a protoplanetary disk dust layer, *ApJ*, **608**, 1050

Gautier, T. N. et al 2007, Far infrared properties of M dwarfs, *ApJ*, **667**, 527

Golimowski, D. A. et al 2006, Hubble Space Telescope ACS multiband coronagraphic imaging of the debris disk around β Pictoris, *AJ*, **131**, 3109

Gomes, R., H. F. Levison, K. Tsiganis, & A. Morbidelli 2005, Origin of the cataclysmic Late Heavy Bombardment period of the terrestrial planets, *Nature*, **435**, 466

Gorlova, N. et al 2004, New debris-disk candidates: 24 micron stellar excesses at 100 Million years, *ApJS*, **154**, 448

Gorlova, N. et al 2006, *Spitzer* 24 μm survey of debris disks in the Pleiades, *ApJ*, **649**, 1028

Gorlova, N., Z. Balog, G. H. Rieke, J. Muzerolle, K. Y. L. Su, V. D. Ivanov, & E. T. Young 2007, Debris disks in NGC 2547, *ApJ*, **670**, 516

Grasdalen, G. L., S. E. Strom, K. M. Strom, R. W. Capps, D. Thompson, & M. Castelaz 1984, High spatial resolution IR observations of young stellar objects – A possible disk surrounding HL Tauri, *ApJL*, **283**, L57

Greaves, J. S., M. C. Wyatt, W. S. Holland, & W. R. F. Dent 2004, The debris disc around τ Ceti: a massive analogue to the Kuiper Belt, *MNRAS*, **351L**, 54

Greaves, J. S. et al 2005, Structure in the ϵ Eridani debris disk, *ApJL*, **619**, L187

Grigorieva, A., P. Artymowicz, & Ph. Thébault 2007, Collisional dust avalanches in debris discs, *A&A*, **461**, 537

Grogan, K., S. F. Dermott, & D. D. Durda 2001, The size-frequency distribution of the zodiacal cloud: Evidence from the Solar System dust bands, *Icarus*, **152**, 251

Gullbring, E., L. Hartmann, C. Briceño, & N. Calvet 1998, Disk accretion rates for T Tauri stars, *ApJ*, **492**, 323

Habing, H. et al 2001, Incidence and survival of remnant disks around main-sequence stars, *A&A*, **365**, 545

Haghighipour, N. & A. P. Boss 2003, On pressure gradients and rapid migration of solids in a nonuniform solar nebula, *ApJ*, **583**, 996

Haisch, K. E., E. A. Lada, & C. J. Lada 2001, Disk frequencies and lifetimes in young clusters, *ApJL*, **553**, L153

Haisch, K. E., R. Jayawardhana, & A. Alves 2005, Constraints on inner disk evolution timescales: A disk census of the η Chamaeleontis young cluster, *ApJL*, **627**, L57

Haisch, K. E., M. Barsony, M. E. Ressler, & T. P. Greene 2006, Mid-infrared observations of class I/flat-spectrum systems in six nearby molecular clouds, *AJ*, **132**, 2675

Hartigan, P., K. M. Strom, & S. E. Strom 1994, Are wide pre-main-sequence binaries coeval?, *ApJ*, **427**, 961

Hartigan, P. & S. J. Kenyon 2003, A spectroscopic survey of subarcsecond binaries in the Taurus-Auriga dark cloud with the Hubble Space Telescope, *ApJ*, **583**, 334

Hayashi, C. 1981, Formation of the planets, in Fundamental Problems in the Theory of Stellar Evolution, (D. Sugimoto et al. eds.), IAU Symp. 93 (Reidel: Dordrecht), pp 113–126

Hernandez, J. et al 2007, *Spitzer* Space Telescope study of disks in the young σ Orionis cluster, *ApJ*, **662**, 1067

Hildebrand, R. H. 1983, The determination of cloud masses and dust characteristics from submillimetre thermal emission, *QJRAS*, **24**, 267

Hillenbrand, L. A. 2002, Young circumstellar disks and their evolution: A review, astro-ph/0210520v1, **recommended review**

Hillenbrand, L. A. 2005, Observational constraints on dust disk lifetimes: Implications for planet formation, astro-ph/0511083, **recommended review**

Hogerheijde, M. R. 2001, From infall to rotation around young stellar objects: A transitional phase with a 2000 AU radius contracting disk?, *ApJ*, **553**, 618

Holland, W. S. et al 1998, Submillimetre images of dusty debris around nearby stars, *Nature*, **392**, 788

Holland, W. S. et al 2003, Submillimeter observations of an asymmetric dust disk around Fomalhaut, *ApJ*, **582**, 1141

Kalas, P., J. R. Graham, & M. Clampin 2005, A planetary system as the origin of structure in Fomalhaut's dust belt, *Nature*, **435**, 1067

Karmann, C., H. Beust, & J. Klinger 2003, The physico-chemical history of falling evaporating bodies around β Pictoris: The sublimation of refractory material, *A&A*, **409**, 347

Kenyon, S. J. & B. C. Bromley 2004, Detecting the dusty debris of terrestrial planet formation, *ApJL*, **602**, L133

Kenyon, S. J. & B. C. Bromley 2006, Terrestrial planet formation. I. The transition from oligarchic growth to chaotic growth, *AJ*, **131**, 1837

Kenyon, S. J. & L. Hartmann 1987, Spectral energy distributions of T Tauri stars – Disk flaring and limits on accretion, *ApJ*, **323**, 714

Kim, J. S. et al 2005, Formation and evolution of planetary systems: Cold outer disks associated with Sun-like stars, *ApJ*, **632**, 659

Kitamura, Y., M. Momose, S. Yokogawa, R. Kawabe, M. Tamura, & S. Ida 2002, Investigation of the physical properties of protoplanetary disks around T Tauri stars by a 1 arcsecond imaging survey: Evolution and diversity of the disks in their accretion stage, *ApJ*, **581**, 357

Krist, J. E. et al 2005, Hubble Space Telescope ACS images of the GG Tauri circumbinary disk, *AJ*, **130**, 2778

Krivov, A. V., I. Mann, & N. A. Krivova 2000, Size distributions of dust in circumstellar debris disks, *A&A*, **362**, 1127

Krivov, A. V., M. Queck, T. Löhne, & M. Sremcević 2007, On the nature of clumps in debris disks, *A&A*, **462**, 199

Lada, C. J. et al 2006, *Spitzer* observations of IC 348: The disk population at 2–3 Million years, *AJ*, **131**, 1574

Leinhardt, Z. M., D. C. Richardson, & T. Quinn 2000, Direct N-body simulations of rubble pile collisions, *Icarus*, **146**, 133

Liou, J.-C. & H. A. Zook 1999, Signatures of the giant planets imprinted on the Edgeworth-Kuiper Belt dust disk, *AJ*, **118**, 580, **recommended for a discussion of planetary disk clearing**

Lovis, C. et al 2006, An extrasolar planetary system with three Neptune-mass planets, *Nature*, **441**, 305

Mariñas, N., C. M. Telesco, R. S. Fisher, C. Packham, & J. T. Radomski 2006, Mid-infrared imaging of the Herbig Ae star AB Aurigae: Extended emission on several scales, *ApJ*, **653**, 1353

Marsh, K. A., C. D. Dowell, T. Velusamy, K. Grogan, & C. A. Beichman 2006, Images of the Vega dust ring at 350 and 450 μm: New clues to the trapping of multiple-sized dust particles in planetary resonances, *ApJL*, **646**, L77

Marshall, J. R., T. B. Sauke, & J. N. Cuzzi 2005, Microgravity studies of aggregation in particulate clouds, *Geophys. Res. Letters*, **32**, Cite ID 11202

Megeath, S. T., L. Hartmann, K. L. Luhman, & G. G. Fazio 2005, *Spitzer*/IRAC Photometry of the η Chameleontis Association, *ApJL*, **634**, L113

Metchev, S. A., J. A. Eisner, L. A. Hillenbrand, & S. Wolf 2005, Adaptive optics imaging of the AU Microscopii circumstellar disk: Evidence for dynamical evolution, *ApJ*, **622**, 451

Moerchen, M. M., C. M. Telesco, C. Packham, & T. J. J. Kehoe 2007, Mid-infrared resolution of a 3 AU radius debris disk around ζ Leporis, *ApJL*, **655**, L109

Moro-Martín, A. & R. Malhotra 2005, Dust outflows and inner gaps generated by massive planets in debris disks, *ApJ*, **633**, 1150

Moro-Martín, A., S. Wolf, & R. Malhotra, 2005, Signatures of planets in spatially unresolved debris disks, *ApJ*, **621**, 1079

Moro-Martín, A. et al 2007, Are debris disks and massive planets correlated?, *ApJ*, **658**, 1312

Muzerolle, J., C. Calvet, L. Hartmann, & P. D'Alessio 2003, Unveiling the inner disk structure of T Tauri stars, *ApJL*, **597**, L149

Muzerolle, J., P. D'Alessio, N. Calvet, L. Hartmann 2004, Magnetospheres and disk accretion in Herbig Ae/Be stars, *ApJ*, **617**, 406

Nagasawa, M., D. N. C. Lin, & E. Thommes 2005, Dynamical Shake-up of planetary systems. I. Embryo trapping and induced collisions by the sweeping secular resonance and embryo-disk tidal interaction, *ApJ*, **635**, 578

Nagasawa, M., E. W. Thommes, S. J. Kenyon, B. C. Bromley, & D. N. C. Lin 2007, The diverse origins of terrestrial-planet systems, in *Protostars & Planets V*, pp 639–654, **recommended review**

Nesvorný, D., W. F. Bottke, L. Dones, & H. F. Levison 2002, The recent breakup of an asteroid in the main-belt region, *Nature*, **417**, 720

Nesvorný, D., W. F. Bottke, H. F. Levison, & L. Dones 2003, Recent origin of the Solar System dust bands, *ApJ*, **591**, 486

Okamoto, Y. et al. 2004, An early extrasolar planetary system revealed by planetesimal belts in β Pictoris, *Nature*, **431**, 660

Oliveira, J. M., R. D. Jeffries, J. Th. van Loon, & M. T. Rushton 2006, Circumstellar discs in the young σ Orionis cluster, *MNRAS*, **369**, 272

Padgett, D. L. et al 1999, Hubble Space Telescope/NICMOS imaging of disks and envelopes around very young stars, AJ, **117**, 1490

Padgett, D. L. et al 2006, The *Spitzer* c2d survey of weak-line T Tauri stars. I. Initial results, *ApJ*, **645**, 1283

Plavchan, P., M. Jura, & S. J. Lipscy 2005, Where are the M dwarf disks older than 10 Million years?, *ApJ*, **631**, 1161

Poppe, T., J. Blum, & T. Henning 2000a, Analogous experiments on the stickiness of micron-sized preplanetary dust, *ApJ*, **533**, 454

Poppe, T., J. Blum, & T. Henning 2000b, Experiments on collisional grain charging of micron-sized preplanetary dust, *ApJ*, **533**, 472

Prato, L. & M. Simon 1997, Are both stars in a classic T Tauri binary classic T Tauri stars?, *ApJ*, **474**, 455

Qi, C., J. E. Kessler, D. W. Koerner, A. I. Sargent, & G. A. Blake 2003, Continuum and CO/HCO+ emission from the disk around the T Tauri star LkCa 15, *ApJ*, **597**, 986

Rhee, J. H., I. Song, B. Zuckerman, & M. McElwain 2007, Characterization of dusty debris disks: the IRAS and Hipparcos catalogs, *ApJ*, **660**, 1556

Rice, W. K. M., G. Lodato, J. E. Pringle, P. J. Armitage, & I. A. Bonnell 2004, Accelerated planetesimal growth in self-gravitating protoplanetary discs, *MNRAS*, **355**, 543

Rieke, G. H. et al 2005, Decay of planetary debris disks, *ApJ*, **620**, 1010

Roberge, Aki, A. J. Weinberger, & E. M. Malumuth 2005, Spatially resolved spectroscopy and coronagraphic imaging of the TW Hydrae circumstellar disk, *ApJ*, **622**, 1171

Semenov, D., Ya. Pavlyuchenkov, K. Schreyer, T. Henning, C. Dullemond, & A. Bacmann 2005, Millimeter observations and modeling of the AB Aurigae system, *ApJ*, **621**, 853

Sicilia-Aguilar, A. et al 2006, Disk evolution in Cep OB2: Results from the *Spitzer* Space Telescope, *ApJ*, **638**, 897

Siegler, N. et al 2007, *Spitzer* 24 μm observations of open cluster IC 2391 and debris disk evolution of FGK stars, *ApJ*, **654**, 580

Silverstone, M. D. et al 2006, Formation and evolution of planetary systems (FEPS): Primordial warm dust evolution from 3 to 30 Myr around Sun-like stars, *ApJ*, **639**, 1138

Song, I., B. Zuckerman, A. J. Weinberger, & E. E. Becklin 2005, Extreme collisions between planetesimals as the origin of warm dust around a Sun-like star, *Nature*, **436**, 363

Spangler, C., A. I. Sargent, M. D. Silverstone, E. E. Becklin, & B. Zuckerman 2001, Dusty debris around solar-type stars: Temporal disk evolution, *ApJ*, **555**, 932

Stapelfeldt, K. R., J. E. Krist, F. Menard, J. Bouvier, D. L. Padgett, & C. J. Burrows 1998, An edge-on circumstellar disk in the young binary system HK Tauri, *ApJL*, **502**, L65

Stapelfeldt, K. R. et al 2004, First look at the Fomalhaut debris disk with the *Spitzer* Space Telescope, *ApJS*, **154**, 458

Su, K. Y. L. et al 2005, The Vega debris disk: A surprise from *Spitzer*, *ApJ*, **628**, 487

Su, K. Y. L. et al 2006, Debris disk evolution around A stars, *ApJ*, **653**, 675

Tanga, P., S. J. Weidenschilling, P. Michel, & D. C. Richardson 2004, Gravitational instability and clustering in a disk of planetesimals, *A&A*, **427**, 1105

Telesco, C. M. et al 2005, Mid-infrared images of β Pictoris and the possible role of planetesimal collisions in the central disk, *Nature*, **433**, 133

Terebey, S., F. H. Shu, & P. M. Cassen 1984, The collapse of the cores of slowly rotating isothermal clouds, *ApJ*, **286**, 529

Trilling, D. E. et al 2007a, Debris disks around F, G, and K stars, *ApJ*, in press

Trilling, D. E. et al 2007b, Debris disks in main-sequence binary systems, *ApJ*, **658**, 1289

Wahhaj, Z., D. W. Koerner, D. E. Backman, M. W. Werner, E. Serabyn, M. E. Ressler, & D. C. Lis 2005, Radial distribution of dust grains around HR 4796A, *ApJ*, **618**, 385

Wahhaj, Z., D. W. Koerner, & A. I. Sargent 2007, High-resolution imaging of the dust disk around 49 Ceti, *ApJ*, **661**, 368

Watson, A. M., K. R. Stapelfeldt, K. Wood, & F. Ménard 2007, Multiwavelength imaging of young stellar object disks: Toward an understanding of disk structure and dust evolution, in Protostars & Planets V, pp 523–538, **recommended review**

Weidenschilling, S. J. 1977, The distribution of mass in the planetary system and solar nebula, Astrophy. Sp. Sci., **51**, 153

Weidenschilling, S. J., D. Spaute, D. R. Davis, F. Marzari, & K. Ohtsuki 1997, Accretional evolution of a planetesimal swarm, *Icarus*, **128**, 429

Wilner, D. J., M. J. Holman, M. J. Kuchner, & P. T. P. Ho 2002, Structure in the dusty debris around Vega, *ApJL*, **569**, L115

Wyatt, M. C. et al 2007a, Steady-state evolution of debris disks around A stars, *ApJ*, **663**, 365

Wyatt, M. C., R. C. Smith, J. S. Greaves, C. A. Beichman, G. Bryden, & C. M. Lisse 2007b, Transience of hot dust around Sun-like stars, *ApJ*, **658**, 569

5 The Brown Dwarf – Exoplanet Connection

I. Neill Reid and Stanimir A. Metchev

Summary. Brown dwarfs form like stars but, with masses below 0.075 solar masses, or 1.5×10^{29} kg, they fail to ignite core hydrogen fusion. Lacking a central energy source, they cool and fade on timescales that are rapid by astronomical standards. Consequently, the observed characteristics of old, cold brown dwarfs provide insight into the expected properties of gas-giant exoplanets. This review focuses on brown dwarfs as companions to main-sequence and evolved stars. Following a brief historical introduction, we consider the different techniques used to identify very low mass companions of stars and discuss the advantages and challenges associated with each method. We summarise results from observational programs, particularly those regarding companion frequency as a function of mass and separation, including discussion of the so-called 'brown dwarf desert'. We consider the implications of those results for brown dwarf and planetary formation mechanisms. Finally, we outline future surveys for low mass companions, particularly direct imaging programs that will have sufficient sensitivity to detect objects of planetary mass.

5.1 Introduction

The existence of brown dwarfs was first hypothesised 45 years ago, when Kumar (1963) pointed out that hydrogen degeneracy halts the collapse of low-mass objects before their cores reach the critical temperature ($T_{crit} \sim 3\times10^7$ K) for self-sustaining hydrogen fusion. Originally called "black dwarfs", the term "brown dwarf" was introduced by Tarter (1976), but it remained a definition looking for an example for another two decades. Throughout the 1980s, a variety of survey programs were initiated, probing to increasingly lower luminosities, lower masses and, since brown dwarfs are cool, to longer wavelengths. Those programs turned up a number of candidates: VB8B (McCarthy, Probst & Low, 1986), which proved to be an optical artefact (Perrier & Marriotti, 1987); G29-38B (Zuckerman & Backlin, 1987), which proved to be circumstellar dust (Greenstein, 1988); HD 114762B (Latham et al, 1989), a confirmed radial velocity companion, whose mass has oscillated depending on the assumed orbital inclination; and GD 165B (Becklin & Zuckerman, 1988), a resolved companion of a white dwarf with extremely red colours and an unusual spectrum. However, while the last two candidates were undoubtedly real, neither could marshal irrefutable evidence that the mass is substellar.

The key turning point in brown dwarf astronomy came with the identification of Gl 229B, a very low luminosity companion to an early-type M dwarf within 6 parsecs of the Sun (Nakajima et al, 1995). Not only is Gl 229B ten magnitudes fainter than the primary star at 2.2μm, but its spectrum is radically different from any previously known star-like object, with strong bands of methane absorption, reminiscent of the Solar System gas giants (Oppenheimer et al, 1995). Those features clearly demonstrated that Gl 229B has an effective temperature less than $\sim$1200K, well below that of any hydrogen-burning star. Thus, the detection of Gl 229B was the first unambiguous identification of a sub-stellar mass brown dwarf[1].

The decade since the coronation of Gl 229B has seen an avalanche of discoveries, most made through photometric searches using either the 2-Micron All-Sly Survey (2MASS: Kirkpatrick et al, 1999; Burgasser et al, 2001; Cruz et al, 2006) or the deep, optical/far-red Sloan Digitial Sky Survey (SDSS: Geballe et al, 2002). Those discoveries have led to the definition of two new spectral classes, L dwarfs (like GD 165B) and T dwarfs (like Gl 229B), with temperatures from $\sim$2100K to $\sim$700K. The lower extreme of this temperature range overlaps with the upper temperature range of exoplanets, particularly hot Jupiters, like 51 Pegb. Consequently, observations of brown dwarfs can provide insight into the expected photometric and spectroscopic properties of exoplanets.

The overwhelming majority of known L and T dwarfs[2] are isolated field objects, but some are in binary systems. The latter include the first three plausible candidates: HD 114729B, GD 165B and Gl 229B. This is not entirely surprising, since one of the most effective means of finding objects with intrinsically low luminosities is to use the "street lamp" approach: search for companions to stars that are already known to lie close to the Sun - *vide* the archetypical M8 dwarf, VB10 (van Biesbroeck, 1944). As in stellar binary systems, the overall frequency and separation distribution of these substellar companions are important factors that can influence the extent and stability of circumstellar disks, and the potential for planet formation.

Brown dwarf astronomy can therefore inform exoplanet studies in two ways: by providing insight into their atmospheric characteristics; and by constraining formation scenarios. This article discusses insights gleaned from recent investigations in these areas.

[1]Interestingly, the announcements of the discovery of Gl 229B, the first unequivocal brown dwarf, and 51 Pegb, the first exoplanet discovered around a main sequence star, were both made during a special session at the 1995 Cool Stars meeting in Florence. Even more interestingly, neither discovery was mentioned in the conference summary - perhaps reflecting the fact that some attendees chose to go shopping during the special session.

[2]Late-M, L and T dwarfs have also been dubbed *ultracool* dwarfs.

5.2 Intrinsic Properties of Brown Dwarfs

5.2.1 Brown Dwarf Evolution

Stars and brown dwarfs form through the collapse of gaseous material (mainly hydrogen) in giant molecular clouds. As the gas collapses, energy is released leading to increased temperatures in the protostellar core. At the same time, the core density, ρ_C, rises and the material becomes partially degenerate. As the protostar collapses further, degeneracy acts as an energy sink, slowing and eventually curtailing the rise in core temperature, T_C. Hydrostatic equilibrium is achieved, and the collapse stops, when the sum of the normal gas pressure and degeneracy pressure is sufficient to balance the gravitational potential. The significant role played by degeneracy in low mass objects ensures that very low-mass stars, brown dwarfs and gas giant exoplanets all have diameters similar to that of Jupiter, R_{Jup}.

The energy released during collapse, and hence $T_C(max)$, depends primarily on the protostellar mass, M. Theoretical models predict that objects with solar abundance and $M > 0.078$ solar masses ($M_\odot$) become stars; their cores are partially degenerate, but the energy released during collapse is sufficient to achieve central temperatures $T_C(max) > T_{crit}$, leading to sustained hydrogen fusion. Objects with solar abundance and total mass $M < 0.074 M_\odot$ never reach core temperatures that exceed T_{crit}; these objects are brown dwarfs. Protostars with slightly higher masses, in the range $0.074 < \frac{M}{M_\odot} < 0.078$, achieve temporary core fusion (for $\sim 10^9$ to $> 10^{10}$ years) before degeneracy absorbs sufficient energy to push the temperature below T_{crit}. The latter objects are known as transition objects, spending their initial years as stars, but ending their lives as brown dwarfs.

Chemical abundance has a secondary role in defining M_{BDcrit}, the mass threshold that separates brown dwarfs and stars (Burrows et al, 2001). Higher helium abundance, Y, leads to a higher mean molecular weight, smaller radii and higher T_C and ρ_C for the same mass; consequently, M_{BDcrit} decreases with increasing Y. Lower metallicity leads to smaller atmospheric opacities, which in turn produces shallower temperature gradients and higher luminosities (less efficient energy retention), and correspondingly lower T_C; thus, decreasing [M/H] leads to increased M_{BDcrit} and higher luminosities and temperatures at the H-burning limit. As a specific example, at zero metallicity, $M_{BDcrit} \sim 0.092 M_\odot$, and the H-burning limit lies at $T_{eff} \sim 3600$K or ~ 1900K hotter than the brown dwarf limit at solar abundances (Burrows et al, 2001).

Lacking a long-lived central energy source, brown dwarfs fade and cool on astronomically rapid timescales. The weak dependence of radius on gravity (mass) and temperature leads to all brown dwarfs following similar trajectories in the (L, T_{eff}) plane. The *rate* of evolution depends on mass, with higher mass brown dwarfs cooling less rapidly than lower mass objects. Burrows et al (2001) give the approximate relations

$$L \propto \tau^{-1.3} M^{2.64} \tag{5.1}$$

$$T_{eff} \propto \tau^{-0.32} M^{0.83} \tag{5.2}$$

$$R \propto g^{-0.18} T_{eff}^{0.11} \approx R_{Jup} \tag{5.3}$$

where g is gravity and τ, age.

Main-sequence stars obey a mass-luminosity relation: higher mass stars are more luminous, with $L \propto M^{4.5}$ for $M > 1 M_\odot$ and $L \propto M^{2.5}$ at lower masses. The similarity in brown dwarf evolutionary tracks means that we cannot estimate the mass of a brown dwarf unless we know its age: in essence, all brown dwarfs have the same luminosity – just at different stages in their careers. However, the longer cooling times for higher-mass brown dwarfs make it likely that most field dwarfs have masses between $0.06 M_\odot$ and the hydrogen-burning limit.

Figures 5.1 and 5.2 illustrate the evolution of solar-abundance low-mass stars and brown dwarfs; the predictions are based on theoretical models computed by the

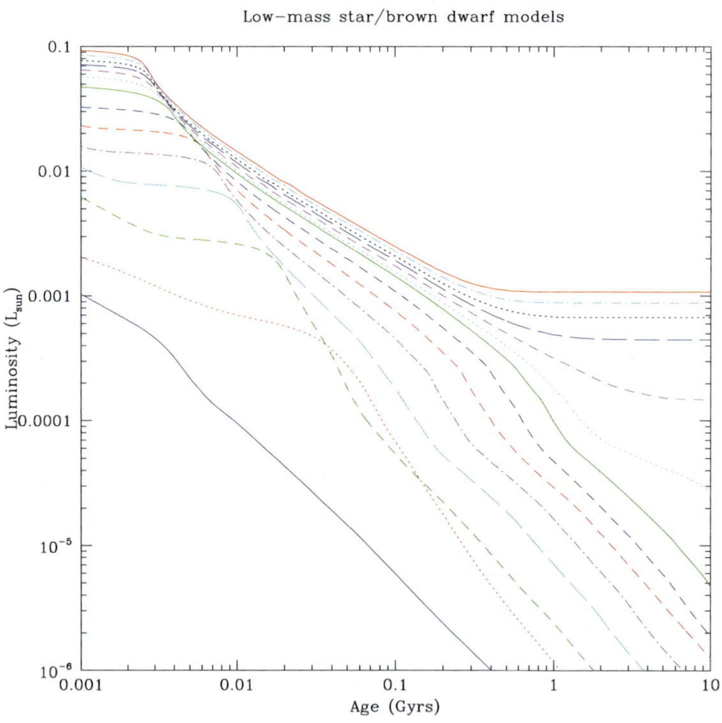

Fig. 5.1. Luminosity evolution of brown dwarfs: the evolutionary tracks are from the model calculations by Burrows et al (1993, 1997). The masses range from 0.10 $M_\odot$ (uppermost solid line) to 0.009 $M_\odot$ (lowest solid line); the latter object is not capable of sustaining deuterium burning, and therefore fades more rapidly over the initial 10^8 years. The higher mass objects (stars) achieve equilibrium, and constant luminosity, after $\sim$300 Myrs; the dotted cyan line and solid green line ($0.075 M_\odot$ and 0.070 $M_\odot$, respectively) are transition objects that are only capable of sustaining hydrogen fusion for a few billion years.

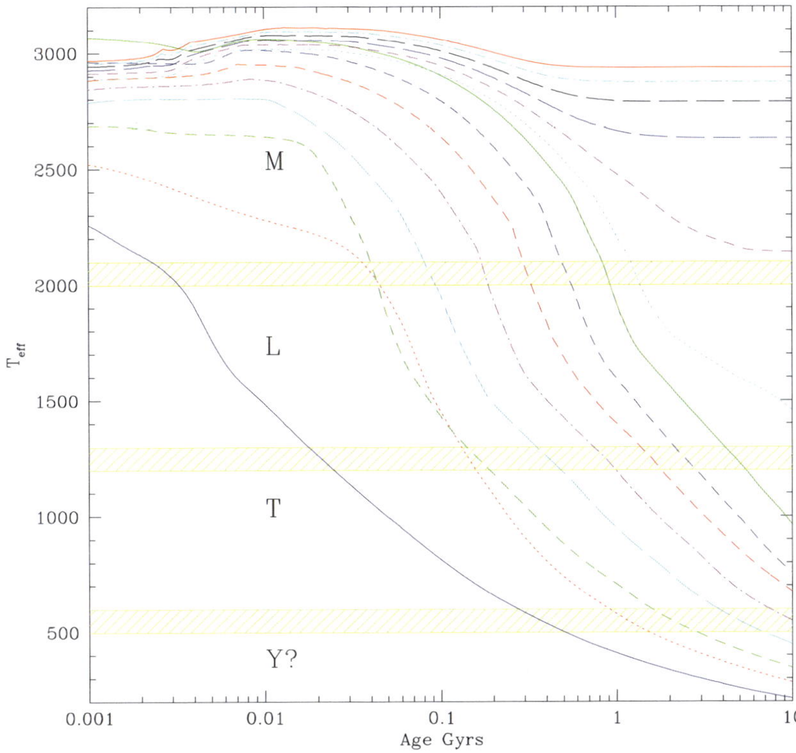

Fig. 5.2. Temperature evolution of brown dwarfs – the models are identical to those shown in Fig. 5.1. The horizontal hatched lines mark the temperatures at the transitions between spectral types M, L, T and (more speculatively) the yet-to-be discovered class Y.

Tucson group (Burrows et al., 1993; 1997). Figure 5.1 shows that a brown dwarf with $L \sim 10^{-4} L_\odot$ could be a 1 Gyr-year old transition object, or a 10^7-year old planetary-mass brown dwarf. There are spectral indicators that can be employed as crude age/mass discriminators for isolated brown dwarfs, as discussed in the following section.

Three features of the tracks plotted in Figs 5.1 and 5.2 deserve comment: first, the slow decline in T_{eff} and L for ages $\tau < 10^7$ years and masses exceeding $\sim 0.013 M_\odot$ is a consequence of fusion of primordial deuterium, which require $T_C > 2 \times 10^5$K (Salpeter, 1954); second, the $0.075 M_\odot$ dwarf is a transition object, and the shallower slope between $\sim 10^9$ and 10^{10} years reflects the presence of temporary hydrogen fusion; and, finally, the increasing separation between the $M > 0.08 M_\odot$ models and lower-mass models at $\tau > few \times 10^9$ years marks the division between fusion-supported stars and passively cooling brown dwarfs.

5.2.2 Observed Characteristics

As a brown dwarf ages and cools, the spectral energy distribution goes through significant changes. The initial temperature ($\sim 3,000$K) corresponds to a mid-type M dwarf, with a spectrum dominated by TiO, VO and metal hydride (MgH, CaH) absorption at optical wavelengths, and water bands in the near infrared. As the surface temperature falls below 2,500K, silicate dust particles condense in the atmosphere, removing TiO and, eventually, VO as significant opacity sources. Metal hydrides (MgH, CaH, FeH) and alkaline absorption lines (Na, K, Cs, Rb) become the most prominent features in the optical and far red, replacing TiO and VO. These objects are L dwarfs (Kirkpatrick et al, 1999), with temperatures cooler than $\sim 2,000$K. As the temperature cools below $\sim 1,700$K, methane forms in the outer atmosphere, becoming a prominent source of near-IR absorption at temperatures below $\sim 1,300$K; these are T dwarfs (Burgasser et al, 2002). At temperatures below ~ 500K, ammonia (NH_3) is predicted to make a significant contribution to the near- and mid-IR spectrum (Kirkpatrick, 2005). While no brown dwarf this cool has yet been identified, they have already been assigned a new spectral class, type Y. The spectral changes are illustrated in Figs 5.3 and 5.4 using representative spectra.

The flux distribution of a 2000K black body peaks at $\sim 1.5\mu m$. M, L and T dwarfs are far from black bodies, but the bulk of the energy is emitted at near-infrared wavelengths. The presence of strong absorption bands due to water and,

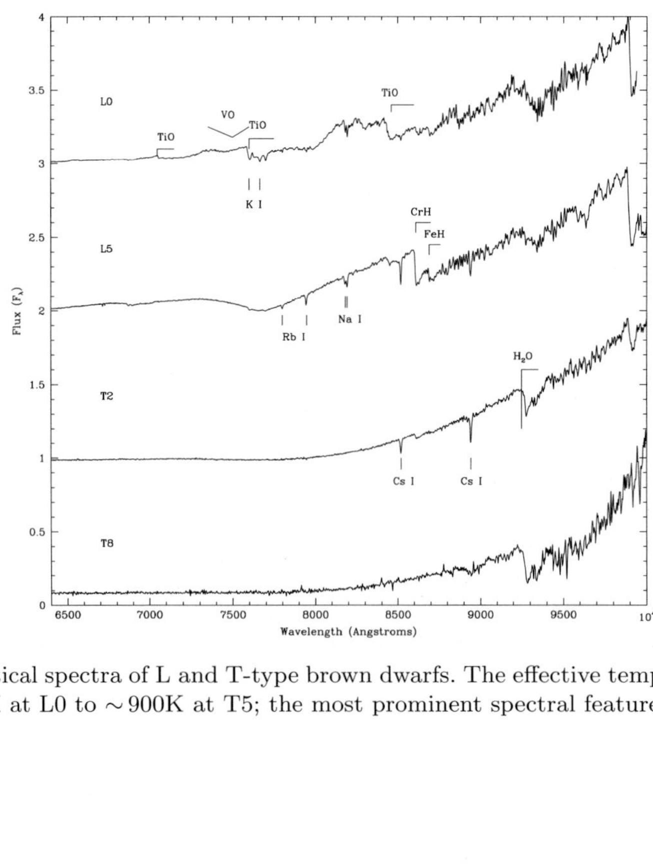

Fig. 5.3. Optical spectra of L and T-type brown dwarfs. The effective temperature ranges from ~ 2100K at L0 to ~ 900K at T5; the most prominent spectral features are labelled.

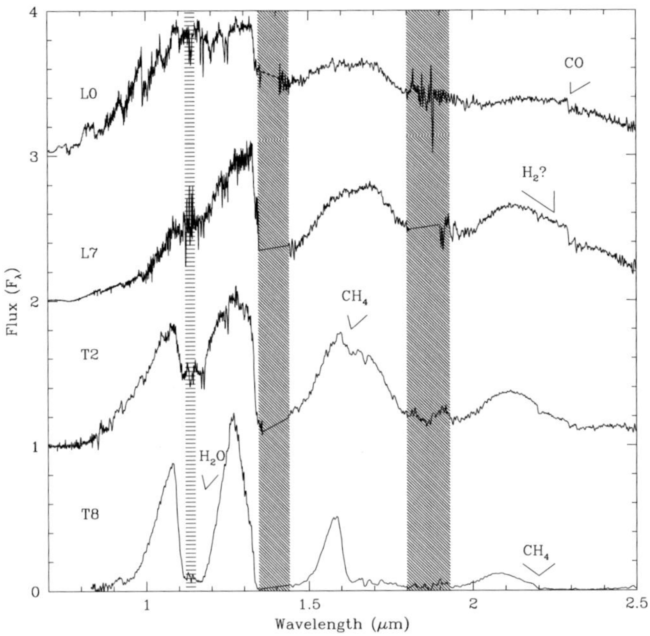

Fig. 5.4. Near-infrared spectra of L and T-type brown dwarfs. As in Fig. 5.3, the most prominent spectral features are labelled. Note, in particular, the onset of methane absorption bands that define spectral type T.

at cooler temperatures, methane favours emission in low opacity windows in the spectrum, notably the 1.2μm J band. As a guide, typical mid-type L dwarfs are ~ 10 magnitudes (a factor of 10^4) brighter at M_J than at visual wavelengths; mid-type T dwarfs, like Gl 229B, are at least ~ 13 magnitudes brighter at M_J than M_V.

Figure 5.5 shows the empirical (M_J, spectral type) and (M_J, (J-K)) colour-magnitude distributions outlined by low-mass stars and brown dwarfs with accurate parallax measurements. Late-type L dwarfs are exceptionally red at near-infrared wavelengths, rivaled only by carbon stars. The rapid blueward evolution in (J-K) between spectral types L and T reflects the onset of methane absorption, and the consequent suppression of flux in the 2.2μm K passband. The lowest luminosity T dwarfs currently known have $M_J \sim 17$, 14 magnitudes (or a factor of 400,000) fainter than the Sun at that wavelength. At visual wavelengths, T dwarfs are even fainter, with $M_V \sim 30$, or 10^{10} less luminous than the Sun.

The coolest T dwarfs known have $T_{eff} \sim 700$K, while the hottest 'hot Jupiters' are predicted to have temperatures T_{eff} from 1200 to 1500K. Stellar irradiation is likely to affect the temperature/density atmosphere profiles for the latter objects. Nonetheless, there are likely to be substantial similarities in the spectral appearances of late-L and T dwarfs, and gas giants in sub-Mercurian orbits around solar-type stars.

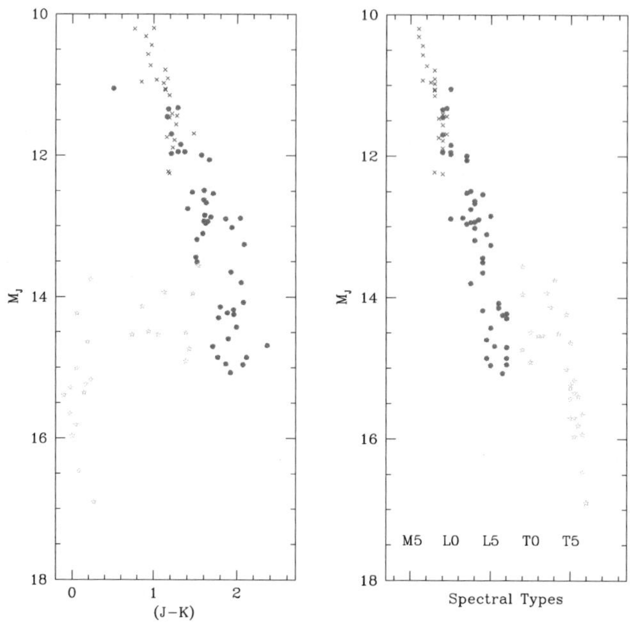

Fig. 5.5. The (M_J, (J-K)) and (M_J, spectral type) diagrams defined by low mass stars and brown dwarfs. M dwarfs are plotted as crosses, L dwarfs as solid points and T dwarfs as 5-point stars. M and then L dwarfs become progressively redder in (J-K) as the luminosity and temperature decrease. The onset of methane absorption at spectral type T reduces the flux emitted in the 2.2μm K band and leads to blue colours, comparable with hot A stars (the optical-IR colours for T dwarfs are much redder than A stars). The flux emitted in the J band is enhanced in early-type T dwarfs, leading to significant overlap in M_J with the later-type L dwarfs.

We can anticipate some spectral differences, however. Most known hot Jupiters have masses less than ~ 2 Jupiter masses[3], M_{Jup}, while most field brown dwarfs have masses between 60 to $75 M_{Jup}$. Since all of these dwarfs have the same radius, $\sim 1 R_{Jup}$, the surface gravities differ by factors of 30 to 60. Recent observations of field brown dwarfs have identified a small number with systematically different spectral features: specifically, the objects have unusually strong VO absorption, weaker alkaline atomic lines and weaker hydride bands, and they tend to be redder than average in (J-K) (Kirkpatrick, 2008; Cruz et al, 2008). Spectroscopically, these dwarfs show a greater resemblance to cool giants, strongly suggesting that the anomalous features are indicative of lower gravities, lower masses and relatively young ages. To date, most of these unusual objects are L dwarfs, lying at the upper temperature limit for even hot Jupiters. However, these observations are starting to provide clues about the likely appearance of gas giant exoplanets.

[3] $1 M_{Jup} = 0.954 \times 10^{-3} M_{\odot}$.

5.2.3 Classifying Brown Dwarfs and Exoplanets

Brown dwarfs have observational properties that overlap with low-mass stars at one extreme and with exoplanets at the other. Brown dwarfs are formally distinguished from stars on the basis that thcy cannot support long-term sustained core hydrogen fusion. Recently, there have been suggestions that similar criteria should be used to separate brown dwarfs and exoplanets; specifically, Basri (2000) and Oppenheimer et al (2000) have proposed setting the break at the mass threshold for deuterium burning, or $\sim 0.013 M_{\odot}$ for solar abundance objects (Grossman & Graboske, 1973; Saumon et al, 1996).

In principle, the deuterium-burning divisor appears to offer the advantage of a unified classification scheme. However, it is based on a property that is not directly observable, save in systems with multiple objects in known orbits. On the other hand, one might argue that the same arguments apply to hydrogen fusion: while the presence of primordial lithium is a key test for low mass brown dwarfs (Rebolo et al, 1992), in most cases the status of an object as a brown dwarf or a star is based on its observed temperature and luminosity, and there can be ambiguities in classification.

The Basri/Oppenheimer proposal, however, sets aside the traditional approach of classifying objects based on how they form: stars and brown dwarfs through core collapse in molecular clouds (Bodenheimer et al, 1980; Padoan & Nordlund, 2004); planets through core accretion (Lissauer & Stewart, 1993) or gravitational instabilities (Boss, 2002) within circumstellar disks. This difference in origin is likely to lead to significant differences in chemical composition, as can be seen in the Solar System, where the non-solar composition of Jupiter and Saturn is radically different from the ice giants, Uranus and Neptune. Corresponding changes in the emergent spectral energy distribution may be subtle, but the higher mean molecular weights lead to significant variations in the mean density and the mass-radius relation. The latter effects are already evident in the range of intrinsic properties deduced for transiting exoplanets (Fortney, Marley & Barnes, 2007).

Thus, formation within a disk does have observable consequences. Indeed, there are likely to be fewer ambiguities under this classification system than relying on an arbitrary, and essentially unmeasureable, mass limit that applies to all very low-mass objects. Moreover, identifying isolated $< 0.013 M_{\odot}$ objects as "planets" or even "exoplanets" is likely to add further unnecessary post-Pluto confusion about that term among the lay public. We prefer to preserve the traditional terminology, and distinguish brown dwarfs and planets on the basis of their formation mechanism. We therefore choose to classify objects such as the recently discovered 2MASSW J1207334–393254B (2M1207–39B), AB Pic B, GQ Lup B and SCRF 1845B as low mass brown dwarfs, not exoplanets.

5.3 Observational Techniques for Identifying Low-mass Companions

A wide variety of observational techniques have been employed to search for very low-mass companions of main-sequence and evolved stars. Here, we place these methods in an historical context, discuss recent results, and consider their relative merits. We recognise that approximately 15% of multiple star systems in the field have more than two components. In most cases, those systems are hierarchical in structure; for example, α Centaui consists of a close pair of solar-type stars, coupled with a third, low-mass component at much wider separation. From the perspective of planet formation, this is effectively a binary combined with a single star. We therefore focus our discussion on binary systems.

5.3.1 Direct Imaging Surveys

Giovanni Battista Riccioli is usually credited with the discovery of the first binary star, resolving ζ Ursae Majoris into its two wide components, Mizar and Alcor. However, the confirmation that close stellar pairs were physically-associated binary systems was actually a later by-product of the quest to measure stellar parallax. Absolute astrometry imposes severe observational requirements. A potential means of circumventing some of these obstacles was originally proposed by Galileo (1630) and restated, over a century later, by James Bradley (1747): target pairs of unequal-magnitude stars that lie in close proximity; if the two stars have similar luminosities, the fainter star lies at a larger distance, and the angular separation should exhibit an annual variation due to larger parallax of the brighter (nearer) star. The program had wait for large-scale implementation until the end of the eighteenth century, when William Herschel instituted an extensive program of double star observations with his 20-foot reflector. However, rather than resulting in the measurement of stellar parallaxes, this program revealed that many close pairs exhibited secular motions consistent with orbital motion (Herschel, 1803). Herschel was the first to refer to these physically associated stars as "binary systems".

Herschel's measurements were made by eye, and visual observations played a major role in binary-star astrometry until the mid-twentieth century. The advent of astronomical photography in the late-nineteenth century provided a means of probing to fainter magnitudes and lower flux ratio ($\frac{F_2}{F_1}$) systems. Photographic proper motions surveys, notably the Lowell survey (Giclas et al, 1958) and Luyten's surveys with the Palomar 48-inch Schmidt (Luyten, 1980; 1981), proved highly effective at identifying wide binaries, although halation rings around bright stars limited the sensitivity at moderate angular separations ($\Delta < 60$ arcseconds).

The search for even lower mass, sub-stellar, companions to stars received new impetus in the early 1990's with the introduction of large-area digital detectors that were sensitive in the near-infrared (1–2 μm) part of the spectrum. The late 1990's and early 2000's saw the completion of the first deep, wide-field sky surveys: the Deep Near-Infrared Southern Sky Survey (DENIS; Epchtein et al., 1997) and the

2MASS (Skrutskie et al., 2006). Among the primary science goals for these surveys was the discovery of brown dwarfs[4].

DENIS and 2MASS, along with the optical/far-red Sloan Digital Sky Survey (SDSS; Stoughton, 2002), account for the overwhelming majority of known brown dwarfs. A subset of known brown dwarfs (~5%; Gizis et al., 2002) are wide ($\Delta > 30$ arcseconds) common proper motion companions of nearby main sequence stars. Brown dwarf secondaries can lie over a thousand astronomical units (AU) from their primaries, distances many times greater than the size of our own planetary system[5]. Such large separations also occur in binaries with low mass stellar companions; for example, the M5 dwarf, Proxima Centauri, lies more than 40,000 AU from α Cen AB.

Imaging brown dwarf companions at smaller separations, comparable to the $\leq$ 30 AU planetary region in our own Solar System, is challenging. Even in the near-infrared, where brown dwarfs are at their brightest, they are still $\geq$ 1000 times fainter than Sun-like stars. For the nearest stars, within 10 parsecs of the Sun, 30 AU spans 3 arcseconds, meaning that any sub-stellar companion within such an orbital separation is embedded in the seeing halo of its host star. Detecting extra-solar planets is even more challenging since their near-IR luminosity is another factor of 1000 fainter.

The existence of seeing haloes around stars is a direct consequence of Earth's turbulent atmosphere. As light from a star enters the atmosphere, it is refracted along its path by multiple pockets of air at slightly different temperatures and indices of refraction, leading to a smearing of the stellar image, typically ~ 1 arcsecond at good observing sites. However, recent developments in telescope technology and fast-processing algorithms have given astronomers an edge. A novel technique, aimed at real time correction of atmospheric turbulence has been implemented at many observatories. "Adaptive optics," or AO has been used in remote sensing applications by the U.S. Air Force since the 1970s, and found its way into astronomy in the early 1990s. The technique dramatically sharpens images blurred by the turbulent atmosphere, reducing the apparent angular size of a point source (i.e., stars, brown dwarfs) to the diffraction limit λ/D of a telescope, where λ is the observation wavelength and D is the telescope's diameter (Fig. 5.6). Additional scattered light suppression and contrast enhancement may be achieved with the use of a coronagraph: a specially fabricated opaque or partially transmissive mask that blocks light from the primary star to reveal fainter objects in its vicinity. The latter technique

[4]The interest in brown dwarfs transcended their importance as a link between the realms of stars and planets. Because of their intrinsic faintness and virtually unknown properties as a galactic population, brown dwarfs were prime candidates to solve the problem of the "missing mass," a.k.a. "dark matter," in the Universe. However, it quickly became clear that the rate at which brown dwarfs were being discovered in DENIS and 2MASS fell far short of what was needed to account for a significant fraction of the unseen 99% of the mass in the Universe.

[5]The radius of our planetary system, as set by the semi-major axis of the orbit of the outermost planet, Neptune, is 30 AU.

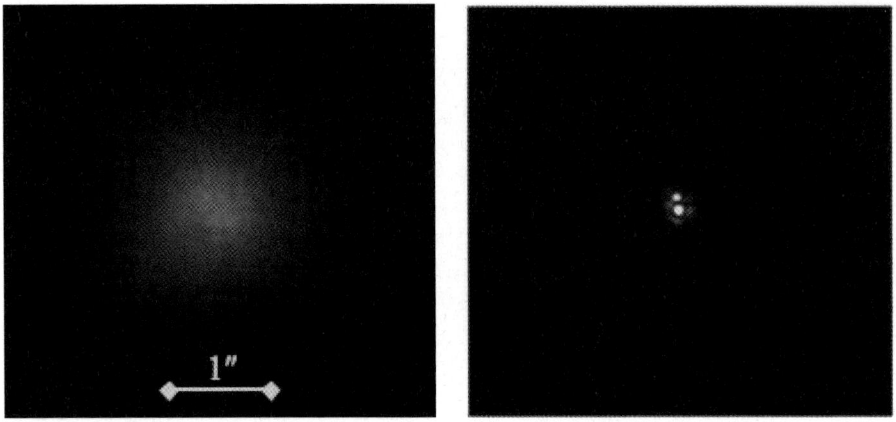

Fig. 5.6. Image of the binary star HD 18940 with the adaptive optics system on the Palomar 5 meter telescope. (a) The adaptive optics system is turned off. The binary is unresolved because of atmospheric turbulence. The scale bar indicates the approximate width of the seeing. (b) The adaptive optics system is turned on. The binary is clearly resolved with an angular separation of 0.167 arcseconds.

was invented by Bernard Lyot, who used a simple circular opaque spot to observe the solar corona in 1930.

The use of adaptive optics and a coronagraph led to the discovery of the first unambiguous brown dwarf Gl 229B (Nakajima et al, 1995) with the 1.5 meter telescope at Palomar Observatory. At an angular separation of 7.8 arcseconds from its stellar primary, Gl 229B is 13 magnitude fainter in the 0.8μm I-band and 10 magnitudess fainter at 2μm. Yet while the companion is easily discerned in the AO-corrected image (Fig. 5.7), it is lost in the glare of the primary in 2MASS seeing-limited images. Brown dwarf companions 10,000 times fainter than their host star can now be detected at separations as small as 0.5 arcseconds using AO on the largest (8 to 10 meters in diameter) ground-based telescopes. Comparable contrast at 1 arcsecond can be achieved with the *Hubble Space Telescope* orbiting above Earth's atmosphere.

Direct imaging allows the most complete characterization of the photospheres of sub-stellar companions, rendering the resolved secondary available for spectroscopic observation. Current observations can resolve separations as small as 5 AU for stars within 10 parsecs, reaching the outer boundaries of the range currently probed by radial velocity surveys for exoplanets. For the moment, image contrast limits potential detections to high mass (5–15 M_{Jup}) exoplanets; however, upgrades to existing systems and progress in coronagraphic techniques will push deeper into the planetary-mass realm, as discussed further in Sect. 5.5.

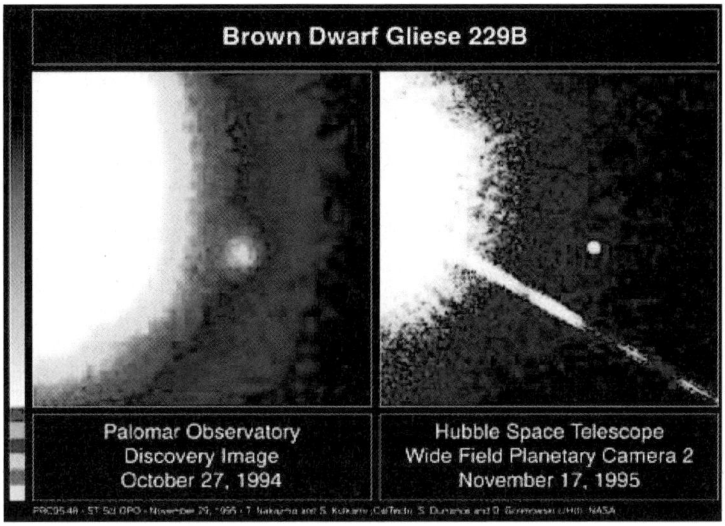

Fig. 5.7. Discovery (*left*) and follow-up (*right*) images of the brown dwarf Gl 229B, taken with an AO system on the Palomar 1.5 meter telescope, and with the Hubble Space Telescope, respectively. Both images are taken at a wavelength of 0.8 microns.

5.3.2 Radial Velocity

The presence of an unseen companion to a star can be inferred from the doubling of the lines in the stellar spectrum or from periodic shifts in the wavelength of spectral lines. Mizar (ζ Ursae Majoris) is the first known *spectroscopic binary*[6], discovered by Henry Draper in 1889. Draper observed that the potassium line in the optical spectrum of ζ UMa periodically appeared double, at intervals of 52 days. He correctly hypothesized that the phenomenon was caused by the fact that ζ UMa is an unresolved binary star with two nearly equal in brightness components. As the stars orbit around the common centre of mass, the individual line-of-sight (radial) velocities change. The potassium line is single when the two stars in the binary lie along our line of sight (zero radial velocity). The spectral lines from the individual components gradually separate due to the Doppler effect as the relative velocities change, reaching the widest separation when there is maximum difference between the blueshift of the approaching star and the redshift of the receding star.

Spectroscopic binaries that show line doubling, such as ζ UMa, are called *double-lined* spectroscopic binaries, SB2. The individual components in these systems have near-equal brightness and mass (mass ratios, $q = M_2/M_1 \sim 1$). Double-lined spectroscopic binaries allow direct measurement of the orbits of both components, and hence of the individual component masses. Binary systems with more disparate component masses ($q \leq 0.5$) and luminosities can still be detected as spectroscopic

[6]Indeed, the first known triple system, given the presence of its visual companion Alcor (Sect. 5.3.1)

binaries, although they lack the characteristic line doubling exhibited by the double-lined systems. In *single-lined* spectroscopic binaries, SB1, only the more luminous (usually more massive) component is visible, and binarity is deduced through its Doppler motion. As discussed by Irwin (this volume), the reflex motion of the primary allows one to deduce the mass ratio, modulo the orbital inclination, i. If we express the semi-amplitude of the velocity variation of the primary as K_1, then

$$K_1 \quad = \quad 28.41 P^{-1/3} \frac{m_2 \sin i}{M_1^{2/3} \sqrt{1-e^2}} \quad \mathrm{ms}^{-1} \tag{5.4}$$

where P is the period in years, M_1 is in solar masses and m_2 in Jupiter masses. Thus, as measured in the ecliptic plane, Jupiter ($P = 11.86$ years) produces a reflex motion of semi-amplitude ~ 12.5 ms^{-1} in the Sun's velocity.

Since SB1 systems have low q (especially planetary systems), accurate measurement of the primary reflex motions requires higher spectral resolution, $R = \frac{\lambda}{\Delta\lambda}$, with $R \geq 30,000$ for exoplanet systems. The first extensive campaigns searching for very low-mass secondaries were initiated in the 1980s, and Latham et al (1989) announced the discovery of the first potential brown dwarf, HD 114752B. With $M_2 \sin i = 0.011 M_\odot \approx 11 M_{Jup}$, HD 114762B falls below the substellar boundary unless $i < 8^o.4$. However, the only means of estimating i is through comparing the observed rotational velocity of HD 114762A, $v_{rot} = 0.8$kms^{-1}, with other F9 dwarfs, v_{rot}=2.3–3 kms^{-1}, leading to ambiguous conclusions. At a distance of 28 parsecs, HD 114762B is within 0.1 arcseconds of its primary, too close and too faint to be resolved even by modern direct imaging techniques.

HD 114762B is most important in that it stood alone until the discovery, 6 years later, of 51 Pegb, the first unequivocal extrasolar planet (Mayor & Queloz, 1995). With $M_2 \sin i \sim 0.47 M_{Jup}$, 51 Pegb falls well below the hydrogen burning limit and squarely in the planetary régime. To this day, nearly 20 years after the announcement of HD 114762b, and with over 200 known exoplanets with masses in the range $0.01 M_{Jup} \leq M_2 \sin i \leq 15 M_{Jup}$, radial velocity surveys have turned up barely two dozen brown dwarfs. This is surprising, since, with their much higher masses, brown dwarfs have a much larger velocity signature than exoplanets. The lack of brown dwarfs in the orbital separation range probed by radial velocity surveys (currently, 0–5 AU) is not due to our inability to detect them. It is a real phenomenon that must arise from the mechanisms of star, brown dwarf, and planet formation. The dearth of close-in brown dwarf companions to stars has given rise to the term "brown dwarf desert".

The precision radial velocity technique (see chapter by Ge, this volume) has by far been the most prolific approach for discovering extrasolar planets. It has given astronomers an unprecedented glimpse into the architectures of other solar systems. Because of the brown dwarf desert, however, most brown dwarf companions have been discovered through direct imaging.

5.3.3 Astrometric Surveys

Astrometric binaries manifest themselves through the presence of systematic irregularities in the tangential motion of the primary star. Sirius B was the first "invisible" companion discovered in this manner, revealed by Friedrich Bessel's analysis of proper motion data for the primary spanning a 6-year period from 1836 to 1842. In this case, the system lies near the Sun, the two components have similar masses and are separated by $\sim$7.5 AU. Consequently, Sirius A exhibits an astrometric "wobble" of several arcseconds, an excursion readily detectable *via* accurate 19th-century visual observations. Brown-dwarf and planetary mass companions produce much smaller reflex motions, and require correspondingly more precise measuring techniques.

The motion produced by a single companion in a circular orbit has a semi-amplitude given by the following relation:

$$A_1 \quad = \quad \frac{m_c}{10.5r}\left(\frac{P}{M_*^{2/3}}\right) \quad \text{arcsec} \tag{5.5}$$

where m_c is the mass of the companion in Jupiter masses, M_* the mass of the primary star in solar masses, r the distance in parsecs, and P the period in years (Gatewood, 1976). Thus, as viewed from a distance of 1 parsec, the Sun has an astrometric wobble of $\sim$0.5 milliarseconds due to Jupiter alone; Saturn contributes an additional 0.27 milliarcsecond. As another example, a $50M_{Jup}$ brown dwarf companion in a Jupiter-like orbit would produce a 2.5 milliarcsecond wobble in a sun-like star as viewed from a distance of 10 parsecs.

Precision astrometry is therefore an alternative to radial velocity in searching for sub-stellar companions to main sequence stars. However, astrometry is most effective in a different régime of orbital parameter space. Astrometric surveys are best suited to detecting sub-stellar secondaries in wide, face-on orbits, since those systems induce larger amplitude (and longer period) reflex motion on the host star. In contrast, radial velocity surveys are optimal for detecting close-in companions in edge-on orbits.

Astrometric searches for sub-stellar companions have had a chequered career, at least until recently. Reuyl & Holmberg (1943) were the first to enter the lists, with the claimed detection of a $\sim 0.01M_\odot$ companion of 70 Ophiuchi (which actually has an unrelated $\sim 1.5M_{Jup}$ planetary companion). Subsequent investigations over the succeeding four decades led to claims of planetary-mass companions around almost a dozen other stars, notably Barnard's star (van de Kamp, 1982). None has survived detailed scrutiny. Recently, however, Pravdo et al (2005) announced the discovery of the brown dwarf Gl 802B, the first and, so far, the only astrometrically discovered sub-stellar companion. The longer orbital periods of the sub-stellar companions sought through astrometric surveys mean that a decade may pass before a large number of brown dwarfs and extrasolar planets are discovered through astrometry.

Although astrometric monitoring has yet to produce many new brown dwarf and planet discoveries, it has proven effective in refining the dynamical masses of known sub-stellar binaries and extrasolar planets. Astrometry allows accurate measurement of the *total* mass and orbital inclination of a binary. If both components are visible, then measurement of the absolute motions allows one to deduce all the orbital parameters directly from the astrometry alone. Alternatively, if only one component is visible (as is the case with planetary companions), the relative astrometry of the host star can be combined with radial velocity data for the same system to uniquely resolve the $\sin i$ ambiguity inherent in measurements of single-lined spectroscopic binaries (Sect. 5.3.2). This allows an exact determination of the mass of the radial velocity planet.

The first sub-stellar objects to have astrometrically determined masses were the components of the binary Gl 569Bab (Lane et al., 2001). Gl 569B is a distant companion of Gl 569A, and its components were originally resolved through AO observations (Martín et al., 2000). It has recently been hypothesized (Simon et al., 2006) that Gl 569Ba is itself a binary (i.e., composed of Gl 569Baa and Gl 569Bab), based on high-resolution spectroscopic observations that suggest two distinct components, as in a double-lined spectroscopic binary (Sect. 5.3.2). Gl 569, a candidate *quadruple* system is thus an excellent example of how different approaches (direct imaging, astrometric, and radial velocity monitoring) depict complementary pieces of the same larger picture. Several other binary brown dwarfs have had their total masses measured astrometrically in recent years, and a number of astrometric campaigns to measure more dynamical masses are currently underway.

Dynamical masses of brown dwarfs are crucial to understanding sub-stellar evolution. Brown dwarf astrometric binaries with known ages and heliocentric distances (e.g., through physical association with a star of a well-determined age and distance, or membership of star clusters or associations) allow us to constrain sub-stellar properties by disentangling the degeneracies between the mass, age, and luminosity of a sub-stellar object (see Sect. 5.2.1).

More recently, astrometric monitoring has been successfully applied to measure the exact dynamical masses both of extrasolar planets (Benedict et al., 2002, 2006) and of brown dwarfs (Reffert et al., 2006) known from radial velocity surveys. The combination of astrometric and radial velocity data thus depict a unique picture of systems with known radial velocity sub-stellar companions.

5.3.4 Photometric Methods: Eclipsing Binaries

Low-mass binary companions can be detected photometrically in three ways: first, through the presence of excess radiation at infrared wavelengths; second, through eclipses of the primary star; and third, through irregularities in the light profile of a microlensing event. In the case of very low-mass/low-luminosity companions, the first of these methods is effective only for white dwarf primaries, and even then can be ambiguous (*vide* G29-38B). Microlensing is discussed extensively by Bennett elsewhere in this volume. Analyses indicate that a handful of microlensing events are likely to be associated with isolated brown dwarfs (e.g. Poindexter et al, 2005), but, so far, no brown dwarf companions have been detected. The latter observation reinforces the brown dwarf desert inferred for solar-type stars from radial velocity measurements (Gaudi, 2005).

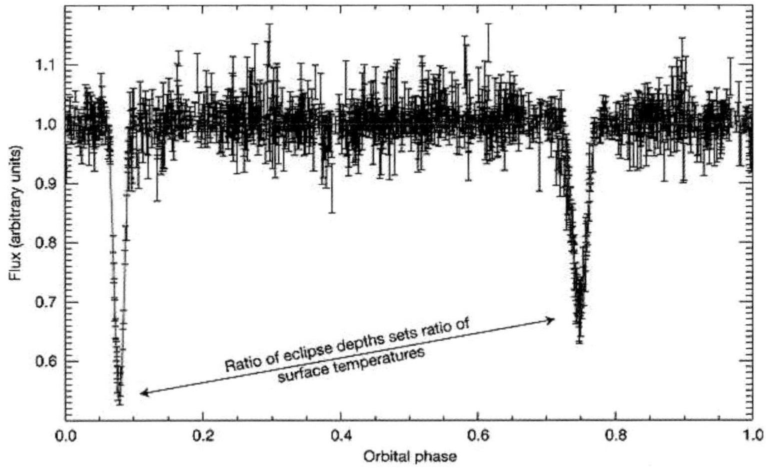

Fig. 5.8. Visual photometry of the eclipsing binary brown dwarf system 2MASS J05352184-0546085 (Stassun et al., 2006). The data are folded on the orbital period of 9.779621 ± 0.000042 days and phased relative to periastron passage. The ratio of eclipse depths provides a direct measure of the ratio of surface temperatures, with the deeper eclipse corresponding to the eclipse of the hotter component.

Eclipsing binaries are binary systems in which the two components orbit each other in a plane that is aligned along the line of sight. The apparent brightness of an eclipsing binary displays a characteristic double-dipped periodic modulation, corresponding to the times when either of the components occults the other (see Fig. 5.8). The dips are generally different in amplitude, their depth depending on the ratio of luminosities of the two components and on the exact viewing geometry.

The first eclipsing binary to be discovered was Algol (β Persei), whose periodicity was noted by Geminiano Montanari in 1667. It was not until more than a century later that an eclipsing mechanism for the variability of Algol was proposed by the British astronomer John Goodricke, for which he was awarded the Royal Astronomical Society's Copley medal in 1783. Edward Pickering proposed a detailed explanation involving two stellar components in 1881. His hypothesis was confirmed in 1889 when the Potsdam astronomer Hermann Vogel discovered Doppler shifts in the spectrum of Algol, confirming variations in the radial velocities of the components. Thus, along with Mizar (Sect. 5.3.2), Algol was one of the first spectroscopic binaries.

Eclipsing binaries are uniquely suited to studying individual binary components. Since the orbital inclination of the binary is known (nearly edge-on), the masses of the components are fully determined in SB2 systems. The ratio of the brightness dips can be used to infer the luminosity ratio, and the duration of the eclipses to estimate their radii. Eclipsing binaries in which at least one of the components is sub-stellar could therefore provide strong constraints on the properties of sub-stellar objects.

Several candidate eclipsing binaries with brown dwarf secondaries have been announced in recent years, though only one, 2MASS J05352184-0546085 (Stassun et al., 2006), has been confirmed. 2MASS J05352184-0546085 is located in the young (1–3 million years) Orion Nebula Cluster star-forming region (distance: ≈ 430 parsecs) and is composed of two brown dwarfs orbiting and eclipsing each other (Fig. 5.8). Given its known age and distance, this eclipsing binary brown dwarf provides the first thorough empirical test of evolutionary models of sub-stellar objects.

Planets crossing in front of their host stars, a.k.a. "transiting planets," are also components of eclipsing binaries, where the primary is the star and the secondary is the planet. The characteristics of these systems are discussed in more detail by Irwin (this volume).

5.3.5 Summary

As we have seen in Sects. 5.3.1–5.3.4, brown dwarf companions have provided an early testing ground for the techniques that are now successfully employed to search for extrasolar planets. Irwin (this volume) has compared the sensitivity of various brown dwarf and planet detection techniques (his Fig. 1.10). Direct imaging is most sensitive at separations beyond 10-100 AU, although the detection limits are luminosity, rather than mass, dependent. The techniques are broadly complementary, with radial velocity, microlensing, and transit (eclipsing) techniques sensitive at small separations, and astrometry and direct imaging sensitive to wider companions. Direct imaging is the only approach where current technology is not capable of detecting Jovian planets around stars on Solar System scales. However, this technique has resulted in the discovery of more brown dwarf companions (more than 30 to date) than all other methods combined. The continued progress in high-contrast imaging technology will undoubtedly result in the direct imaging of planets around other stars. Such detections are expected to usher in a new era of study of exoplanets, one in which we will able to investigate their atmospheres in detail, and address questions about the possibility of the existence of life elsewhere in the Universe. Future prospects for the detection of planets, using direct imaging in particular, are discussed in Sect. 5.5.

5.4 Brown Dwarfs as Companions

In Sect. 5.3 we described the applicability of the various companion detection techniques to finding brown dwarf companion to stars. With the exception of the microlensing technique, these methods have been successful in detecting sub-stellar secondaries. Different techniques are sensitive to different types of systems, and a single method cannot detect all binary systems across a wide span of separations. However, combining several approaches can reveal the broader picture. Thus, radial velocity monitoring probes the closest ($< 1 - 10$ AU) systems, astrometric monitoring is sensitive to wider ($\sim 1 - 100$ AU) systems, and direct imaging resolves the widest ($> 10 - 100$ AU) pairs.

5.4.1 Stellar Binary Systems

Individual binary systems have been known for many years, but serious systematic surveys have been possible only within the last 15 years or so. This was primarily a sampling issue: reliable statistics demand a well-defined sample; the most reliable sample for present purposes is a complete, volume-limited sample; and it is only within the last two decades that such samples have become available. Low luminosity stars have received less observational attention than solar-type stars, but these stars are also more common: approximately 80% of the stars in the Galactic Disk are spectral type K or M dwarfs, while only ∼15% are spectral types F or G. Consequently, there are sufficient numbers of late-type dwarfs in the immediate solar neighborhood to allow the derivation of reliable multiplicity statistics.

Most multiplicity investigations are limited to stars within 20-50 parsecs of the Sun, so we are sampling only a tiny fraction of the total volume of the Galactic disk. This might raise concerns about whether the derived properties are truly representative of the broader stellar populations. The saving grace is that stars (and brown dwarfs) acquire random motions through dynamical interactions with massive objects, such as molecular clouds, as they orbit the Milky Way. The net result is that objects currently near the Sun (Galactic radius ∼8kpc) originated at Galactic radii from 4 to 12 kpc. A local sample *can* provide reliable global characteristics.

The most comprehensive analysis of multiplicity in solar-type stars remains that by Duquennoy & Mayor (1991). They selected an unbiased sample of 164 nearby G dwarfs, which they followed with direct imaging, radial velocity, and astrometry for 13 years. The salient result from their study was that more than half of Sun-like stars in the solar neighborhood are binaries, $f_{bin} > 57\%$. They also found that the period distribution is log-normal with a peak near 150 years, and tails extending from under 1 day to ∼10 million years (Fig. 5.9). This corresponds to a distribution of orbital semi-major axes extending from 0.01 AU to 50,000 AU, and peaking near 30 AU, comparable to the size of the Solar System. Finally, Duquennoy & Mayor (1991) found that the distribution of the mass ratio $q = M_2/M_1$ in binary systems rises toward lower mass ratios and attains a broad peak near $q = 0.2$ although their data are largely insensitive to lower mass ratio binaries. That is, most Sun-like primaries have companions that are a fifth of their mass, or approximately 200 times the mass of Jupiter.

The nearest late-type dwarfs have been subjected to extensive scrutiny for spectroscopic and resolved companions. Most have high proper motion, $\mu > 0.2$ arcsec yr^{-1}, allowing ready identification of wide stellar companions from photographic surveys. Those photographic data have been supplemented by deep optical and infrared imaging using both conventional CCD cameras (e.g. Simons et al, 1996) and AO techniques (e.g. Close et al, 2003). Fischer & Marcy (1992) were the first to apply the comprehensive approach of Duquennoy & Mayor (1991) to M dwarfs. They derived a binary fraction $f_{bin} \sim 42\%$, with a period distribution broadly consistent with G dwarf binaries. Since these are lower mass binaries, this implies that the semi-major axis distribution is more compressed; that is, less massive binaries tend to be tighter. The mass ratio distribution for the M dwarf sample is approximately

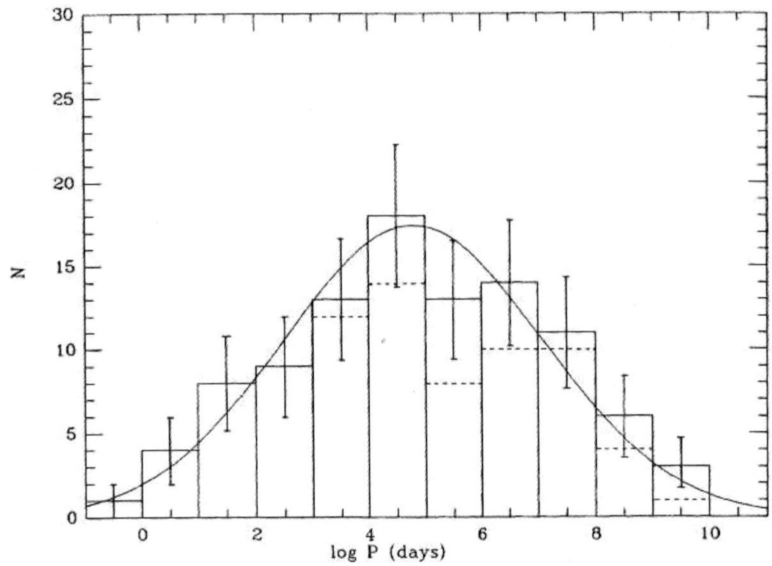

Fig. 5.9. Distribution of orbital periods for Sun-like binary stars from the study of Duquennoy & Mayor (1991).

flat between 0.4 and 1.0, indicating a higher proportion of equal-mass binaries than in high-mass stars.

The Fischer & Marcy analysis centred on early- and mid-type M dwarfs. Reid & Gizis (1997) extended the investigation to later type dwarfs, concentrating on stars within 8 parsecs of the Sun. The main result from that analysis is that the binary fraction decreases further, to $f_{bin} \sim 30\%$, both due to the inclusion of cooler M dwarfs and the formal restriction to a trigonometric-parallax defined, volume-limited sample. There is also a stronger tendency towards equal mass-ratio systems among the 8-parsec binaries than in the Fischer & Marcy sample. In general, the proportion of low mass ratio systems declines with decreasing mass. Thus, $q = M_2/M_1 < 0.2$ systems are most common among B and A (2.5–20 $M_\odot$) stars, where they occur in 35–40% of all binaries (Tokovinin et al., 1999; Shatsky & Tokovinin, 2002; Kouwenhoven et al., 2005). G and K dwarfs (0.5–1.2 $M_\odot$ stars) have a smaller fraction of $q < 0.2$ binaries (10–20%; Duquennoy & Mayor, 1991), and late-M stars ($< 0.2 M_\odot$) essentially have none (Burgasser et al., 2007, and references therein).

None of these investigations was highly sensitive to brown dwarf companions, since all were undertaken before the era of high-contrast imaging, precision radial velocity and precision astrometry. However, the results suggest several important trends. Namely, (1) binarity is more common among higher mass stars, (2) higher-mass binaries extend to wider separations, and (3) low-mass ratio binaries are more common among binaries with higher mass primaries. All of these trends are relevant to studying brown dwarf companions to stars, since binaries in which one of the components is sub-stellar are simply very low mass ratio ($q < 0.1 - 0.2$) binaries. In the following subsections we summarize the results from searches for brown dwarfs around high-mass and low-mass stars.

5.4.2 Solar-Type Stars

Extending the work of Duquennoy & Mayor (1991) to detect brown dwarf companions has been challenging on two fronts. For one, detecting sub-stellar secondaries required revolutionary changes in all of the popular techniques used for companion searches: radial velocity, direct imaging, and astrometry. Some technological challenges were addressed by the mid-1990's, and 1995 saw the announcement of the first bona-fide brown dwarf and planetary companions (Sect. 5.1). However, it was soon realized that brown dwarf companions to stars are extremely rare at small separations (Sect. 5.3.2). At ≤ 3 AU orbital separations from Sun-like stars, brown dwarfs are more than 10 times less common than extrasolar planets (Marcy & Butler, 2000).

The dearth of close-in brown dwarfs did not deter attempts to image sub-stellar objects at wider (>100 AU) separations. Following the imaging program that discovered Gl 299B (Oppenheimer et al, 2001), a slew of other surveys followed, using the *Hubble Space Telescope* (Brandner et al., 2000; Lowrance et al., 2005) and ground-based AO systems (Lloyd, 2002; McCarthy & Zuckerman, 2004; Carson et al., 2005). These surveys targeted nearby ($\leq$50 pc) and/or young (<300 million years) stars, deemed optimal for the direct imaging of sub-stellar companions. The emphasis on proximity and youth is central to direct imaging searches. Nearby stars offer a double advantage: they allow the detection of companions that are fainter and at smaller physical separations. In addition, young stars allow the detection of lower-mass companions because brown dwarfs are more luminous at young ages (Sect. 5.2.1).

In the late 1990s, high-contrast direct imaging emerged as the foremost technique for searching for wide sub-stellar companions to stars. The most exciting aspect of the technique was the opportunity to characterise sub-stellar companions through direct spectroscopic analysis. Indeed, the holy grail of direct imaging is the detection of an extrasolar planet, which will offer the unprecedented opportunity to study an exoplanetary atmosphere in detail.

By the year 2003, every $\geq$3 meter telescope equipped with an AO system had multiple on-going direct imaging searches. However, despite concentrated efforts, the pace of discovery remained slow. Meanwhile, the rate of radial velocity planet discoveries grew exponentially. By the end of 2003, more than 100 exoplanets were known from radial velocity, compared to less than 20 brown dwarfs discovered through direct imaging, despite the fact that the first exoplanet and the first brown dwarf were announced simultaneously eight years prior. The two largest direct imaging surveys for sub-stellar companions (Oppenheimer et al., 2001; McCarthy, 2001) produced only one brown dwarf, Gl 229B, in a combined sample of more than 340 stars and a combined separation range of 10–1200 AU. This stands in stark contrast to the frequency of RV extra-solar planets within 3 AU (5–15%; Marcy & Butler, 2000; Fischer et al., 2002) and the frequency of stellar companions to stars at 10–1200 AU

(9–13% for M–G-dwarf primaries; Fischer & Marcy, 1992; Duquennoy & Mayor, 1991). This led to suggestions that the brown dwarf desert extends to separations of ~1000 AU McCarthy & Zuckerman (2004)

A different picture has begun to emerge in the last few years. Novel, technologically advanced AO systems installed on 5–10 meter class telescopes (Palomar, the Monolithic Mirror Telescope [MMT], Gemini, the Very Large Telescope [VLT], Keck), combined with ingenious differential imaging techniques (Marois et al., 2005, 2006; Biller et al., 2006), provide unprecedented sensitivity to sub-stellar companions at separations of tens of AU. Furthermore, better understanding of stellar ages and the discovery of several co-moving groups of young stars in the solar neighborhood (Zuckerman & Song, 2004, and references therein) have allowed improved target star selection. The pace of sub-stellar companion discoveries has accordingly picked up, and almost 30 new sub-stellar companions to stars have been imaged over the past 3.5 years, doubling the catalogue.

We now possess sufficient observational evidence from high-contrast imaging surveys to claim that brown dwarfs at 10–1000 AU orbital separations are a factor of 5–8 more common (2–4%; Neuhäuser & Guenther, 2004; Metchev, 2006) than in the 0–3 AU radial velocity brown dwarf desert ($\leq$ 0.5%; Marcy & Butler, 2000). At wider (> 1000 AU) separations, Gizis et al. (2001) found that the frequency of sub-stellar companions to stars in 2MASS (a seeing-limited, shallow, all-sky survey) is fully consistent ($\sim$18%) with the frequency of stellar secondaries. That is, the frequency of sub-stellar companions increases with semi-major axis (see Sect. 5.4.4).

Besides probing the low mass end of the companion mass function, the direct imaging effort is also motivated by the desire to study sub-stellar properties at low effective temperatures, 500–2000 K, in anticipation of future imaging of giant extrasolar planets with similar effective temperatures. Because of their association with stars of known ages and distances, fundamental properties of sub-stellar companions can be constrained more closely than those of isolated free-floating brown dwarfs. A first glimpse at the relevance of companion imaging surveys to exoplanet studies emerged last year with the discovery of the coolest young sub-stellar objects, HD 203030B and HN PegB (Metchev & Hillenbrand, 2006; Luhman et al., 2007). Both objects have effective temperatures between 1100–1300 K, and, because of their relative youth ($\approx$ 0.3 billion years), are still undergoing contraction. Their radii are expected to be similar to those of the comparably hot short-period extrasolar Jupiters that are puffed up because of proximity to their host stars. Although the physical conditions in the hot Jupiter atmospheres are likely to be affected by irradiation from their host stars, we expect to gain a rudimentary picture of the chemistry in these hot planetary atmospheres by studying the much more easily accessible young brown dwarf companions.

5.4.3 Low Mass Binaries

Brown dwarfs also exist as companions of late-type main sequence stars. Gl 229B, the archetypical T dwarf, lies at a separation of $\sim$50 AU from an M0.5 primary.

Since that discovery, the nearest stars have been subjected to intense scrutiny (Simons et al, 1997; Oppenheimer et al, 2001), but only two have yielded brown dwarf companions: Gl 570D is an extremely cool T8 dwarf, lying 1525 AU from Gl 570ABC, a K5/M1/M2 triple (Burgasser et al, 2000); and ϵ Indi, a K5 dwarf, is accompanied at a separation of 1459 AU by a T1/T5 binary, ϵ Indi Bab (Scholz et al, 2003; McCaughrean et al, 2006). Other binaries within 10-20 parsecs are known, including Gl 802B (Sect. 5.3.3); G196-3B, a resolved L-dwarf companion of a young ($\sim 10^7$ year-old) M dwarf (Rebolo et al); and LP 216-75/2M0951+35, a Pleiades-age M4.5 dwarf with a $\sim 0.02 M_\odot$ wide (Δ ~450 AU) L6 companion (Reid & Walkowicz, 2006). The implication, however, is that luminous (L/T) brown dwarf companions of late-type dwarfs (particularly M dwarfs) are rare.

The qualifier "luminous" in the last sentence of the previous paragraph highlights one of the key observational issues: brown dwarfs cool and fade with age (Figs. 5.1 and 5.2), making it difficult to confidently assess their presence among local field stars, which have typical ages of several Gyrs. Radial velocity measurements can set some limits on brown dwarfs at small separations. Approximately 100 field M dwarfs (mainly early types, M0–M3) are included in the radial velocity planet-search programs described by Irwin (this volume). So far, four systems (Gl 436, Gl 581, Gl 849 and Gl 876) are known to harbour planetary companions. Only Gl 849b has a minimum mass close to that of Jupiter ($M_2 \sin(i) = 0.82 M_{Jup}$, Butler et al, 2006); the remaining planets have $M_2 \sin i < 0.1 M_{Jup}$. While many nearby M dwarfs still lack adequate observations, no brown dwarf companions have been discovered to date.

An alternative means of circumventing the visibility issue is to survey a much younger population. The nearest star forming regions, however, are several hundred parsecs distant, setting strong constraints on our ability to detect faint companions. Recently, a number of young ($\sim 10^7$ years) stellar associations have been identified, with sparser membership but lying at distances of 40–60 parsecs from the Sun (Zuckerman & Song, 2004). Principal among those is the TW Hydrae Association (TWA), which has at least 28 stellar/brown dwarf systems as members. Nine are binaries, three are triples and one is a quadruple system, giving an overall multiplicity of 43%. The companions include two brown dwarfs (TWA 5B and 2M1207-39B), both resolved systems.

The 2M1207-39AB system deserves particular comment. The primary is itself a brown dwarf, whose spectral type of M8 (Gizis, 2002) implies a mass of 25–$45 M_{Jup}$ for an age of $\sim 10^7$ years. The companion, discovered by Chauvin et al (2004), is ~ 8 magnitudes fainter with an inferred mass of 4–$6 M_{Jup}$, corresponding to a mass ratio of $q \sim 0.2$. At a separation of ≥ 40 AU, the companion lies well beyond the plausible extent of 2M1207-39A's protoplanetary disk. This indicates that both components probably formed like stars, from the gravitational collapse of molecular gas, rather than like planets, from the accretion of rocks and gas in a circumstellar protoplanetary disk. 2M1207-39B is therefore the lowest mass brown dwarf yet identified.

2M1207-39AB also stands apart from most other very low-mass binaries. Follow-up observations of 2MASS, DENIS and SDSS sources have provided extensive catalogues of nearby L and T dwarfs (Cruz et al, 2006; Burgasser et al, 2006). Those datasets have been surveyed for binary companions, although current observations are largely limited to high resolution imaging with either the Hubble Space Telescope (e.g. Reid et al, 2006; Burgasser et al, 2006) or ground-based AO systems (e.g. Close et al, 2003). With angular resolutions better than 0.1 arcseconds, those observations can resolve binaries with separations of a few AU, corresponding to periods of 10-50 years for massive brown dwarfs. More than 90 L dwarfs, including 72 within 20 parsecs, and $\sim$25 T dwarfs have been observed to date. Spectroscopic observations are sparser (e.g. Reid et al, 2002), and are complicated by the broad absorption features of L dwarfs that limit the accuracy of the measured velocities.

The results from recent investigations have been summarised by Burgasser et al (2007). In brief, there are three main conclusions: first, the multiplicity fraction for ultracool dwarfs is probably less than 20%, with $\sim$12% in resolved systems and no more than $\sim$6% in spectroscopic binaries. Second, there is a clear preference for equal-mass systems. Fig. 5.10 (from Reid et al, 2006) illustrates this point, where we compare the point-spread function distribution for HST near-infrared images

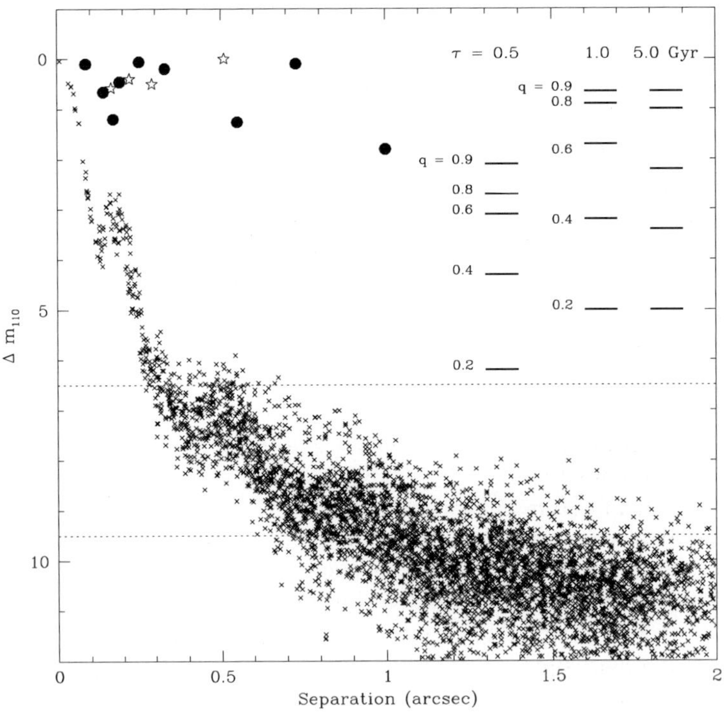

Fig. 5.10. The HST NICMOS J-band (F110W) point-spread function, plotted in relative magnitudes, matched against the peak brightness of known companions of ultracool dwarfs. The dotted lines mark the effective detection limits. It is clear that there is a substantial area of discovery space that is accessible, but unoccupied.

against the peak brightness of known L dwarf binaries. All of the systems have nearly equal luminosities, which, for brown dwarfs, implies nearly equal masses, yet there is a vast expanse of discovery space for lower luminosity companions. Finally, the component separation distribution for brown dwarf binaries peaks at 3-10 AU, with few binaries having separations exceeding 15 AU. All of the wide binaries, including 2M1207-39AB, GG TauBab and DENIS 0551-4434AB, are younger than 10^8 years, raising the possibility that an evolutionary effect, perhaps dynamical stripping, might play a role in modifying the population.

5.4.4 Summary

In summarising these results, we must emphasise that brown dwarf binaries should be considered as part of a continuum, stretching from O-type systems through solar-type stars to ultracool brown-dwarf/brown-dwarf binaries, rather than as a distinct category unto themselves. Taking that perspective, there are three broad characteristics of binary systems:

- The overall multiplicity fraction decreases from early to late spectral types (where we use the spectral type of the primary to characterise the system);
- The proportion of equal-mass systems increases at later spectral types; and
- The distribution of component separations becomes more compressed at later spectral types.

The data underpinning the last conclusion are shown in Fig. 5.11 (adapted from Reid & Walkowicz, 2006), which plots the total system mass, $M_t = M_1 + M_2$, and the mass ratio, q, as a function of component separation, Δ. Reid et al (2001) originally pointed out that the maximum separation of low-mass binaries appeared to scale logarithmically with M_t (i.e. $\log(\Delta_{max}) \propto M_t$). Burgasser et al (2003) subsequently demonstrated that the outer envelope is better matched by $\Delta_{max} \propto M_t^2$ at $M_t \geq 0.3 M_\odot$. As noted in Sect. 5.4.3, the handful of low-mass binaries that violate these limits are generally younger than $\sim 10^8$ years.

Fig. 5.12 offers some clues as to how we might understand these results. Here, we plot the mass/separation diagram for stellar, brown dwarf and planetary companions to a volume-complete sample of nearby solar-type stars. This clearly illustrates the high frequency of planetary companions and dearth of brown dwarf companions at small separations ($\leq$ 10AU). More importantly, it shows that the brown dwarf desert extends well into the stellar mass régime, with only a handful of companions with $M_2 \sin i < 0.7 M_\odot$. This is in stark contrast to the mass distribution of wide companions (>100 AU), which extends to brown dwarf masses and, indeed, closely matches the mass distribution of single stars.

Rephrasing these results, Fig. 5.12 shows that there is a clear preference for near-equal mass systems at small separations among solar-type stars. The inner region presumably reflects the effects of competitive accretion: big stars don't let small stars form close to them. Wide ($\Delta >$ 100AU) systems form through the gravitational association of independent protostellar cores, and therefore span a greater range of mass ratios.

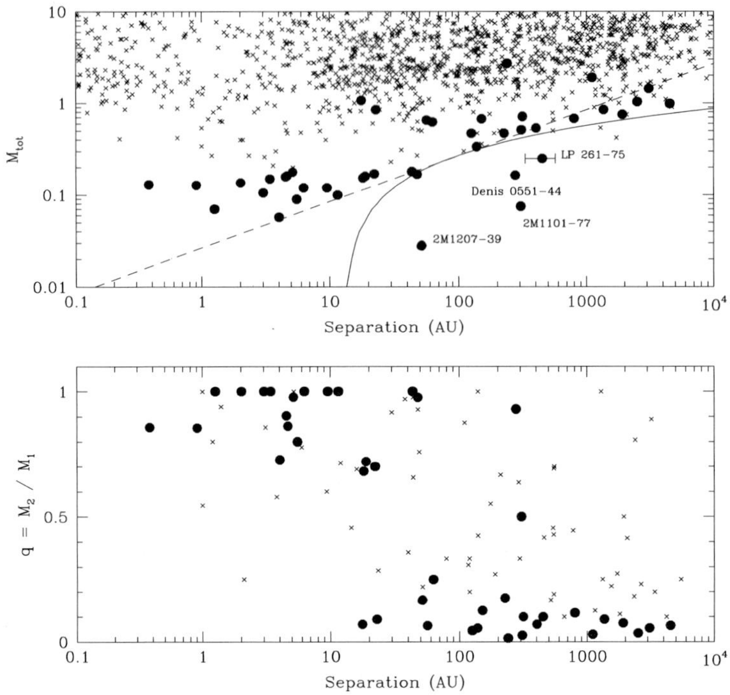

Fig. 5.11. Total system masses and mass ratios as a function of separation (adapted from Reid & Walkowicz, 2006): crosses mark stellar binaries; solid points are systems with ultracool companions; the dashed line and solid line in the mass/separation diagram are from Burgasser et al (2003).

Suppose that this dichotomy holds over the full mass range of primary stars. It seems likely that the boundary between the inner and outer regions will scale with the gravitational potential, i.e. the mass of the primary. Subsequent dynamical interactions are likely to be more effective at disrupting wide low-mass binaries. Removing those systems decreases the total binary fraction, and preferentially eliminates low-q systems, leading to a higher proportion of equal-mass systems and the mass/separation diagram shown in Fig. 5.11.

Close binaries are clearly a "bad thing" (to quote Sellars & Yateman) for the formation and survival of planetary systems, since gravitational interactions are liable to truncate and even disrupt the protostellar disk. Taken at face value, the decrease in the fraction of stellar or brown dwarf companions with decreasing primary mass suggests that the environment around low-mass stars and brown dwarfs may be more suitable for planet formation. However, if post-formation disruption plays a significant role in constricting the separation distribution plotted in Fig. 5.11, then the present-day multiplicity fraction may lead to an overestimate of the number of low-mass dwarfs with undisturbed planetary systems. From the brown dwarf perspective, Figs. 5.11 and 5.12 suggest that higher mass ($M > 0.7M_{\odot}$) AFGK stars

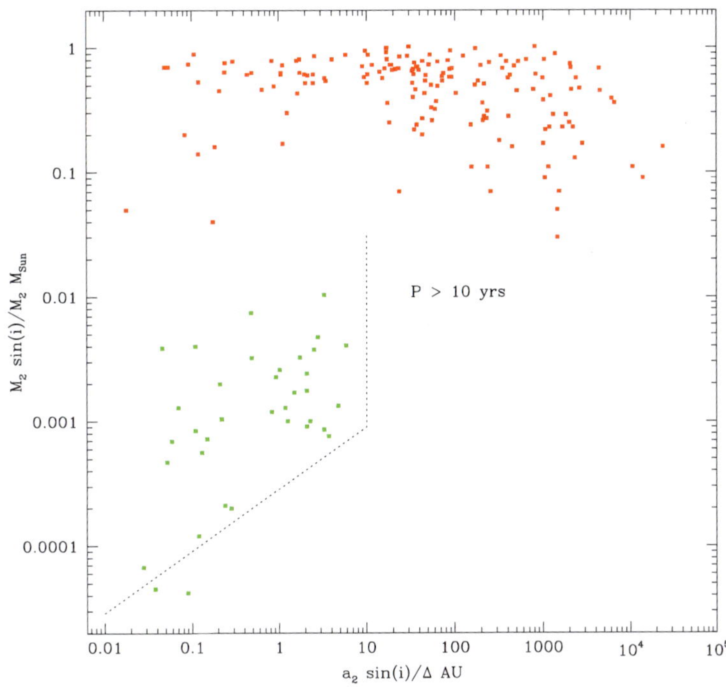

Fig. 5.12. The mass/separation diagram for known stellar, brown dwarf and planetary companions of the volume-complete sample of 479 solar-type stars ($4 < M_v < 7$) within 25 parsecs of the Sun. Notice that there are only a handful of companions with $0.01 < \frac{M}{M_\odot} < 0.5$ and $\delta < 10$AU – *the* brown dwarf desert extends through the M dwarf régime.

are the best targets for direct imaging surveys, since they offer the best prospects for harbouring sub-stellar companions at wide separations.

5.5 Future Work

A major portion of the brown dwarf effort in the future will aimed at discovering sub-stellar objects that are cooler and fainter than the ones known to date. As discussed in Sect. 5.2.1, these dwarfs are likely to bear close similarities to giant planets found in our own Solar System and around other stars. In particular, the effective temperatures of extremely cool brown dwarfs may be sufficiently low (<500 K) to allow the formation of ammonia and water molecules in their atmospheres, as in the atmospheres of the solar system giant planets. Direct imaging and spectroscopy of brown dwarfs will thus remain at the forefront of exoplanet research. Planned future studies will follow several main approaches: (1) direct detection of extra-solar planets during secondary eclipses in transiting systems; (2) direct imaging of planetary-mass companions to stars through enhanced high-contrast imaging capabilities; (3) wide-field infrared imaging surveys in search for very cool and low-mass nearby brown dwarfs; (4) precise astrometric and spectroscopic monitoring of binary brown dwarfs and exoplanets to determine dynamical sub-stellar masses; and (5) characterization of the atmospheres and of the evolutionary states of known brown dwarfs through detailed spectroscopy and theoretical modeling.

5.5.1 Direct Detection of Transiting Planets

Direct detection of planetary atmospheres in secondary eclipses became feasible with the launch of the mid-infrared Spitzer Space Telescope in 2003. The high precision of the Spitzer measurements and the high brightness of extrasolar planets in the mid-infared recently allowed Richardson et al. (2007) to record the first spectrum of an extrasolar planet: the mid-infrared spectrum during secondary eclipse, when the planet is hidden, is subtracted from the spectrum outside eclipse (star plus planet); the difference is the planetary spectrum. The Spitzer mission is nearing the end of its fully operational lifetime (the cryogens are expected to run out in 2008), but larger mid-infrared-optimized space telescopes are planned for the future, notably the James Webb Space Telescope (JWST; expected launch date: 2013). JWST (diameter 6.5 m) will continue and enhance the exoplanet science done with Spitzer (0.85 m).

In the meantime, the recently (2006) launched COnvection, ROtation, and planetary Transits (COROT; 0.2 m) space mission and the upcoming Kepler space mission (0.95 m; to be launched in 2008) are expected to find many more transiting planet systems. Both satellites work at optical wavelengths. COROT will survey a few thousand bright stars (Bordé, Rouan & Léger, 2003), and has already detected a Jupiter mass transiting planet (CoRoT-Exo-1 b; $M \sim 1.3 M_{Jup}$, $P \sim 1.5$ days). Kepler is designed with the specific goal of discovering Earth-sized exoplanets (Borucki et al, 2003). The future of the planet transit technique is thus wide open, as we have just scratched the surface of the immense scientific potential of this method.

5.5.2 High Contrast Imaging

High-contrast direct *imaging* efforts are aimed at spatially resolving planets from their host stars. The future of the high-contrast imaging approach is promising and exciting, as the AO technology developed in the early 1990's is starting to mature. Multiple efforts are underway to improve and enhance the available instrumentation to allow the direct imaging of photons from exoplanets. Both ground- *and* space-based telescopes in the future will rely on AO to achieve optimum imaging contrast and stability.

In the near term, upcoming next-generation high-contrast imaging instruments include the PALM-3000 AO system for the Palomar 5 meter telescope, the Near-Infrared Coronagraphic Imager (NICI) on the Gemini South 8 meter telescope, the Gemini Planet Imager (GPI) on the 8 meter Gemini North telescope, the Spectro-Polarimetric High-contrast Exoplanet Research (SPHERE) instrument on the 8-meter Very Large Telescope (VLT), and the Next Generation Adaptive Optics System on the Keck 10 meter telescope. All of these instruments are expected to be able to detect 5–10 Jupiter-mass planets within 30 AU of stars within 30 parsecs from the Sun. NICI is already undergoing integration at the telescope, while the remaining systems are expected to see "first light" by 2012.

A separate approach to high-contrast imaging involves the detection of polarized emission from the day side of an extrasolar planet. Light reflected from the day side of a planet in an inclined ($0° < i < 90°$) orbit with respect to the line of sight is expected to be polarized through Thomson scattering in the planetary atmosphere. Phase variations in the degree of system polarization can be used to infer the reflected luminosity of the planet. This concept is already integrated in the design for SPHERE on the VLT. In the more distant future (beyond 2020), ground-based high-contrast imaging will take advantage of the unprecedented 20–60 meter apertures of the next generation of extremely large telescopes, such as the Giant Segmented Mirror Telescope (GSMT; $>$ 20 meters), the Thirty-Meter Telescope (TMT; 30 meters), and the European Extremely Large Telescope (E-ELT; 30–60 meters). The factor of 2–6 larger apertures of these future telescopes may allow detection of planets at 2–6 times smaller angular separations from their stars, potentially probing the Jovian region ($\sim$5 AU) for Jupiter-mass planets around stars within 30 parsecs from the Sun.

In parallel with the ground-based effort, there is a wide range of space-based programs at various stages of development. At present, JWST is the only mission slated for launch before 2015, and will be capable of detecting super-Jupiters at $>$ 5 AU separations from stars within 30 pc. Plans are also underway for more ambitious programs, including the Terrestrial Planet Finder Coronagraph (TPF-C; 8×4 meter, visible light) and Interferometer (TPF-I; four 4-meter formation-flying mirrors; mid-infrared), and the European Space Agency's (ESA) Darwin interferometer (three $\geq$3-meter formation-flying mirrors; mid-infrared). TPF-C, TPF-I, and Darwin will all aim for the direct imaging of Earth-like planets in the habitable zone ($\sim$300 K; $\sim$1 AU) of nearby Sun-like stars.

5.5.3 Wide Field Imaging Surveys

Sensitive wide-field imaging surveys in the future will detect the largest numbers of new sub-stellar objects. Indeed, the vast majority of known brown dwarfs were discovered using the present generation of wide-area imaging surveys: DENIS, 2MASS, and SDSS (Sect. 5.3.1), where they are seen as isolated free-floating objects. However, those surveys have revealed only a small fraction of the expected sub-stellar objects in the solar neighborhood, and barely hinted at their diversity. The census of sub-stellar objects in the solar neighborhood remains woefully incomplete, especially at the cool end of the brown dwarf sequence (spectral type T8).

A second generation of sensitive wide-field surveys is now under way, with the UK Infrared Digital Sky Survey (UKIDSS) in operation on the 3.8 meter UK Infrared Telescope (UKIRT) since 2005, and the Panoramic Survey Telescope & Rapid Response System (Pan-STARRS) survey expected to commence on the first of three 1.8 meter telescopes in 2007. UKIDSS is 3 magnitudes more sensitive than 2MASS, and will therefore be correspondingly more effective at detecting very cool brown

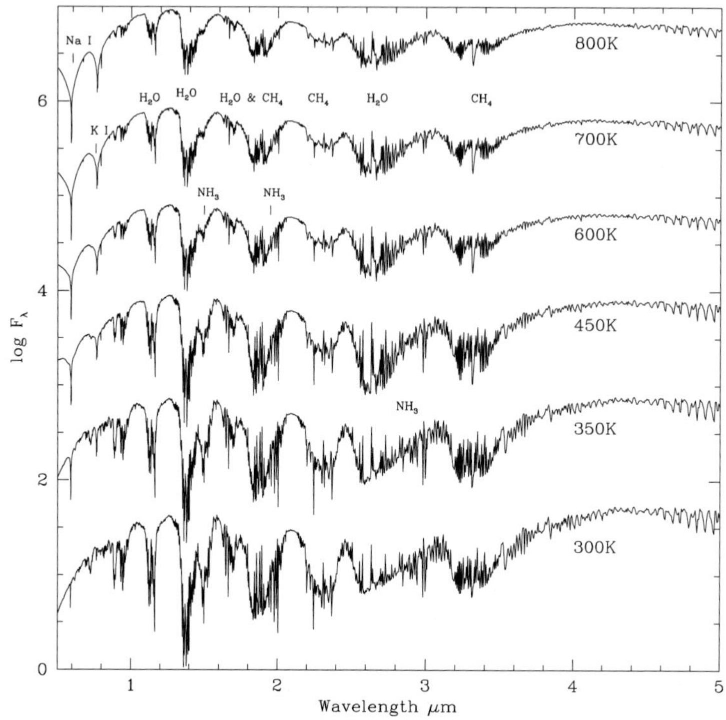

Fig. 5.13. The expected spectroscopic characteristics of cool brown dwarfs in the wavelength range covered by JWST's Near InfraRed Spectrograph (NIRSPEC) (from the model calculations by Burrows, Sudarsky & Lunine, 2003). The spectra are labeled by temperature and the main features identified. Note the appearance of ammonia at $\sim$600K and the weakening of the alkaline lines (e.g. Na I) at $T_{eff} < 450$K.

dwarfs, with effective temperatures 500–700 K. Kendall et al (2007) recently announced the first T dwarf discoveries from UKIDSS data.

Even cooler brown dwarfs should be detectable with the planned 8.4 meter Large Synoptic Survey Telescope (LSST; commissioning expected circa 2012), which will scan the entire sky every 3 days. LSST will be able to detect very faint, fast-moving, nearby brown dwarfs, which may have evaded previous shallower surveys performed on smaller telescopes. The search for cool brown dwarfs is also a primary scientific goal of the upcoming Wide-Field Infrared Survey Explorer (WISE; mid-infrared; 0.4 m) space mission, an all-sky 3μm survey due for launch in 2009. All of these programs aim to push detections to temperatures below 600K, where, as discussed in Sect. 5.2.1, the presence of significant absorption by ammonia will lead to sufficient changes in the spectroscopic appearance to warrant their classification in a new spectral type, Y. Figure 5.13 shows the expected appearance of late-T and Y dwarfs at near- and mid-infrared wavelengths.

5.5.4 Radial Velocity and Astrometric Surveys

Tried and true classical techniques such as radial velocity and astrometric monitoring also continue to improve in sensitivity. The High Accuracy Radial velocity Planet Searcher (HARPS) spectrograph on the European Southern Observatory (ESO) 3.6 m telescope has demonstrated precision levels well below 1 m sec^{-1}, and was recently used to identify the lowest mass radial velocity planet discovered to date, Gl 581c with $M \sin i = 5M_{\rm Earth}$ (Udry et al., 2007). With the use of larger telescopes in the future, the precision of the radial velocity technique is expected to further improve.

Precision astrometry is also leaving its footprints, having provided the first dynamical mass measurements of extrasolar planets (Sect. 5.3.3). Ongoing ground-based projects include the twin 10 m Keck interferometer and the four 8 m (+ three 1.8 m auxiliary) telescopes of the VLT Interferometer (VLTI), which offer the potential of both sub-milliarcsecond astrometry and direct detection of close companions, through on-axis nulling of the light from the primary star (Hinz et al, 1999). Space-based systems are also planned, with technical work continuing on NASA's Space Interferometry Mission SIM-PlanetQuest (Unwin et al, 2008), which will achieve microarcsecond precision, permitting the potential detection of terrestrial planets around nearby stars.

Besides detecting new extrasolar planets, astrometric and radial velocity monitoring of known brown dwarfs is also expected to produce many important results in the near future. A fundamental parameter of any self-gravitating object is its mass. For brown dwarfs, mass is an elusive quantity that is degenerately linked to its other parameters: effective temperature, luminosity, composition, and age(Sect. 5.2.1). To date, barely a handful of sub-stellar objects have had their masses measured dynamically, through monitoring of their orbital motions (Sects. 5.3.3, 5.3.4). Several astrometric, radial velocity, and photometric monitoring campaigns are currently under way (e.g., Stassun et al., 2006; Konopacky et al., 2007; Burgasser et al., 2007, and references therein) that promise to radically increase the number of such systems in the near future. These measurements are urgently needed to calibrate present and future editions of sub-stellar evolutionary models (e.g., Burrows et al., 1997; Chabrier et al., 2000; Baraffe et al., 2003).

5.5.5 Brown Dwarf Atmospheres

Several hundred brown dwarfs are currently known (see J. D. Kirkpatrick's Dwarf Archive, http://www.dwarfarchives.org), and their detailed characterization has now begun in earnest. An important step in that direction has been the Brown Dwarf Spectroscopic Survey (McLean et al., 2003, 2007), which has compiled medium- and high-resolution spectroscopic data on brown dwarfs covering the full L and T dwarf sequences. These near-infrared data complement the extensive optical data obtained in the course of the 2MASS follow-up program.

As discussed in Sect. 5.2.2, nearby brown dwarfs are now known to span a range of surface gravities and metallicities that lead to significant variations in their spectroscopic appearance even within the same spectral type (e.g. Kirkpatrick, 2005, 2008, and references therein). Both surface gravity and metallicity are linked to age. For example, low-surface gravity indicates that the object either has a low mass or an unusually large radius. In either case, this implies youth: if the object is a low-mass brown dwarf, then it has been caught during its relatively brief luminous phase; if higher mass, then the large radius implies that it is still contracting onto the main sequence.

Conversely, metal deficiency in any object is usually an indication of old age, as the object likely formed when the inter-stellar medium was not enriched to its present level of heavy-element content. Hence, brown dwarfs with peculiar surface gravities and metallicities can be used as tracers for the history of sub-stellar object formation in our galaxy. Peculiar brown dwarfs will also provide important benchmarks for extrasolar giant planets that orbit stars of various ages and metallicities. The availability of a larger brown dwarf sample and more powerful spectroscopic instruments on future ground- and space-based telescopes will undoubtedly reveal more surprises and provide more insight into the characteristics of these very low mass objects.

5.6 Summary and Conclusions

Brown dwarfs and exoplanets originate through different formation mechanisms. However, they have similar atmospheric characteristics, and the distribution of brown dwarfs as secondary components in binary systems affects the prevalence of planetary systems. We have briefly discussed the basic properties of brown dwarfs, and summarised the techniques employed to search for them as companions of main sequence stars. After a slow start, astronomers are piecing together a consistent picture of the formation and evolution of stellar/brown dwarf binaries, and the consequent implications for the survival of exoplanetary systems. With the development of new instrumentation and new techniques, the next decade is likely to see the detection of planetary-mass brown dwarfs at planetary temperatures, and perhaps even the direct detection of exoplanets around the nearest stars.

References

Baraffe, I., Chabrier, G., Barman, T. S., Allard, F., & Hauschildt, P. H., 2003, Evolutionary Models for Cool Brown Dwarfs and Extrasolar Giant Planets. The Case of Hd 209458, *Astr. Ap.*, **402**, 701

Basri, G., 2000, Observations of Brown Dwarfs, *Ann. Rev. Astr. Ap.*, **38**, 485.

Becklin, E. E., & Zuckerman, B., 1988, A low-temperature companion to a white dwarf star, *Nature*, **336**, 656

Benedict, G. F., et al., 2002, A Mass for the Extrasolar Planet Gliese 876b Determined from Hubble Space Telescope Fine Guidance Sensor 3 Astrometry and High-Precision Radial Velocities, *ApJ*, **581**, L115

Benedict, G. F., et al., 2006, The Extrasolar Planet ε Eridani b: Orbit and Mass, *AJ*, **132**, 2206

Biller, B. A., Close, L. M., Lenzen, R., Brandner, W., McCarthy, D., Nielsen, E., Kellner, S., & Hartung, M., 2006, Suppressing Speckle Noise for Simultaneous Differential Extrasolar Planet Imaging (SDI) at the VLT and MMT, In: *Proceedings of IAU Colloquium 200: Direct Imaging of Exoplanets: Science & Techniques*, ed. by C. Aime and F. Vakili, Cambridge Univ. Press, pp 571-576

Bodenheimer, P., Tohline, J. E., & Black, D. C., 1980, Criteria for fragmentation in a collapsing rotating cloud, *ApJ*, **242**, 209

Bordé, P., Rouan, D., & Léger, A., 2003, Exoplanet detection capability of the COROT space mission, *Astr. Ap.*, **405**, 1137

Borucki, W. J., et al., 2003, The Kepler Mission: Finding the Sizes, Orbits and Frequencies of Earth-size and Larger Extrasolar Planets, In: *ASP Conf. Ser. 294: Scientific Frontiers in Research on Extrasolar Planets*, ed. by D. Deming and S. Seager, ASP, San Francisco, pp 427-440

Boss, A. P., 2002, Stellar Metallicity and the Formation of Extrasolar Gas Giant Planets, *ApJ*, **567**, L149

Brandner, W., et al., 2000, Timescales of Disk Evolution and Planet Formation: HST, Adaptive Optics, and ISO Observations of Weak-Line and Post-T Tauri Stars, *AJ*, **120**, 950

Burgasser, A. J., et al., 2000, Discovery of a Brown Dwarf Companion to Gliese 570ABC: A 2MASS T Dwarf Significantly Cooler than Gliese 229B, *ApJ*, **531**, L57.

Burgasser, A. J., et al., 2002, The Spectra of T Dwarfs. I. Near-Infrared Data and Spectral Classification, *ApJ*, **564**, 421

Burgasser, A. J., Kirkpatrick, J. D., Reid, I. N., Brown, M. E., Miskey, C. L., & Gizis, J. E., 2003, Binarity in Brown Dwarfs: T Dwarf Binaries Discovered with the Hubble Space Telescope WFPC2, *ApJ*, **586**, 512

Burgasser, A. J., Kirkpatrick, J. D., Cruz, K. L., Reid, I. N., Leggett, S. K., Liebert, J., Burrows, A., & Brown, M. E., 2006, Hubble Space Telescope NICMOS Observations of T Dwarfs: Brown Dwarf Multiplicity and New Probes of the L/T Transition, *ApJS*, **166**, 585

Burgasser, A. J., Reid, I. N., Siegler, N., Close, L., Allen, P., Lowrance, P., & Gizis, J. 2007, Not Alone: Tracing the Origins of Very-Low-Mass Stars and Brown Dwarfs Through Multiplicity Studies, In: *Protostars and Planets V*, ed. by B. Reipurth, D. Jewitt, and K. Keil, University of Arizona Press, Tucson, pp. 427-441

Burrows, A., Hubbard, W. B., Saumon, D., & Lunine, J. I. 1993, An expanded set of brown dwarf and very low mass star models, *ApJ*, **406**, 158

Burrows, A., et al. 1997, A Nongray Theory of Extrasolar Giant Planets and Brown Dwarfs, *ApJ*, **491**, 856

Burrows, A., Hubbard, W. B., Lunine, J. I., & Liebert, J. 2001, The theory of brown dwarfs and extrasolar giant planets, *Reviews of Modern Physics*, **73**, 719

Burrows, A., Sudarsky, D., & Lunine, J. I. 2003, Beyond the T Dwarfs: Theoretical Spectra, Colors, and Detectability of the Coolest Brown Dwarfs, *ApJ*, **596**, 587

Butler, R. P., Johnson, J. A., Marcy, G. W., Wright, J. T., Vogt, S. S., & Fischer, D. A. 2006, A Long-Period Jupiter-Mass Planet Orbiting the Nearby M Dwarf GJ 849. *PASP*, **118**, 1685

Carson, J. C., Eikenberry, S. S., Brandl, B. R., Wilson, J. C., & Hayward, T. L. 2005, The Cornell High-Order Adaptive Optics Survey for Brown Dwarfs in Stellar Systems. I. Observations, Data Reduction, and Detection Analyses, *AJ*, **130**, 1212

Chabrier, G., Baraffe, I., Allard, F., & Hauschildt, P. 2000, Evolutionary Models for Very Low-Mass Stars and Brown Dwarfs with Dusty Atmospheres, *ApJ*, **542**, 464

Chauvin, G., Lagrange, A.-M., Dumas, C., Zuckerman, B., Mouillet, D., Song, I., Beuzit, J.-L., & Lowrance, P. 2004, A giant planet candidate near a young brown dwarf. Direct VLT/NACO observations using IR wavefront sensing, *Astr. Ap.*, **425**, L29

Close, L. M., Siegler, N., Freed, M., & Biller, B. 2003, Detection of Nine M8.0-L0.5 Binaries: The Very Low Mass Binary Population and Its Implications for Brown Dwarf and Very Low Mass Star Formation, *ApJ*, **587**, 407

Cruz, K. L., et al. 2007, Meeting the Cool Neighbors. IX. The Luminosity Function of M7-L8 Ultracool Dwarfs in the Field, *AJ*, **133**, 439

Cruz, K. L., et al. 2008, In: *Proceedings of the 14th Cambridge Workshop on Cool Stars, Stellar Systems, and the Sun*, ed. by G. van Belle, in press

Duquennoy, A., & Mayor, M. 1991, Multiplicity among solar-type stars in the solar neighbourhood. II – Distribution of the orbital elements in an unbiased sample, *Astr. Ap.*, **248**, 485

Epchtein, N., et al. 1997, The deep near-infrared southern sky survey (DENIS), *The Messenger*, **87**, 27

Fischer, D. A., & Marcy, G. W. 1992, Multiplicity among M dwarfs, *ApJ*, **396**, 178

Fischer, D. A., Marcy, G. W., Butler, R. P., Vogt, S. S., Walp, B., & Apps, K. 2002, Planetary Companions to HD 136118, HD 50554, and HD 106252, *PASP*, **114**, 529

Gizis, J. E., Kirkpatrick, J. D., Burgasser, A., Reid, I. N., Monet, D. G., Liebert, J., & Wilson, J. C. 2001, Substellar Companions to Main-Sequence Stars: No Brown Dwarf Desert at Wide Separations, *ApJ*, **551**, L163

Gizis, J. E., Reid, I. N., & Hawley, S. L. 2002, The Palomar/MSU Nearby Star Spectroscopic Survey. III. Chromospheric Activity, M Dwarf Ages, and the Local Star Formation History, *AJ*, **123**, 3356

Galileo, G., 1630, Dialogo sopra i due massimi sistemi del mondo

Gatewood, G. 1976, On the astrometric detection of neighboring planetary systems, *Icarus*, **27**, 1

Gaudi, B. S. 2003, Microlensing Constraints on Low-Mass Companions, In: *Proceedings of IAU Symposium 211: Brown Dwarfs*, ed. E. Martín, Astronomical Society of the Pacific, San Francisco p. 305

Geballe, T. R., et al. 2002, Toward Spectral Classification of L and T Dwarfs: Infrared and Optical Spectroscopy and Analysis, *ApJ*, **564**, 466

Greenstein, J. L. 1988, The companion of the white dwarf G29-38 as a brown dwarf, *AJ*, **95**, 1494

Grossman, A. S., & Graboske, H. C. 1973, Evolution of Low-Mass Stars. V. Minimum Mass for the Deuterium Main Sequence, *ApJ*, **180**, 195

Herschel, W. 1803, *Philosophical Transactions Series I*, **93**, 339

Hinz, P. M., Angel, J. R. P., Woolf, N. J., Hoffmann, W. F., & McCarthy, D. W., Jr. 1999, Imaging Extra-solar Systems from the Ground: The MMT and LBT Nulling Interferometers, In: Working on the Fringe: Optical and IR Interferometry from Ground and Space. Proceedings from ASP Conference Vol. 194, eds. S. Unwin and R. Stachnik, 194, p. 401

Kendall, T. R., et al. 2007, Two T dwarfs from the UKIDSS early data release, *Astr. Ap.*, **466**, 1059

Kirkpatrick, J. D., et al. 1999, Dwarfs Cooler than "M": The Definition of Spectral Type "L" Using Discoveries from the 2 Micron All-Sky Survey (2MASS), *ApJ*, **519**, 802

Kirkpatrick, J. D. 2005, New Spectral Types L and T, *Ann. Rev. Astr. Ap.*, **43**, 195

Kirkpatrick, J. D. 2008, In: *Proceedings of the 14th Cambridge Workshop on Cool Stars, Stellar Systems, and the Sun*, ed. by G. van Belle, in press

Konopacky, Q. M., Ghez, A. M., Rice, E. L., & Duchene, G. 2007, New Very Low Mass Binaries in the Taurus Star-forming Region, *ApJ*, **663**, 394

Kouwenhoven, M. B. N., Brown, A. G. A., Zinnecker, H., Kaper, L., & Portegies Zwart, S. F. 2005, The primordial binary population. I. A near-infrared adaptive optics search for close visual companions to A star members of Scorpius OB2, *Astr. Ap.*, **430**, 137

Kumar, S. S. 1963, The Structure of Stars of Very Low Mass, *ApJ*, **137**, 1121

Lane, B. F., Zapatero Osorio, M. R., Britton, M. C., Martín, E. L., & Kulkarni, S. R. 2001, The Orbit of the Brown Dwarf Binary Gliese 569B, *ApJ*, **560**, 390

Latham, D. W., Stefanik, R. P., Mazeh, T., Mayor, M., & Burki, G. 1989, The unseen companion of HD114762 – A probable brown dwarf, *Nature*, **339**, 38

Lissauer, J. J., & Stewart, G. R. 1993, Growth of planets from planetesimals, In: Protostars and Planets III, University of Arizona Press, Tucson, Arizona, pp. 1061-1088

Lloyd, J. P. 2002, The detection and characterization of low mass companions to sun-like stars, Ph.D. Thesis, University of California, Berkeley, California

Lowrance, P. J., et al. 2005, An Infrared Coronagraphic Survey for Substellar Companions, *AJ*, **130**, 1845

Luhman, K. L., et al. 2007, Discovery of Two T Dwarf Companions with the Spitzer Space Telescope, *ApJ*, **654**, 570

Luyten, W.J. 1979, Catalogue of stars with proper motions exceeding 0"5 annually (LHS), Univ. of Minnesota Publ., Minneapolis, Minnesota

Luyten, W.J. 1980, Catalogue of stars with proper motions exceeding 0"2 annually (NLTT), Univ. of Minnesota Publ., Minneapolis, Minnesota

McCarthy, C. 2001, An infrared coronagraphic search for substellar companions to young nearby stars, Ph.D. Thesis, University of California, Los Angeles, California

McCarthy, C., & Zuckerman, B. 2004, The Brown Dwarf Desert at 75-1200 AU, *AJ*, **127**, 2871

McCarthy, D. W., Jr., Probst, R. G., & Low, F. J. 1985, Infrared detection of a close cool companion to Van Biesbroeck 8, *ApJ*, **290**, L9

McLean, I. S., McGovern, M. R., Burgasser, A. J., Kirkpatrick, J. D., Prato, L., & Kim, S. S. 2003, The NIRSPEC Brown Dwarf Spectroscopic Survey. I. Low-Resolution Near-Infrared Spectra, *ApJ*, **596**, 561

McLean, I. S., Prato, L., McGovern, M. R., Burgasser, A. J., Kirkpatrick, J. D., Rice, E. L., & Kim, S. S. 2007, The NIRSPEC Brown Dwarf Spectroscopic Survey. II. High-Resolution J-Band Spectra of M, L, and T Dwarfs, *ApJ*, **658**, 1217

Marcy, G. W., & Butler, R. P. 2000, Planets Orbiting Other Suns, *PASP*, **112**, 137

Marois, C., Doyon, R., Nadeau, D., Racine, R., Riopel, M., Vall'ee, P., & Lafrenière, D. 2005, TRIDENT: An Infrared Differential Imaging Camera Optimized for the Detection of Methanated Substellar Companions, *PASP*, **117**, 745

Marois, C., Lafrenière, D., Doyon, R., Macintosh, B., & Nadeau, D. 2006, Angular Differential Imaging: A Powerful High-Contrast Imaging Technique, *ApJ*, **641**, 556

Martín, E. L., Koresko, C. D., Kulkarni, S. R., Lane, B. F., & Wizinowich, P. L. 2000, The Discovery of a Companion to the Very Cool Dwarf Gliese 569B with the Keck Adaptive Optics Facility, *ApJ*, **529**, L37

Mayor, M., & Queloz, D. 1995, A Jupiter-Mass Companion to a Solar-Type Star, *Nature*, **378**, 355

Metchev, S. A. 2006, Brown dwarf companions to young solar analogs: An adaptive optics survey using Palomar and Keck, Ph.D. Thesis, California Institute of Technology, Pasadena, California

Metchev, S. A., & Hillenbrand, L. A. 2006, HD 203030B: An Unusually Cool Young Substellar Companion near the L/T Transition, *ApJ*, **651**, 1166

Nakajima, T., Oppenheimer, B. R., Kulkarni, S. R., Golimowski, D. A., Matthews, K., & Durrance, S. T. 1995, Discovery of a Cool Brown Dwarf, *Nature*, **378**, 463

Neuhäuser, R., & Guenther, E. W. 2004, Infrared spectroscopy of a brown dwarf companion candidate near the young star GSC 08047-00232 in Horologium, *Astr. Ap.*, **420**, 647

Oppenheimer, B. R., Kulkarni, S. R., Matthews, K., & Nakajima, T. 1995, Infrared Spectrum of the Cool Brown Dwarf GL:229B, *Science*, **270**, 1478

Oppenheimer, B. R., Kulkarni, S. R., & Stauffer, J. R. 2000, Brown Dwarfs, In: Protostars and Planets IV, University of Arizona Press, Tucson, Arizona, p. 1313

Oppenheimer, B. R., Golimowski, D. A., Kulkarni, S. R., Matthews, K., Nakajima, T., Creech-Eakman, M., & Durrance, S. T. 2001, A Coronagraphic Survey for Companions of Stars within 8 Parsecs, *AJ*, **121**, 2189

Padoan, P., & Nordlund, Å. 2004, The "Mysterious" Origin of Brown Dwarfs, *ApJ*, **617**, 559

Perrier, C., & Mariotti, J.-M. 1987, On the binary nature of Van Biesbroeck 8, *ApJ*, **312**, L27

Poindexter, S., Afonso, C., Bennett, D. P., Glicenstein, J.-F., Gould, A., Szymánski, M. K., & Udalski, A. 2005, Systematic Analysis of 22 Microlensing Parallax Candidates, *ApJ*, **633**, 914

Pravdo, S. H., Shaklan, S. B., & Lloyd, J. 2005, Astrometric Discovery of GJ 802b: In the Brown Dwarf Oasis?, *ApJ*, **630**, 528

Rebolo, R., Martín, E. L., & Magazzu, A. 1992, Spectroscopy of a brown dwarf candidate in the Alpha Persei open cluster, *ApJ*, **389**, L83

Reffert, S., Quirrenbach, A., Mitchell, D. S., Albrecht, S., Hekker, S., Fischer, D. A., Marcy, G. W., & Butler, R. P. 2006, Precise Radial Velocities of Giant Stars. II. Pollux and Its Planetary Companion, *ApJ*, **652**, 661

Reid, I. N., Cruz, K. L., & Allen, P. R. 2007, Meeting the Cool Neighbors. XI. Beyond the NLTT Catalog, *AJ*, **133**, 2825

Reid, I. N., & Gizis, J. E. 1997, Low-Mass Binaries and the Stellar Luminosity Function, *AJ*, **113**, 2246

Reid, I. N., Kirkpatrick, J. D., Liebert, J., Gizis, J. E., Dahn, C. C., & Monet, D. G. 2002, High-Resolution Spectroscopy of Ultracool M Dwarfs, *AJ*, **124**, 519

Reid, I. N., Gizis, J. E., Kirkpatrick, J. D., & Koerner, D. W. 2001, A Search for L Dwarf Binary Systems, *AJ*, **121**, 489

Reid, I. N., Lewitus, E., Burgasser, A. J., & Cruz, K. L. 2006, 2MASS J22521073-1730134: A Resolved L/T Binary at 14 Parsecs, *ApJ*, **639**, 1114

Reid, I. N., & Walkowicz, L. M. 2006, LP 261-75/2MASSW J09510549+3558021: A Young, Wide M4.5/L6 Binary, *PASP*, **118**, 671

Richardson, L. J., Deming, D., Horning, K., Seager, S., & Harrington, J. 2007, A spectrum of an extrasolar planet, *Nature*, **445**, 892

Salpeter, E. E. 1954, Reactions of Light Nuclei and Young Contracting Stars, In: Memoires de Societe Royale des Sciences de Liege, 1, p.116

Saumon, D., Hubbard, W. B., Burrows, A., Guillot, T., Lunine, J. I., & Chabrier, G. 1996, A Theory of Extrasolar Giant Planets, *ApJ*, **460**, 993

Scholz, R.-D., McCaughrean, M. J., Lodieu, N., & Kuhlbrodt, B. 2003, ε Indi B: A new benchmark T dwarf, *Astr. Ap.*, **398**, L29

Shatsky, N., & Tokovinin, A. 2002, The mass ratio distribution of B-type visual binaries in the Sco OB2 association, *Astr. Ap.*, **382**, 92

Simon, M., Bender, C., & Prato, L. 2006, The Gl569 Multiple System, *ApJ*, **644**, 1183

Simons, D. A., Henry, T. J., & Kirkpatrick, J. D. 1996, The Solar Neighborhood. III. A Near-Infrared Search for Widely Separated Low-Mass Binaries, *AJ* , **112**, 2238

Skrutskie, M. F., et al. 2006, The Two Micron All Sky Survey (2MASS), *AJ*, **131**, 1163

Stassun, K. G., Mathieu, R. D., & Valenti, J. A. 2006, Discovery of two young brown dwarfs in an eclipsing binary system, *Nature*, **440**, 311

Stoughton, C. e. et al. 2002, Sloan Digital Sky Survey: Early Data Release, *AJ*, **123**, 485

Tarter, J. C. 1976, Brown Dwarfs, Lilliputian Stars, Giant Planets and Missing Mass Problems, *Bulletin of the American Astronomical Society*, **8**, 517

Tokovinin, A. A., Chalabaev, A., Shatsky, N. I., & Beuzit, J. L. 1999, A near IR adaptive optics search for faint companions to early-type multiple stars, *Astr. Ap.*, **346**, 481

Udry, S., Bonfils, X., Delfosse, X., Forveille, T., Mayor, M., Perrier, C., Bouchy, F., Lovis, C., Pepe, F., Queloz, D., & Bertaux, J. 2007, The HARPS search for southern extra-solar planets. XI. Super-Earths (5 and 8 $M_\oplus$) in a 3-planet system, *Astr. Ap.*, **469**, 43

van Biesbroeck, G. 1944, The star of lowest known luminosity, *AJ*, **51**, 61

van de Kamp, P. 1982, The planetary system of Barnard's star, *Vistas in Astronomy*, **26**, 141

Unwin, S. C. et al., 2008, Taking the Measure of the Universe: Precision Astrometry with SIM PlanetQuest, *PASP*, in press

Zuckerman, B., & Becklin, E. E. 1987, Excess infrared radiation from a white dwarf – an orbiting brown dwarf?, *Nature*, **330**, 138

Zuckerman, B., & Song, I. 2004, Young Stars Near the Sun, *Ann. Rev. Astr. Ap.*, **42**, 685

6 Close-Orbiting Exoplanets: Formation, Migration Mechanisms and Properties

Hugh R.A. Jones, James S. Jenkins & John R. Barnes

Summary. The discovery of a signal from a putative extra-solar planet around the nearby Sun-like star 51 Peg by Mayor & Queloz and the rapid confirmation by Marcy & Butler was the main starting point for the field of extrasolar planets. More than a decade later, '51 Peg - type planets' or 'hot Jupiters' are frequently discovered and characterised by a variety of methods. Developments in experimental capabilities means that so called hot Saturn's and hot Neptune's have also been discovered. The wide range of properties of close-orbiting planets has stimulated a plethora of physical models to explain their properties. They provide the sharpest test for theories of formation, e.g., gravitational instability versus core-accretion, the role of stellar metallicity in determining planetary core mass and how an irradiating star influences planetary contraction and migration, e.g., type I, type II and delayed. With the continuous development of experiments close-orbiting terrestrial-mass extra-solar planets are the exciting new frontier in astrophysics and will test a wide range of theoretical predictions.

6.1 Introduction

It was the renaissance philosopher Giordano Bruno who first suggested there might be other worlds orbiting the stars of the night sky. Bruno's heretical philosophising came to a fiery end, when in 1600 he is believed to have been burned at the stake. However, his musings set the stage for one of astronomy's 'Holy Grails' - the search for planets around other stars. Bruno's death at the hands of the Holy Inquistion was followed by fruitless searches over the following 395 years. In 1991 the first extrasolar planet (exoplanet) based on its low-mass was discovered in close-orbit around a pulsar using timing measurements (Wolzcan & Frail 1992). A further three planets have now been discovered around PSR 1257+12 and even a possible comet, as well as a planet around PSR B1620-26 (Backer, Foster & Sallmen 1993; Sigurdsson et al. 2003). While these are landmark discoveries, the planets' location, next to a stellar remnant and perhaps forming after its collapse, probably helps little in understanding our own Solar System. Nonetheless it gives the impression that planet formation is robust and fuels our idea that planets are common throughout the universe.

In 1995 Mayor & Queloz (1995) announced the detection of the first exoplanet around a Sun-like star. The radial velocity of the G2V star 51 Pegasi was used to infer the presence of a Jupiter mass planet in a 4.2 day orbit. The discovery was quickly confirmed independently (Marcy & Butler 1996) and also corroborated by Doppler evidence for Jupiter mass planets in close-orbit around a number of other nearby stars. Our knowledge of exoplanets has been fuelled by the growth in the sheer number and also by the broad range of parameter space now populated. However, close-orbiting planets characterised with a combination of precise radial velocity measurements and transit photometry have played a key role. In these systems we can determine the mass and radius of the planet, which in turn yields constraints on its physical structure and bulk composition. The transiting geometry also permits the study of the planetary atmosphere without the need to spatially isolate the light from the planet from that of the star. This technique is known as transit spectroscopy or sometimes occultation spectroscopy and has allowed for photometric and spectroscopic measurements of exoplanets to be made.

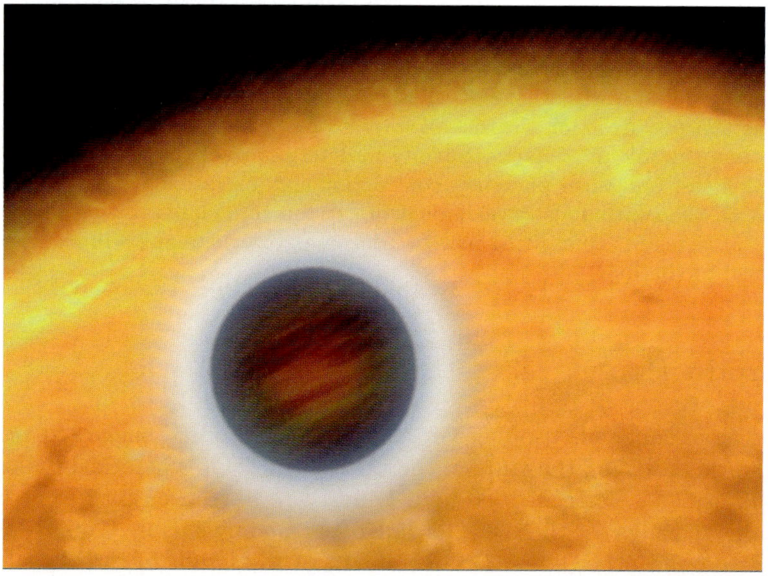

Fig. 6.1. In an artist's impression, a "hot Jupiter" in tight orbit about its parent star, is seen to puff up under the intense heat and its outer gases boil off into space. Image courtesy NASA, ESA, and G. Bacon (STScI).

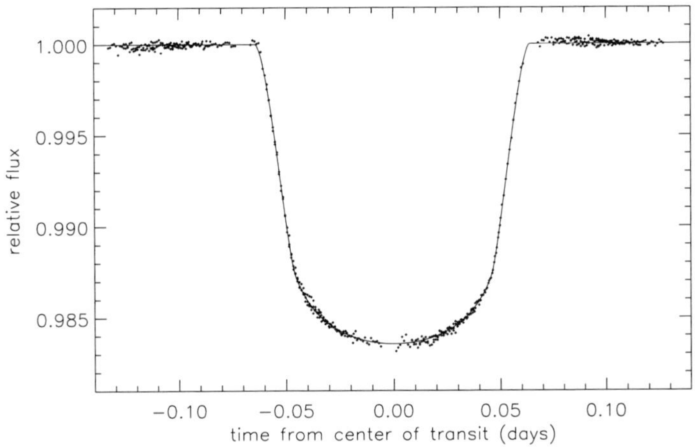

Fig. 6.2. Light curve from Brown et al. (2001) obtained by observing four transits of the planet of HD209458 using the STIS spectrograph on the Hubble Space Telescope. The folded light curve can be fitted within observational errors using a model consisting of an opaque circular planet transiting a limb-darkened stellar disk. In this way the planetary radius is estimated as 1.347 ± 0.060 R_{JUP}, the orbital inclination 86.6 ± 0.14, the stellar radius 1.146 ± 0.050 $R_{\odot}$. Satellites or rings orbiting the planet would, if large enough, be apparent from distortions of the light curve or from irregularities in the transit timings. No evidence is found for either satellites or rings, with upper limits on satellite radius and mass of 1.2 R_{JUP} and 3 M_{JUP}, respectively. Opaque rings, if present, must be smaller than 1.8 planetary radii in radial extent.

6.2 51 Pegasi as a Prototypical Close-Orbiting Exoplanet

While the discovery of 51 Pegasi was a landmark, as with the earlier discovery of planets around pulsars, it was met with scepticism in part because of the difficulties of the measurement but as much because 51 Pegasi b seemed to have nothing in common with our Jupiter (except mass). Interpretation of its orbit required rediscovery of the robust concept of inward planetary migration driven by tidal interactions with the protoplanetary disk (Goldreich & Tremaine 1980; Lin & Papaloizou 1986; Ward & Hourigan 1989). Thus, while such a large mass planet could not form in the glare of radiation from its sun, it was entirely plausible that it had migrated into position through the disk of material around 51 Pegasi (Lin, Bodenheimer & Richardson 1996). Although 51 Pegasi-like objects dominated early discoveries, other types of planets are considerably more common. Marcy et al. (2005) find that the occurrence of "hot Jupiters" within 0.1 au of FGK stars is 1.2 %. The 51 Pegasi class were found first because they are easiest to find. Relatively heavy and close to their stars they exert the largest force on their hosts and are thus by far the easiest to detect by the radial velocity method. In addition to being favoured by radial velocity surveys, the bias is even stronger in transit surveys. All known transiting exoplanets have periods less than a week.

Although the exoplanets discussed in this review were found using a range of techniques discussed in Chapter 1, it is important to emphasize that they are not discovered from a single well documented and quantified methodology. The compilation relies on a number of different ongoing surveys operating with different samples, sensitivities, instruments, scheduling, strategies and referencing techniques (e.g., Schneider et al. 2007).

6.3 Transit Discovery of Close-Orbiting Planets

Around 1 in 3000 stars (e.g. Horne 2003) are expected to have a planet in orbit around them which moves into the line of sight between the star and the Earth. The drop in brightness of a star can be detected as a cool planet transits across it and blocks some of the light. The first discovery of a transiting exoplanet was made by monitoring the known radial velocity discovered exoplanet HD209458 (Brown et al. 2001; Charbonneau et al. 2001). This validation of the radial velocity technique put the field of exoplanets on an indisputable empirical footing. The transit technique is now well-established as a primary discovery technique and has provided more than 20 exoplanets. The nearby examples of these are particularly valuable quarry. When a planet transits, its orbital inclination can be accurately measured which enables the mass of the planet to be determined from the $M \sin i$ measurement provided by a complementary radial velocity measurements. The planetary radius can be directly obtained by scaling the depth of the transit event to the radius of the star. With the mass and radius, quantities such as average density and surface gravity can be determined.

6.4 Orbital Characteristics of Close-Orbiting Planets

The term close-orbiting planets is not well-defined. Indeed this is perhaps not surprising given that the term planet, whilst having an ancient definition of 'wandering star' is itself somewhat controversial (e.g., Gingerich 2006). It is illuminating to examine what the current census of exoplanets reveals. All inferences and plots will be made by reference to the exoplanets catalogue (Butler et al. 2006, updated 2007 May 3 and further augumented by parameters for H43691b, HD132406b, GJ317b, TrES-3, HD147506b and GJ436b taken from Schneider et al. 2007). This catalogue provides the best available parameters for exoplanets though it is a compilation from many surveys and is drawn from an inhomogeneous sample.

Since exoplanets are found at a wide range of semi-major axes it is instructive to examine the overall distribution. Figure 6.3 (and a version for close orbiting planets Fig. 6.4) suggests that exoplanets show key differences with semimajor axis. Two separate features are apparent in the exoplanet period distribution. A peak of short-period exoplanets is seen in the 51 Pegasi-type objects, then a dearth, followed by a sharp rise and plateau in the number of exoplanets toward longer semimajor axes. It can be seen that when a crude attempt to remove the selection effect due to mass

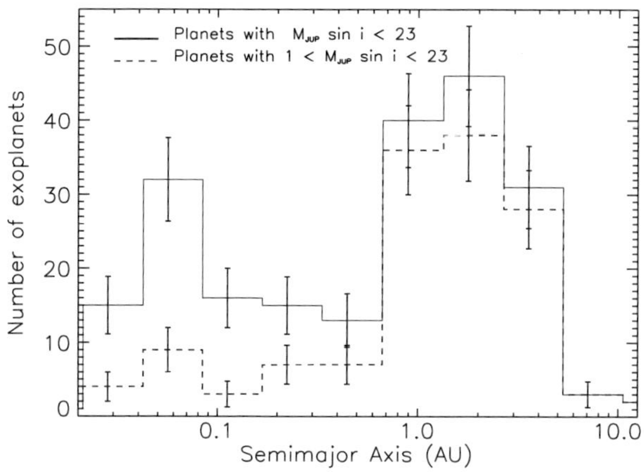

Fig. 6.3. The semimajor axis distribution for exoplanets. The solid line represents the entire exoplanet sample. The dashed line represents all planets above 1 M_{JUP} sin i. This was chosen as the limiting mass which is relatively unbiased throughout most of the range of semimajor axes over which exoplanets have been detected. For example, HD 154345b with 1.01 M_{JUP} sin i is found at 4.36 au around a V = 6.8, [Fe/H]=-0.1 dex star. It is thus not so bright or metal-rich that its detection is not expected to have resulted from extra data points and thus to have its exoplanetary detectability substantially enhanced (e.g. Cumming 2004). The error bars are from $\sqrt{(number)}$ statistics and are only indicative. While a peak at small values of semi-major axes may arise from incompleteness there is evidence for a sharp change in the distribution, presumably due to migration, between 0.5 and 1 au.

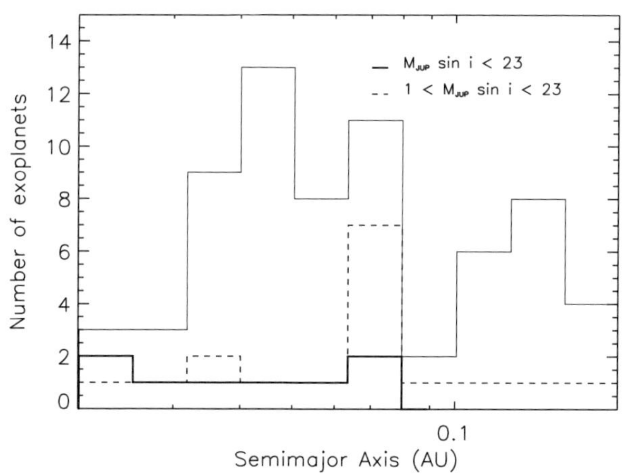

Fig. 6.4. The semimajor axis distribution for close-orbiting exoplanets. The solid line represents the entire exoplanet sample. The thick solid line shows the sample of transit planets. The small semimajor axis side of the plot is dominated by transit discoveries and the right-hand side by radial velocity discoveries. The dashed line represents all planets above 1 M_{JUP} sin i. The error bars are from $\sqrt{(number)}$ statistics and are only indicative. There is evidence for a significant peak in the distribution for the 1 M_{JUP} sin i and above sample at around 0.07 au which maybe symptomatic of mass dependent migration, however, in order to have confidence it is necessary to better understand the relatively sharp selection biases for transit and radial velocity surveys which exist within this plot range.

is made that the short-period peak largely disappears. The distribution we are then left with is clearly distinguished into short- and long-period at around 0.5 au.

6.4.1 Exoplanetary Mass Function

Figure 6.5 shows a mass histogram representing all exoplanets as well as solid lines showing sub-samples of short, intermediate and long period exoplanets. All indicate a rise in the number of exoplanets per unit mass towards lower masses despite the selection bias toward finding heavier exoplanets. This overall distribution is similar to the mass function based on subsamples of the long-period and intermediate-period exoplanets. Also over-plotted on this graph for comparison is a dotted line for $dN/dM \propto M^{-1.1}$ (following Butler et al. 2006). The sensitivity of the radial velocity planets below 1 M_{JUP} sin i is limited (see Fig. 6.3 caption) and so the turnover that both mass functions exhibit below 1 M_{JUP} sin i is expected. Authors making corrections for incompleteness find steeper relationships, e.g., $dN/dM \propto M^{-1.6}$ Lineweaver, Grether & Hidas (2003). Cumming et al. (quoted in Butler et al. 2006) finds $dN/dM \propto M^{-1.1}$ for M <0.6 M_{JUP} sin i and $dN/dM \propto M^{-1.9}$ for M>0.6 M_{JUP} sin i. However, until an approach such as that in Cumming et al. (2003), including a full treatment of eccentricity which incorporates a knowledge of detection sensitivities for a single sample, is made such results

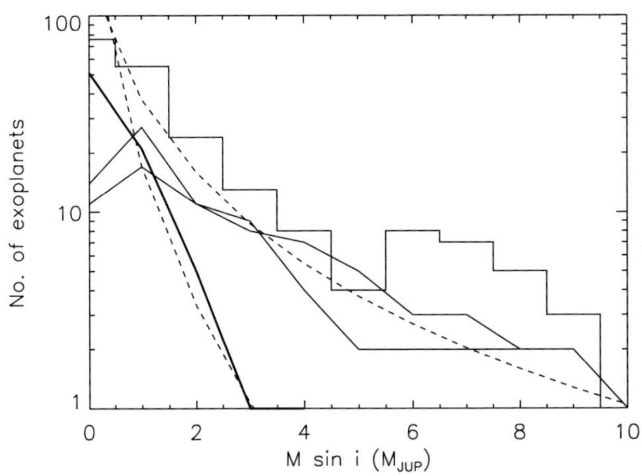

Fig. 6.5. The plot shows various representations for the number of exoplanets per unit mass. By the dotted lines it may depicit the apparently different mass functions prevalant for close-orbiting planets and others. The two solid lines which approximately follow the shape of the histogram are smoothed sub-samples of exoplanets with intermediate (0.24-1.52) and large (1.6-11.4) values of au. The dashed line among them is the power law for the number of exoplanets scaling as $dN/dM \propto M^{-1.1}$. To the left of the plot is a steeper solid line which is a smoothed version of the small (0.0177-0.238) au sample as well as a dotted line showing $dN/dM \propto M^{-3.1}$ a tantalising indication that the close-orbiting exoplanet mass function may be steeper.

are problematic. Nonetheless, it is noteworthy that the short-period mass function appears to have a somewhat different form. In Fig. 6.5 it is plotted with a thicker line and is overplotted with a normalised curve for $dN/dM \propto M^{-3}$.

6.4.2 Exoplanetary Eccentricities

The upper plot in Fig. 6.6 shows that exoplanets can have a wide range of eccentricities at all semi-major axes. This is contrary to our Solar System where planets have orbits close to circular (<0.21). The binned plot in the lower plot of Fig. 6.6 indicates the expected trend that planets with small semi-major axes should have small eccentricities due to tidal circularisation of close-in planets (e.g. Halbwachs, Mayor & Udry 2005). However the clear turning point at periastron distances of less than 0.1 au (e.g., Rasio & Ford 1996) is not seen. For periastron distances of 0.1 au

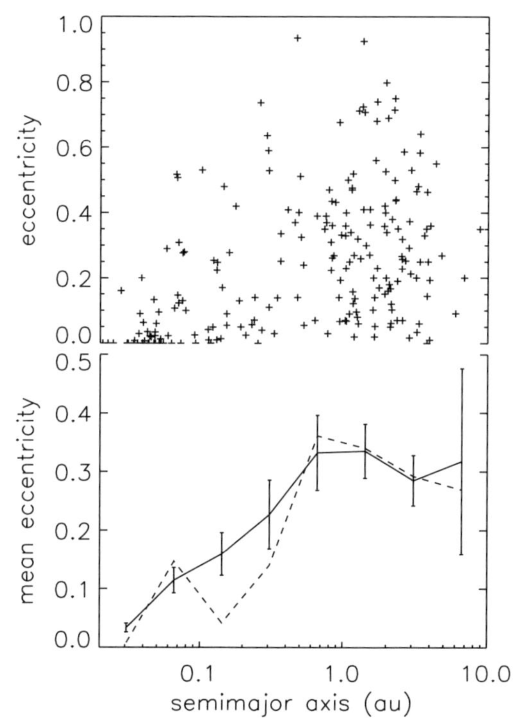

Fig. 6.6. The upper plot shows a scatter plot for eccentricities against semimajor axis for exoplanets. The lower plot shows mean eccentricities plotted against semimajor axis for exoplanets. Mean eccentricities are shown binned in the solid line (and median in dotted since there are cases where exoplanet eccentricity fits are set to zero when there is relatively little radial velocity data). The error bars are from $\sqrt{(number)}$ statistics and are only indicative. While a wide range of eccentricity values can be found across the range of measured semi-major axes it is evident that circularisation processes are evident below 1 au.

the tidal circularisation timescale is already likely to be many Gyr, for inferred Q values (based on planets having semi-major axes with less than 0.1 au, where Q is the specific dissipation function). The circularisation timescale, $t_{\rm circ}$, is a sensitive function of the radius of the planet, $R_{\rm planet}$ ($t_{\rm circ} \propto R_{\rm planet}^{-5}$, Goldreich & Soter 1966) and supposedly occurs due to dissipation in the planet, not the star (e.g., Gu et al. 2004). Perhaps, the larger planet radii during contraction, in the first 10 Myr or so, helps to shorten the circularisation time. Thus the rising mean eccentricities versus semi-major axis apparent in Fig. 6.6 from 0.1 to around 0.5 au may be explained by the decreasing effectiveness of the tidal circularisation mechanism.

One might expect that the orbital eccentricities of smaller planets would respond more to eccentricity boosting mechanisms, however, the lower plot in Fig. 6.7 suggests the opposite. While an attempt has been made to correct for the bias against finding high eccentricity planets (by only looking at the more 'complete' sample of higher mass planet) this is a crude correction (e.g. Cumming 2004). Ribas & Miralda-Escude (2007) hypothesize that there are two populations of gaseous planets: the low-mass population which form by gas accretion onto a rock-ice core in a circumstellar disk and is more abundant at high metallicities, and a high-mass population which forms directly by fragmentation of a pre-stellar cloud. Planets

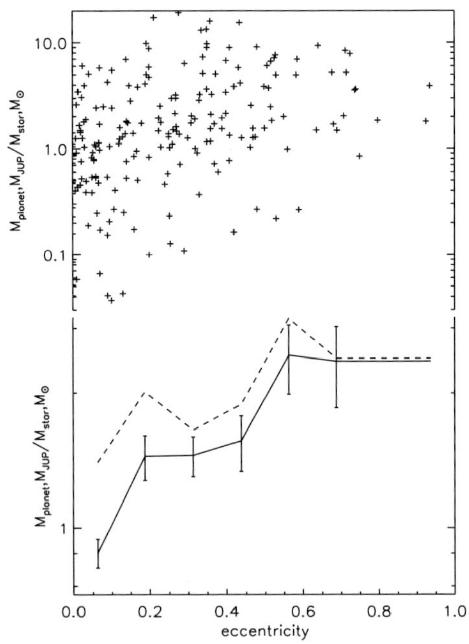

Fig. 6.7. The upper plot shows a scatter plot for $M_{{\rm planet}_{\rm Jup}}/M_{{\rm star}_\odot}$ against eccentricities for exoplanets. The lower plot shows median $M_{{\rm planet}_{\rm Jup}}/M_{{\rm star}_\odot}$ values and suggests that for a given mass host star a larger planet will have a higher eccentricity. The dashed line represents all planets above 1 $\rm M_{JUP}$ sin i. The error bars are from $\sqrt{(number)}$ statistics and are only indicative.

of the first population form in initially circular orbits and grow their eccentricities later, and may have a mass upper limit from the total mass of the disk that can be accreted by the core. The second population may have a mass lower limit resulting from opacity-limited fragmentation. This would roughly divide the two populations in mass, although they would likely overlap over some mass range. If most objects in the second population form before the pre-stellar cloud becomes highly opaque, they would have to be initially located in orbits larger than ~30 AU, and would need to migrate to the much smaller orbits in which they are observed. The higher mean orbital eccentricity of the second population might be caused by the larger required intervals of radial migration, and the brown dwarf desert might be due to the inability of high-mass brown dwarfs to migrate inwards sufficiently in radius.

6.4.3 The Parent Stars of Close-Orbiting Exoplanets are Metal-Rich

An important characteristic of a star is the fraction of 'metals' (elements heavier than helium) it contains – the metallicity (usually measured as [Fe/H]). Gonzalez (1997) found exoplanet host stars to be metal-rich. This conclusion has been confirmed by many authors with different samples, methodologies and spectral synthesis codes. Fig. 6.8 shows just how metal-rich all exoplanet primaries are. Although close-orbiting exoplanets show the highest [Fe/H], there is no strong trend with semi-major axis. All points are 0.15 dex above the solar; whereas the Sun and other solar type dwarf stars in the solar neighbourhood have an average metallicity of 0 or even slightly less (Reid 2002). This result, representing the only link between the presence of planets and a stellar photospheric feature, has been given two main explanations. The first, is based on the classical view that giant planets are formed by runaway accretion of gas on to a 'planetesimal' having up to 10 Earth masses. In such a case, we can expect that the higher the proportion of dust to gas in the primordial cloud (i.e. the proportion of metals), and consequently in the resulting protoplanetary disc, the more rapidly and easily may planetesimals, and subsequently the observed giant planets be built.

Opposing this view, it had been proposed that the observed metallicity 'excess' may be related to 'pollution' of the convective envelope of the star by the infall of planets and/or planetesimals (e.g., Hole et al. 2001). So far evidence for this has only been found in one of the many systems investigated. While such pollution may play a small role, giant stars with planets all have high metallicities and show no evidence for pollution (e.g., Santos et al. 2003). These are crucial systems because they are well mixed and not subject to the mixing and surface uncertainties that limit our confidence in measuring 'pollution' in Sun-like stars. Besides the [Fe/H] differences, there is currently some debate about possible anomalies concerning other elements. But the relatively low number of exoplanets known, and possible systematics with respect to the samples, do not permit firm conclusions to be reached (Santos et al. 2003). So far, it seems that the higher metallicity of most planet-harbouring stars arises because high metallicity environments have a higher probability of planet formation (e.g., Livio & Pringle 2003). Thus it appears that like the early detections of 51 Pegasi-type exoplanets we are finding a surfeit of exoplanets around metal-

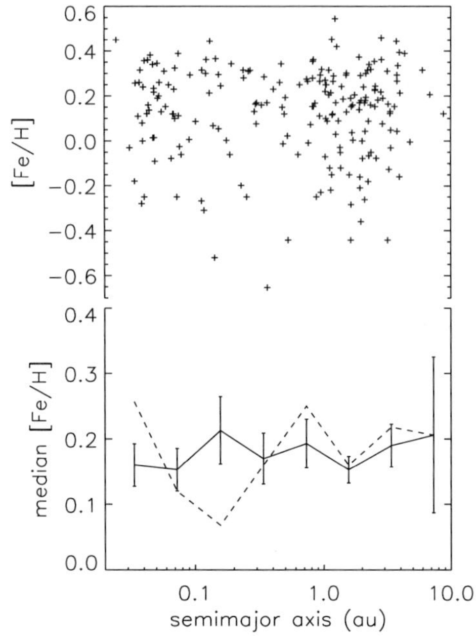

Fig. 6.8. Median spectroscopic [Fe/H] values for the primaries of exoplanets plotted as a function of semimajor axis. The metal-rich *Nature* of exoplanets is evident from this plot, with no binned data approaching solar metallicity. The dashed line represents all planets above 1 M_{JUP} sin *i*. The error bars are from $\sqrt{(number)}$ statistics and are only indicative. No relationship between semi-major axis and metallicity is apparent and thus any metallicity dependence of migration can be expected to be relatively modest.

rich stars because they are easier to detect. The observed lack of [Fe/H] correlation with semi-major axis or any theoretical expectation of a link (e.g., Pringle et al. 2003) indicates metallicity does not play a key role in exoplanet migration. Note that global trends are difficult to quantify due the addition of metal-rich stars to most major planet search campaigns.

6.5 Migration and Formation of Exoplanets

The distribution of exoplanets with semimajor axes in Figs. 6.4 and 6.5 suggest that exoplanets show key differences with period. The short-period peak has been an important motivation in the development of migration theories for exoplanets (Lin et al. 1996; Murray et al. 1998; Trilling et al. 1998; Ward 1997; Armitage et al. 2002; Trilling, Lunine & Benz 2003). This peak can be explained by a stopping mechanism such as Lindblad resonances between the planet and disk (Kuchner & Lecar 2002) and photoevaporation (Matsuyama, Johnstone & Murray 2003) limiting exoplanet migration to periods of less than 3 days. Once there exists a sample of short-period exoplanets derived from a wider range of primary masses it should

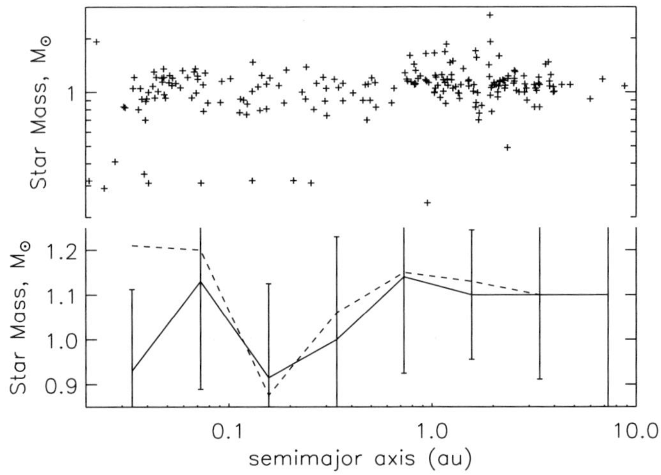

Fig. 6.9. The plot shows median $M_{star_\odot}$ values for exoplanet host stars against semimajor axis for their exoplanets. The upper plot indicates a relative lack of stars with masses between 0.4 and 0.7 $M_\odot$ around which planets have been found. This is not unexpected since surveys target a greater fraction of 'Solar mass' stars and more recently have also focussed telescope time on the lowest-mass stars observable with current experiments. The lower plot indicates median $M_{star_\odot}$ values, the dashed line represents the sub sample of all planet hosts with above 1 M_{JUP} sin i. No strong relationship between host mass and semi-major axis, which might be a proxy for mass-dependent migration, is indicated. The error bars are from $\sqrt{(number)}$ statistics and are only indicative.

be possible to distinguish between different stopping mechanisms by comparing the orbital characteristics of the exoplanet with those of the primary. For example, Kuchner & Lecar predict that planets around early A-type main sequence stars will collect at a radius much further from the star (0.3 au) than the radius where solar-type stars collect. Further data is required in order to robustly test this hypothesis, e.g., Fig. 6.9. The rise in the number of exoplanets towards longer periods is becoming more apparent as more planets are discovered and is very well reproduced by exoplanet migration scenarios which envisage planets migrating inwards (e.g., Armitage et al. 2002; Trilling et al. 2003) as well as outwards (Masset & Papaloizou 2003).

6.5.1 Planet Formation

The favoured model of forming the cohort of currently discovered planets is by core accretion (Laughlin, Bodenheimer & Adams 2004) this is backed up by both observations (Fischer & Valenti 2005) and theory (Rafikov 2005). Core accretion forms planets by the build-up of particles in the stellar disk; the disk being leftover from the formation of the parent star. Initially the small dust grains in the disk begin to coagulate until they form larger structures whose gravity can affect their

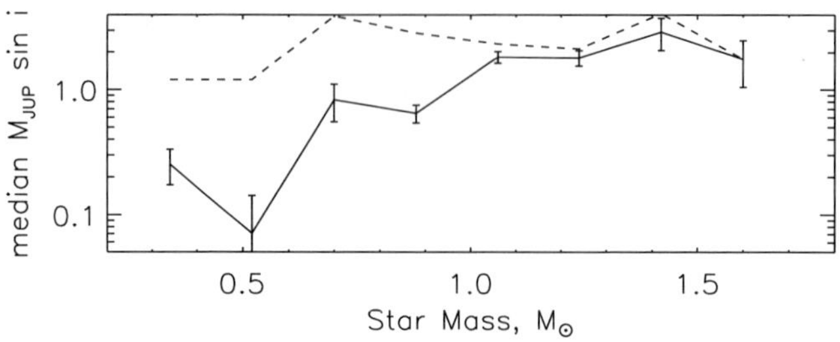

Fig. 6.10. The plot shows median M_{JUP} versus $M_{\mathrm{star}_\odot}$. The solid line of binned data suggests a relationship between host mass and planet mass, however, the dashed line of all planets above 1 $\mathrm{M_{JUP}}$ sin i does not support this conclusion for a more 'complete' sample. The error bars are from $\sqrt{(number)}$ statistics and are only indicative.

surroundings. The simplest method of planet formation occurs when a core grows by the accretion of a number of planetesimals at a fixed orbital radius from the star (Safranov 1969). Initially this core will be surrounded by a near-hydrostatic gaseous envelope, with the majority of the luminosity being generated by the continuous planetesimal accretion. The growth of the core will continue until a critical core mass is reached, at which point there is no stable core-envelope solution and the planetary envelope will collapse onto the protoplanetary core (Mizuno 1980; Pollack et al. 1996). The mass of the accreted planetesimals and accreted gas is almost equally distributed and as the envelope contracts the gas accretion rate increases, which increases the energy losses through radiation giving rise to a runaway accretion process. This formation scenario describes the types of planets discovered thus far but the picture is slightly different for terrestrial planets, where the core does not grow large enough for the surrounding gas to collapse onto it, and hence a significant planetary atmosphere must be generated by different processes like volcanism.

One of the long standing problems with the core accretion model relates to the timescale of formation of the planetary embryos. To generate a core massive enough for the surrounding disk gas to collapse onto requires timescales on the order of 10^7 years (Hayashi, Nakazawa & Adachi 1977), however the observational upper limits for the lifetimes of disks forming planets is in the range 10^5 -10^7 years. Thus the planetesimals will have had insufficient time to accrete enough gas to create the gas giant planets we observe today. More recently models of formation and the environment of the disk can somewhat account for this discrepancy. Alibert et al. (2005a) has shown that by including disk evolution, migration and gap formation into their models, the formation timescales for planetesimal cores become more compatible with typical disk lifetimes. This speed-up is mostly due to migration, which ensures there is no depletion of the feeding zones of forming planets, a problem for in situ models. They have also shown that these models are consistent with the wealth of information garnered on Jupiter and Saturn on their core masses and total amount of heavy elements in their envelopes (Alibert et al. 2005). However, another problem arises due to the fast migration timescales of small planetesimal cores.

An alternative model for planetary formation is the disk instability mechanism which is based on the formation and survival of self-gravitating clumps of gas and dust in a marginally gravitationally unstable protoplanetary disk (Boss 1997; reviewed by Durisen et al. 2007). In order for a disk instability to succeed, the disk must be able to cool its midplane as the clumps form allowing them to continue to contract to higher densities. The clumps must be able to survive indefinitely in the face of Keplerian shear, tidal forces and internal thermal pressure. Different resolution and sized calculations together with different boundary conditions and treatments of inner disk heating lead different authors (e.g. Boss 2007, Boley et al. 2007, Mayer et al. 2007) to almost rule out or even promote the disk instability mechanism. Thus further work is required before the importance of the disk instability formation mechanism can be established.

6.5.2 Migration and Evolution

When planetary migration and formation were first modelled together (e.g., Trilling et al. 1998, Rafikov 2002, Lufkin et al. 2004) it was discovered that beyond a critical core mass for the forming planet, a gap in the protoplanetary disk is created. The critical mass is dependent on the mass and metallicity of the disk and hence it doesn't have a singular value, but has been shown to reside in a region between around 10 - 30 M_{Earth}. Planets below the critical mass undergo Type I migration with no gap created, whereas planets that accrete enough disk material to cross the critical mass undergo migration in a gap (Type II migration).

The two types of migration both have different characteristics and hence both need to be fully understood to explain the distribution of planetary systems discovered thus far. However, both types are caused by the same processes, those being tidal interactions between the planet and the protoplanetary disk. Other processes are involved with migration, such as star-planet interactions and planetary mass-loss, however Trilling et al. (1998) has shown that it is the disk-planet torques that dominate. The interactions that induce the planetary migration are located at resonant sites throughout the disk. Linblad and co-rotation resonances cause the planets to migrate through the disk, with Linblad resonances being the most important. The planetesimals migrate by the emission of spiral density waves that travel away from the planet and interact at these Linblad sites. Over a large range of density gradients and disk pressures the outer-Linblad resonances dominate in planetary systems, causing inward migration to occur more frequently than outward migration (Ward 1997). Ward (1986) has also shown that Type I migration is extremely fast, as fast as 10^3 years for Jupiter-type planets. Trilling, Lunine & Benz (2002) have shown that disk lifetimes can reach up to 10^7 years, which poses a problem for planets undergoing Type I migration as at this speed a lot of planetesimals would migrate onto their host stars and be destroyed. When the critical mass has been breached, the planetesimal will open a gap in the disk and the orbital migration will change completely. Type II migration will occur, and even though the processes inducing the migration are similar, their characteristics are different. When a gap has been opened in the disk the resonance sites are located within the gap. This leads to higher-order resonances and therefore the planet undergoes a much slower migration rate. Also Lin & Papaloizou (1986) found that tidal locking can occur in the gap, slowing the migration further as the planets migration follows the disk's viscous flow. The migration timescale is now on the order of 10^6 - 10^7 years, which is comparable to the longest disk lifetimes and helps alleviate the problem of planets spiralling into their host stars. However, for typical planetesimal parameters the timescale of the Type I migration is still considerably shorter than the accretion timescale. Thommes & Murray (2006) and Thommes, Nilsson & Murray (2007) have shown that by using a model with a viscously evolving disk the accretion and migration timescales become comparable and can help to halt the destruction of the forming planetesimals.

Planetary migration not only has implications for the planet that is migrating through the disk, but it also puts limits on the probability of other planets existing in the disk. Smaller planets that have formed in the same disk could be ejected from the system due to the tidal forces induced by a migrating gas giant (Juric & Tremaine 2007). On the other hand, Edgar & Artymowicz (2004) have found that a migrating protoplanet may not inhibit the formation of more planetesimals after migration, and the migrating planet will only pump-up the eccentricities of planetesimals, with more massive planets inducing larger pumping. Recent work from Fogg & Nelson (2007) on the other hand have shown that migrating gas giants can help to replenish the habitable zones of stars with water by the mixing of material from beyond the "snow-line" to the inner regions of the systems. This, coupled with the scattering of planetesimals outwards to create a solid disk where dynamical frictions are strong and terrestrial planet formation can resume, gives rise to water abundant terrestrial planets.

Once the disk of material that formed the planets has cleared ($>10^7$ years) the planets remain relatively fixed in their orbits unless perturbed by an outside influence such as a binary star or another planets gravitational field. The relatively high number of planets in some form of commensurability shows that planet-planet interactions exist long after the initial migration phase. Planets that are not eventually scattered from their orbits will evolve in a radiative fashion. Evolutionary models from Burrows et al. (1997) and Baraffe et al. (2003) have shown that the effective temperatures and luminosities decay rapidly as a function of time as the internal en- ergy leftover from the formation of the planet is lost. A 12 M_J planet orbiting a Sun-like star, after 5 Gyrs of evolution, will be $\sim 10^{-6}$ times fainter than the host. This has repercussions for imaging such objects as the brightness difference between the host star and the planet is so large even after the inclusion of reflected light.

Winn et al. (2007) suggest how orbital migration mechanisms may be empirically constrained. Simulations of inward planet migration via tidal interactions with the protoplanetary disk predict low values of eccentricities, not as large as 0.5 (see, e.g., D'Angelo, Lubow & Bate 2006). While most of the close-orbiting exoplanets have low values of eccentricity there are several with semi-major axes of 0.1 au and eccentricities around 0.5. They presumably acquired their high eccentricities by mechanisms naturally giving rise to large eccentricities: planet-planet scattering (e.g., Chatterjee et al. 2007) or due to the tide of a third body (the Kozai mechanism whereby an orbit undergoes eccentricity/inclination oscillations and ultimately shrinks in semimajor axis due to tidal dissipation, e.g., Fabrycky & Tremaine 2007, Wu et al. 2007). A corollary of either scattering or Kozai migration is that the orbit can be tilted considerably with respect to its initial orbital plane, which was presumably close to the stellar equatorial plane. One can search for such a misalignment by exploiting the spectral distortion observed during a transit due to stellar rotation (the Rossiter-McLaughlin effect), whereby the planet hides part of the rotational velocity field of the stellar photosphere, resulting in an 'anomalous Doppler shift' (see, e.g., Snellen 2004, Ohta et al. 2005). The time sequence of anomalous Doppler shifts depends on the angle between the stellar spin axis and the orbital axis, as projected on the sky. This angle has been measured to be small or consistent with zero in several systems (e.g., Winn et al. 2007). Nonetheless, further refinements to theoretical predictions and measurements for eccentric close-orbiting planets which transit have the potential to empirically distinguish between different migration mechanisms.

6.6 Close-Orbiting Planet Atmospheres

The first spectroscopic probe of exoplanet atmospheres was provided by Charbonneau et al.'s (2002) detection of the additional dimming of sodium absorption during transit due to absorption from sodium in the atmosphere of HD209458b. The observed dimming is reasonably well modelled by planetary atmosphere models that incorporate irradiation and allow for sodium to be out of thermal equilibrium (Barman et al. 2002). Vidal-Madjar et al. (2003, 2004) have also detected atomic hydrogen, carbon and oxygen during transits of HD209458b. The large implied physical radii exceeds the Roche limit, leading them to conclude that material is escaping from the planet. However, the minimum escape rate based on the data is low enough to reduce the planetary mass by only 0.1% over the age of the system (confirmed 'empirically' by Melo et al. 2006). Combining this with models for HD209458b, which put the upper atmosphere at a temperature of 10,000 K, this gives strong evidence for atmospheric evaporation. This evaporation confirms the conclusions by Hebrard et al. (2003) and may lead to new types of planets being discovered with hydrogen poor atmospheres or even with no atmospheres at all (Trilling et al. 1998).

Spitzer has detected radiation from several hot Jupiters, over six bandpasses from 3.6 to 24 μm and has detected phase-dependent flux implying significant day-night temperature contrasts on two hot Jupiters (Harrington et al. 2006, Knutson et al. 2007). Further announcements are expected. Such observations are now fuelling new research into the meterology of exoplanets (Cooper & Showman 2005, 2006; Fortney et al. 2006). Spectroscopic observations of stars during primary and secondary eclipse of close-orbiting planets have been particularly revealing. The detection of sodium in the atmosphere of HD209458b (Charbonneau et al. 2002) and detections of water in the far red optical regime have been made on HD209458b by Barman (2007) and on HD189733b by Tinetti et al. (2007). These detections can now be well modelled by a near isothermal vertical profile for the planet's atmosphere. Furthermore, the lack of water vapour and possible silicate features observed by Richardson et al. (2007) and Grillmair et al. (2007) during secondary eclipse is consistent with the expected strong circulation on close-orbiting planets which can flatten the day side temperature gradient. Mid-infrared measurements have been made possible by the exquisite precision made possible by transit spectroscopy and space-borne instrumentation. Ground-based near-infrared spectroscopy has proved more difficult (Richardson et al. 2003) though is expected to soon enable access to near-infrared flux measurements (e.g. Barnes et al. 2007a and Fig. 6.11 from Barnes et al. 2007b), which will allow the resolution to resolve different atomic and molecular species.

Reflected light studies carried out by (Collier Cameron et al. 1999; Charbonneau et al. 1999; Collier Cameron et al. 2002; Leigh et al. 2003a,b) (as well as results from MOST photometry, e.g., Rowe et al. 2006) place albedo upper limits on the atmospheres of CEGPs. These upper limits to the planet/star contrast ratio have established the low reflectivity when compared with the solar system gas giants. High precision polarimeters (e.g. Hough et al. 2006) are now able to measure the polarisation of reflected light from extra-solar planets. Combining this with $M \sin i$ measurements from radial velocity data will provide the orbital inclination of the planet and hence the planet's mass. Detailed modelling will determine the nature of the scattering particles and the geometric albedo as a function of wavelength.

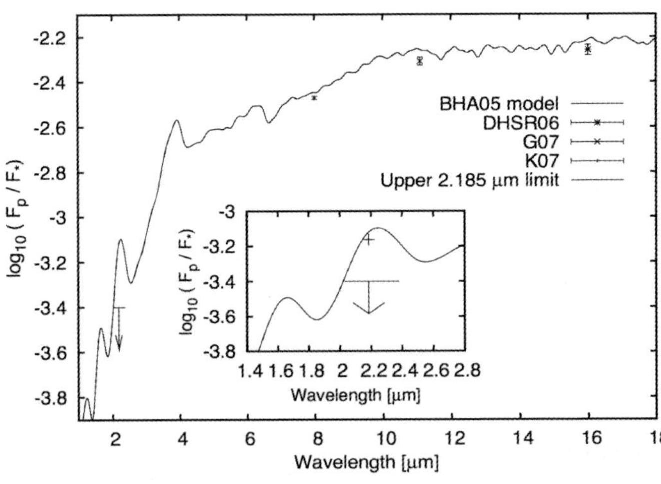

Fig. 6.11. Model planet/star flux ratio for the HD189733 system. *Spitzer* eclipse depth measurements are plotted for 8 μm (Knutson et al. 2007), 11.1 μm (Grillmair et al. 2007) and 16 μm (Deming et al. 2006). A horizontal bar with vertical down-pointing arrow indicates Barnes et al. (2007) 1 σ limit; the width of the horizontal bar represents the wavelength range of the data. The single point plotted in the inset at $\log_{10}\epsilon_0$ = -3.16 represents the model mean flux over the range of the non-detection. The upper limit is at a level of $\log_{10}\epsilon_0 = 0.24$ lower than the model (i.e. ϵ_0(model)/ϵ_0(observed)=1.7).

6.7 Composition

Transiting planets are key because they provide accurate estimates of mass, radius, and, by inference, composition. The position of a planet in the mass-radius diagram is a direct indication of its overall composition, while other factors such as temperature play only a minor role (see e.g. Fortney et al. 2007). Some such trends include trends between planet mass (Mazeh, Zucker & Pont 2005) or gravity (Southworth, Wheatley & Sams 2007) and orbital period for the known transiting planets. The existence of such correlations indicates something about the composition and their physical nature. Hansen & Barman (2007) identify a bimodality in the distribution of transiting planet properties, based on the Safronov number, which essentially measures the efficiency with which a planet scatters other bodies. They expect that this reflects the influence of planet or planetesimal scattering in determining when planetary migration stops. Another possibility is that some planets lose more mass to evaporation than others. If this evaporation process preferentially removes helium from the planet, the consequent reduction in the mean molecular weight may explain why some planets have anomalously large radii.

Fig. 6.12 from Gillon et al. (2007a,b) puts the known transiting planets in the context of the Solar System planets and the range of mass-radius relationships expected for different compositions. Some exoplanet systems appear to have considerably higher radii than expected. In the current paradigm, intermediate-mass planets are composed of an iron/nickel core, a silicate layer, an ice layer (H_2O, CH_4, NH_3), and an H/He envelope. The mass and radius that is measured for the lowest-mass example GJ436b indicate that it is not a low-mass gas giant or a very heavy "super-Earth". The presence of a significant amount of methane and ammonia in addition to water within a pure ice planet could slightly increase the radius above the theoretical value for a pure water ice planet. Nevertheless, a planet composed only of ice (and thus without any rock) is improbable in the current paradigm: all the icy objects in the Solar System have a considerable fraction of rock and have their ice composition dominated by water. GJ436b likely has an H/He envelope and thus appears to be very similar to our Neptune.

6.8 Future

Future space missions such as Darwin and TPF will enable direct imaging and spectroscopy for exoplanet spectra. To gain the full benefit of these missions, commensurate investments in basic atomic and molecular data as well as appropriately irradiated 3D circulation models will need to be made in order that synthetic spectra for exoplanet atmospheres will provide a worthy match. In the meantime very substantial advances are expected. A glimpse of the high quality of the space-based transit missions COROT and Kepler is provided by the first COROT press release announcing its first exoplanet whilst still undergoing system calibration.

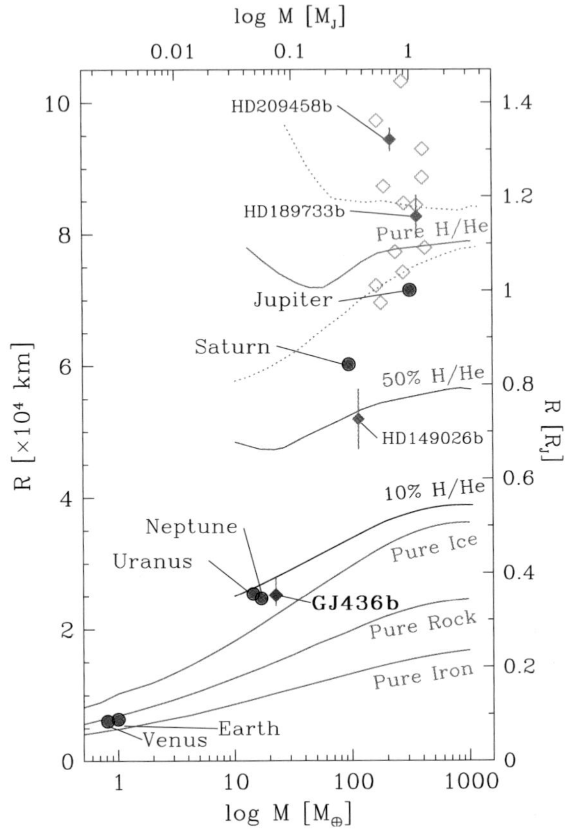

Fig. 6.12. Planetary mass-radius diagram (from Gillon et al. 2007a) comparing the position of Solar System planets, transiting hot Jupiters (diamonds), and GJ 436 b. The lines indicate the position of the Fortney et al. (2007) models for different compositions: pure iron, pure silicate, pure water ice (with thermal profiles from Solar System planets), and models for irradiated planets at 0.1 au from a Solar-type star with a fraction of 10%, 50% and 100% of Hydrogen/Helium. The dotted lines show the models for a cold ($a = 10$ au) and very hot ($a = 0.02$ au) pure H/He gas giant.

6.8.1 The Hunt for Terrestrial Planets

Radial velocity surveys have announced 5 M_{Earth} planets in the solar neighbourhood. However, the radial velocity detection method is extremely mass dependent as its signal is proportional to the ratio of the planet and stellar mass. In order to detect terrestrial planets around stars via the Doppler technique, it is necessary to increase the sensitivity of the detection method and/or decrease the mass of the target primary stars. Both strategies are being pursued. Existing surveys teams expand their allocation of time and construct new dedicated or larger telescopes (e.g. HARPS-N, Automated Planet Finder). Alternatively, a number of groups are con-

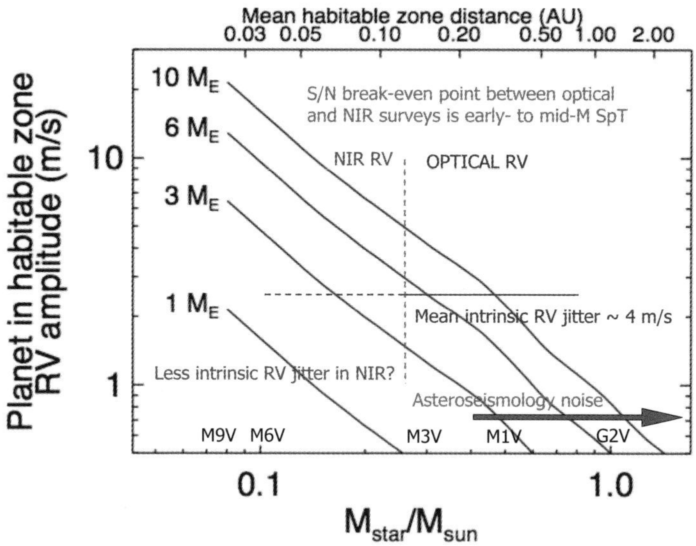

Fig. 6.13. The plot (from John Rayner) indicates how the power of radial velocities maybe further extended by carrying out surveys in the infrared and around lower mass stars. Such surveys have the potential to detect close-orbiting planets down to terrestrial masses.

structing promising high precision infrared spectrographs (e.g., Fig. 6.13, Precision Radial Velocity Spectometer, www.roe.ac.uk/ukatc/prvs)

The impetus of discovery and characterisation means that exoplanet discovery should continue to increase as objects are found from a wide range of techniques. The power of characterisation using several techniques has already been proven for GJ436 and HD209458. A much deeper understanding of exoplanets and our Solar System should become apparent once such data exists for a large sample. In the near term the continuing powerful combination of radial velocity together with transit photometry and timing is very pomising. Astrometric (e.g., CTIO) and interferometric (e.g. Magellan Ridge) imaging measurements should also provide important constraints. Notwithstanding an increasing rate of discovery and characterisation of an even broader spectrum of exoplanets, the next few years should continue to bring dramatic improvements in the realism of exoplanet formation and evolution simulations.

The revolution in exoplanet science is just beginning. At the moment we are close to being sensitive to Earth-mass exoplanets. While we need a better idea of planets in general terms as a function of host star properties, planet mass, composition (gas, ice, rock, metal) and orbit parameters and the intercorrelation of these parameters the recent launch of COROT and impending launch of Kepler means that progress should continue to accelerate. It will be fascinating to see the importance of environment on exoplanets, not only on the major planets that we detect over the next few years but also the minor constituents such as comets and asteroids which are already been constrained with improved transit timings. Substantial research into chemical differentiation will be necessary and allow the serious investigation into the extent to which terrestrial-like planets have non-equilibrium atmospheres.

References

Alibert Y., Mousis O., Mordasini C., Benz W., 2005, New Jupiter and Saturn Formation Models Meet Observations, *ApJ*L, **626**, 57

Armitage P.J., Bonnell I. A., 2002, The Brown Dwarf Desert as a Consequence of Orbital Migration, *MNRAS*, **330**, 11

Armitage P.J., Livio M., Lubow S.H., Pringle J.E., 2002, Predictions for the Frequency and Orbital Radii of Massive Extrasolar Planets, *MNRAS*, **334**, 248

Backer D. C., Foster R. S. Sallmen S., 1993, A Second Companion of the Millisecond Pulsar 1620-26, *Nature*, **365**, 817

Baraffe I., Chabrier G.., Barman T. S., Allard F., Hauschildt P. H., 2003, Evolutionary models for cool brown dwarfs and extrasolar giant planets. The case of HD 209458, *A&A*, **402**, 701

Barman T. et al., 2002, Non-LTE Effects of Na I in the Atmosphere of HD 209458b, *ApJ*, **569**, 51

Barman T., 2007, Identification of Absorption Features in an Extrasolar Planet Atmosphere, *ApJL*, **669**, 191

Barnes J., 2007a, Near-infrared spectroscopic search for the close orbiting planet HD 75289b, *MNRAS*, **379**, 1097

Barnes J., 2007b, 2.2 micron limits on HD189449, *MNRAS*, in press (astroph/07084300)

Boley A.C. et al., 2007, The Internal Energy for Molecular Hydrogen in Gravitationally Unstable Protoplanetary Disks, *ApJ*L, **656**, 89

Boss A.P., 1997, Giant planet formation by gravitational instability, *Science*, **276**, 1836

Boss A.P., 2007, Testing Disk Instability Models for Giant Planet Formation, *ApJL*, **661**, 73

Brown T., Charbonneau D., Gilliland R., Noyes R., Burrows A., 2001, HST Time-Series Photometry of the Transiting Planet of HD 209458, *ApJ*, **552**, 699

Burrows A., 1997, A Nongray Theory of Extrasolar Giant Planets and Brown Dwarfs, *ApJ*, **491**, 856

Butler R.P., et al. 2006, Catalog of Nearby Exoplanets, *ApJ*, **646**, 505

Charbonneau D., Brown T.M., Noyes R.W., Gilliland R.L., 2002, Detection of an Extrasolar Planet Atmosphere, *ApJ*, **568**, 377

Chatterjee, S., Ford, E. B., Rasio, F. A. 2007, Dynamical Outcomes of Planet-Planet Scattering, astroph/0703166

Cooper C.S., Showman A.P., 2005, Dynamic Meteorology at the Photosphere of HD 209458b, *ApJL*, **629**, 45

Cooper C.S., Showman A.P., 2006, Dynamics and Disequilibrium Carbon Chemistry in Hot Jupiter Atmospheres, with Application to HD 209458b, *ApJ*, **649**, 1048

Cumming A., 2004, Detectability of extrasolar planets in radial velocity surveys, *MNRAS*, **354**, 1165

D'Angelo, G., Lubow, S. H., Bate, M. R. 2006, Evolution of Giant Planets in Eccentric Disks, *ApJ*, **652**, 1698

Durisen R.H. et al. 2007, Gravitational Instabilities in Gaseous Protoplanetary Disks and Implications for Giant Planet Formation in Protostars & Planets V, eds, Reipurth B., Jewitt D. & Keil K., Tuscon, Univ. of Arizona Press, p607

Edgar R., Artymowicz P., 2004, Pumping of a planetesimal disc by a rapidly migrating planet, *MNRAS*, **354**, 769

Fabrycky, D., Tremaine, S. 2007, Shrinking binary and planetary orbits by Kozai cycles with tidal friction, *ApJ*, **669**, 1298

Fischer D. A., Valenti J., 2005, The Planet-Metallicity Correlation, *ApJ*, **622**, 1102

Fogg M. J., Nelson R. P., 2007, On the formation of terrestrial planets in hot-Jupiter systems, *A&A*, **461**, 1195

Fortney J.J. et al., 2006, he Influence of Atmospheric Dynamics on the Infrared Spectra and Light Curves of Hot Jupiters, *ApJ*, **652**, 746

Gillon M. et al., 2007a, Detection of transits of the nearby hot Neptune GJ 436 b, *A&A*, **472**, 13

Gillon M. et al., 2007b, Accurate Spitzer infrared radius measurement for the hot Neptune GJ 436b, *A&A*, **471**, 51

Goldreich P., Soter S., 1966, Q in the Solar System, Icarus, **5**, 375

Goldreich P., Tremaine S., 1980, Disk-Satellite Interactions, *ApJ*, **241**, 425

Gonzalez G., 1997, The Stellar Metallicity-Giant Planet Connection, *MNRAS*, **285**, 403

Grillmair C.J. et al. 2007, A Spitzer Spectrum of the Exoplanet HD 189733b, *ApJ*, **658**, L115

Gu P.-G., Bodenheimer P.H., Lin D.N.C., 2004, The Internal Structural Adjustment Due to Tidal Heating of Short-Period Inflated Giant Planets, *ApJ*, **608**, 1076

Harrington J. et al. 2006, The Phase-Dependent Infrared Brightness of the Extrasolar Planet ν Andromedae b, *Science*, **314**, 623

Hough J. et al. 2006, PlanetPol: A Very High Sensitivity Polarimeter, *PASP*, **118**, 1305

Knutson H.A. et al. 2007, A map of the day-night contrast of the extrasolar planet HD 189733b, *Nature*, **447**, 183

Hebrard G., Etangs A. Lecavelier des., Vidal-Madjar A., Desert J. -M., Ferlet R., 2004, Evaporation Rate of Hot Jupiters and Formation of Chthonian Planets, *ASPC*, **321**, 203

Horne K., 2003, Status and Prospects of Planetary Transit Searches: Hot Jupiters Galore, ASP Conference Series 294: Scientific Frontiers in Research on Extrasolar Planets, eds. D. Deming and S. Seager, 361

Hole et al., 2001, Stellar Metallicity Enhancement by Exoplanet Consumption, *AAS*, **199**, 611

Gingerich O., et al., 2006, IAU resolution votes, IAU0603, http://www.iau.org/iau0603.414.0.html

Juric M., Tremaine S., 2007, The Eccentricity Distribution of Extrasolar Planets, astroph/0703160

Kuchner N., Lecar M., 2002, Halting Planet Migration in the Evacuated Centres of Protoplanetary Disks, *ApJ*L, **574**, 87

Laughlin G., Bodenheimer P., Adams F. C., 2004, The Core Accretion Model Predicts Few Jovian-Mass Planets Orbiting Red Dwarfs, *ApJ*L, **612**, L73

Livio M., Pringle J., 2003, Metallicity, Planetary Formation and Migration, *MNRAS*, **346**, 42

Lineweaver C.H., Grether D., Hidas, 2003, What Can Extrasolar Planets Tell Us About Our Solar System?, ASP Conf. Ser., 294, 161

Lin D.N.C., Bodenheimer P., Richardson D.C., 1996, Orbital Migration of the Planetary Companion of 51 Pegasi to its Present Location, *Nature*, **380**, 606

Lin D.N.C., Papaloizou J.C.B., 1986, On the Tidal Interaction Between Protoplanets and the Protoplanetary Disk. III Orbital Migration of Protoplanets, *ApJ*, **309**, 846

Lucas P., Roche, 2002, A search for the infrared spectroscopic signature of hot Jupiter planets, *MNRAS*, **336**, 637

Lufkin G., Quinn T., Wadsley J., Stadel J., Governato F., 2004, Simulations of gaseous disc-embedded planet interaction, *MNRAS*, **347**, 421

Marcy G.W. et al., 2005, Observed Properties of Exoplanets: Masses, Orbits, and Metallicities, Progress in Theoretical Physics, 158, **24**

Marcy G.W., Butler R.P., 1996, First Three Planets, SPIE, **2704**, 46

Masset F.S., Papaloidzou J., 2003, Runaway Migration and the Formation of Hot Jupiters, *ApJ*, **588**, 494

Matsuyama I., Johnstone D., Murray N., 2003, Halting Planet Migration by Photoevaporation from the Central Source, *ApJ*L, **585**, 143

Mayor M., Queloz D., 1995, A Jupiter-Mass Companion to a Solar-Type Star, *Nature*, **378**, 355

Mazeh T., Zucker S., Pont F., 2005, An intriguing correlation between the masses and periods of the transiting planets, *MNRAS*, **356**, 955

Melo C. et al., 2006, On the age of stars harboring transiting planets, *A&A*, **460**, 251

Murray N., Hansen B., Holman M., Tremaine S., 1998, Migrating Planets, *Science*, **279**, 69

Ohta, Y., Taruya, A., Suto, Y. 2005, The Rossiter-McLaughlin Effect and Analytic Radial Velocity Curves for Transiting Extrasolar Planetary Systems, *ApJ*, **622**, 1118

Rafikov R. R., 2002, Planet Migration and Gap Formation by Tidally Induced Shocks, *ApJ*, **572**, 566

Rafikov R. R., 2005, Can Giant Planets Form by Direct Gravitational Instability?, *ApJ*L, **621**, 69

Rasio F.A., Ford E.B, 1996, Dynamical instabilities and the formation of extrasolar planetary systems, *Science*, **274**, 954

Ribas I., Miralda-Escude J., 2007, The eccentricity-mass distribution of exoplanets: signatures of different formation mechanisms?, *A&A*, **464**, 779

Richardson L.J. et al. 2007, A spectrum of an extrasolar planet, *Nature*, **445**, 892

Reid I.N., 2002, On the *Nature* of Stars with Planets, *PASP*, **114**, 306

Safranov V. S., 1969, NASA Tech. Trans. F-677

Santos N.C., Israelian G., Mayor M., Rebolo R., Udry S., 2003, Metallicity, Orbital Parameters, and Space Velocities, *A&A*, **398**, 363

Sigurdsson S., Richer H.B., Hansen B.M., Stairs I.H., Thorsett S.E., 2003, A Young White Dwarf Companion to Pulsar 1620-26, *Science*, **301**, 193

Schneider J., 2007, The Extrasolar Planets Encyclopaedia, http://www.obspm.fr/encycl/encycl.html

Snellen, I, 2004, A new method for probing the atmospheres of transiting exoplanets, *MNRAS*, **353**, 1

Southworth J., Wheatley P.J. Sams G., 2007, A method for the direct determination of the surface gravities of transiting extrasolar planets, *MNRAS*, **379**, 11

Swain M.R., et al. 2007, The mid-infrared spectrum of the transiting exoplanet HD 209458b, astroph/0702593

Thommes E. W., Murray N., 2006, Giant Planet Accretion and Migration: Surviving the Type I Regime, *ApJ*, **644**, 1214

Thommes E. W., Nilsson L., Murray N., 2007, Overcoming Migration during Giant Planet Formation, *ApJL*, **656**, 25

Tinetti G et al., 2007, Water vapour in the atmosphere of a transiting extrasolar planet, *Nature*, 448, 169

Trilling D.E., Benz W., Gulliot T., Lunine J.I., Hubbard W.B., Burrows A., 1998, Orbital Evolution and Migration of Giant Planets: Modelling Extrasolar Planets, *ApJ*, **500**, 428

Trilling D., Lunine J.I., Benz W., 2003, Orbital migration and the frequency of giant planet formation, *A&A*, **394**, 241

Vidal-Madjar A. et al., 2003, An extended upper atmosphere around the extrasolar planet HD209458b, *Nature*, **422**, 143

Vidal-Madjar A. et al., 2004, Detection of Oxygen and Carbon in the Hydrodynamically Escaping Atmosphere of the Extrasolar Planet HD209458b, *ApJL*, **604**, 69

Ward W.R., Hourigan K., 1989, Orbital Migration of Protoplanets,: The Inertial Limit, *ApJ*, **347**, 490

Ward W.R., 1997, Protoplanet Migration by Nebula Tides, *ICAR*, **126**, 261

Winn J.N. et al., 2007, Spin-Orbit Alignment for the Eccentric Exoplanet HD 147506b, *ApJL*, **665**, 167

Wolszczan A., Frail D., 1992, A Planetary System around the Millisecond Pulsar PSR1257+12, *Nature*, **255**, 145

7 Dynamics of Multiple Planet Systems

Rory Barnes

Summary. This chapter discusses the dynamical properties of multiple planet systems. The orbits of these planets evolve due to tidal, resonant, and/or secular (long-term) effects. Basic analytical and numerical techniques can describe these interactions. Multiple planet systems may also evolve chaotically, and some principles of chaos theory are described. Finally, this chapter discusses the current distributions of dynamical properties of exoplanetary systems, possible origins of these distributions, and compares exoplanetary systems to the Solar System.

7.1 Introduction

Dynamics is a general term that encompasses all methods of describing and predicting the motion of systems of fluids and particles. Systems of planets (which, for the purpose of understanding their motion, can be considered particles) can therefore be described, understood, and characterized by equations that have been developed for centuries. These equations predict the positions and velocities of planets as a function of time, their "evolution". The evolution may be periodic (the motion repeats on a given timescale), or chaotic (the motion is non-repeating). The discovery of exoplanets has provided new examples of dynamical behavior not seen among planets in our Solar System. These new systems reveal how our Solar System is similar to, and different from, the general population of planetary systems. By compiling statistics of planetary system properties, we can validate planet formation scenarios. For most planetary systems, the classical dynamics of point particles is a reliable approximation for the motions of planets, although the orbits of a few planets on very short period orbits are affected by general relativity and/or tides. A comprehensive review of planetary dynamics would be too long to present here, so we focus on recent advances in analytical and numerical techniques to analyze, interpret and compare multiple-planet systems.

In this chapter we ignore single planet systems, in which the planets' orbits are simply ellipses. However, the orbits of these planets may provide important evidence about their origin. For example, some have large eccentricities ($e > 0.3$) that may reflect interactions with now-ejected, or currently undetectable, companions (Rasio

& Ford, 1996; Marzari & Weidneschilling, 2002; Bodenheimer, Laughlin & Lin, 2003; Wu & Murray, 2003). Additionally, planets on small orbits (with periods ≤ 11 days), the so-called "hot Jupiters", most likely underwent tidal circularization (Rasio et al., 1996). For a review of the dynamics of planet formation see Papaloizou & Terquem (2006).

7.1.1 Planetary Orbits

Johannes Kepler showed in the 1600s that the path of a planet about a star, the orbit, is an ellipse with the star located at one focus. This ellipse lies in the planet's orbital plane. The shape of this orbit is described by 5 parameters called orbital elements (a sixth parameter defines the position of the planet on the orbit).

The orbital elements that can be identified by radial velocity surveys are the orbital period P, which is related to the semi-major axis a (the average distance of a planet from its star) by Kepler's Third Law, eccentricity e (a measure of the deviation from a perfectly circular orbit), the longitude of periastron ϖ (the angle, measured from the star, between a reference direction and the direction of the planet's point of closest approach). The reference direction is the line between the Earth and the host star. The planet's closest approach distance to the primary, periastron, is $a(1-e)$, and the furthest distance, apoastron, is $a(1+e)$. The true longitude, ν, is the angle between the reference direction and the direction from the sun to the planet. This geometry is shown in Fig. 7.1.

7.1.2 Observational Constraints

The study of exoplanet dynamics is severely hampered by observational uncertainties. Although the detections themselves are robust, the orbital elements have significant uncertainties. The most problematic aspect of the Doppler technique is the mass-inclination degeneracy. If the inclination, the angle between the plane of the orbit and a reference plane, could be determined by a complementary method, such as astrometry or transits, this degeneracy may be broken, and the planetary masses and full three dimensional orbits could be identified. The mass-inclination degeneracy therefore makes many simulations, analyses, and hypotheses unreliable. Generally, in dynamical studies and throughout this chapter, the masses are assumed to be the "minimum mass", the mass if the orbit was exactly edge-on. Statistically, we expect this choice to be reasonably accurate.

The Doppler technique also limits the ranges of planetary masses and orbital radii that may be observed. A planet must be massive enough and close enough to the star for its orbit (about the planet) to be observable. Furthermore, the orbital period must be short enough that at least one complete orbit can be detected. Therefore the observed planets may not be all the planets in a system. The conclusions presented here are subject to revision as additional planets may exist in each system that are either low-mass or orbit at large distances, and these unseen companions may significantly alter the best-fit orbits of the known planets.

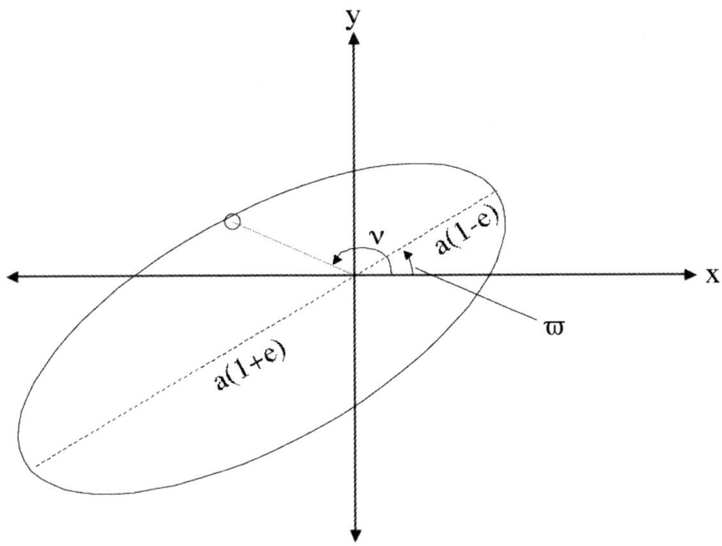

Fig. 7.1. Schematic of an exoplanet orbit, as detected by radial velocity surveys. The planet is denoted by the open circle. The reference direction, positive x, points toward the Earth. The host star is located at the origin, which is also one focus of the elliptical orbit. The major axis of the orbit is denoted by the dashed line, and the relationship between semi-major axis, a, and eccentricity, e, is shown. The true longitude, ν is is the angle between the reference direction and the direction of the planet, as measured from the origin.

Beyond the mass-inclination degeneracy and incompleteness, orbital parameters also suffer from random errors. Stellar jitter (a result of the turbulent surface of stars), the photometric condition of the observations, spectral resolution, etc., all contribute an uncertainty to each individual observation, which in turn produces errors in the orbital parameters. These uncertainties in orbital elements are often large enough to include regions of instability, in which numerical simulations predict at least one planet is ejected from the system (see Sect. 7.4.4) (Barnes & Quinn, 2004). Therefore dynamical studies of individual systems must be interpreted cautiously. The dynamics described here are based on the recent catalog of exoplanets (Butler et al., 2006), unless otherwise cited.

7.2 Review of Orbital Theory

Dynamical analyses may be divided into three categories: analytical, semi-analytical, and N-body. Analytical investigations must make major assumptions, but the resulting equations can be solved with just pencil and paper. Semi-analytical research makes some assumptions to produce analytical equations, but the equations often can only be solved by numerical methods (with computers). N-body methods use computers to calculate the gravitational forces among all the bodies involved and determine their motions, i.e. how the positions and velocities change with time.

Analytical and semi-analytical methods generally ignore short-period changes, which are assumed to average to zero over long timescales. These methods model the long-term orbital evolutions, not the actual positions of planets, and provide insight into how certain systems will behave. Since these methods have individual terms whose numerical values can be calculated, the relative importance of each effect can be quantified. The disadvantage of these methods is they are often only accurate in certain regimes, like low eccentricity. More terms can be added to a semi-analytic description to improve accuracy, but eventually the complexity outweighs the advantage of averaging short-period effects. In these situations it is best to turn to numerical simulations. These N-body integrations are grounded in first principles as they solve the fundamental laws of gravity and motion, i.e. they are "self-consistent". N-body simulations are often used to test the validity of analytical and semi-analytical results. Modern computing power limits the simulation time and/or the number of bodies that can be considered, but for many systems the changes in the shapes of the orbits are periodic, and the equations that describe the motion need only be integrated for 1 period ($\lesssim 10^5$ years) to reveal the dynamics. In summary, analytic methods approximate the long-term motion ($\gtrsim 1$ Gyr) and N-body simulations show the true motion. When used appropriately, these approaches provide powerful insight into the dynamics of planetary systems.

7.2.1 Analytical Methods

The basis for analytical methods lies in consideration of the "disturbing function", the difference between the gravitational potential of a planet due to a star, and that due to a star and one or more additional planets. We will focus on the Fourier series expansion of the disturbing function. Analytic methods can describe two phenomena often seen in planetary systems: resonant and secular evolution. Resonant and secular theory assume certain terms in the series will average to zero, and can therefore be ignored when modeling orbits. Secular theory ignores all terms that depend on the mean motion, n (the orbital frequency if the planet were on a circular orbit), and, in many cases, all terms that are of order 3 or higher, i.e. $e^3 \approx 0$. Resonant theory adds terms that do depend on mean motions, but only those related to the resonance in question. For a detailed description of the disturbing function, secular theory and resonance theory, consult Murray & Dermott (1999).

Secular Theory

To visualize secular theory, imagine the distribution of the planets' masses over a long timescale. The distribution would be that of a ring of matter. Secular theory predicts how the shapes of these rings will change with time, and hence how the orbit changes with time. In this second order approximation, motion out of the plane is decoupled from motion in the plane.

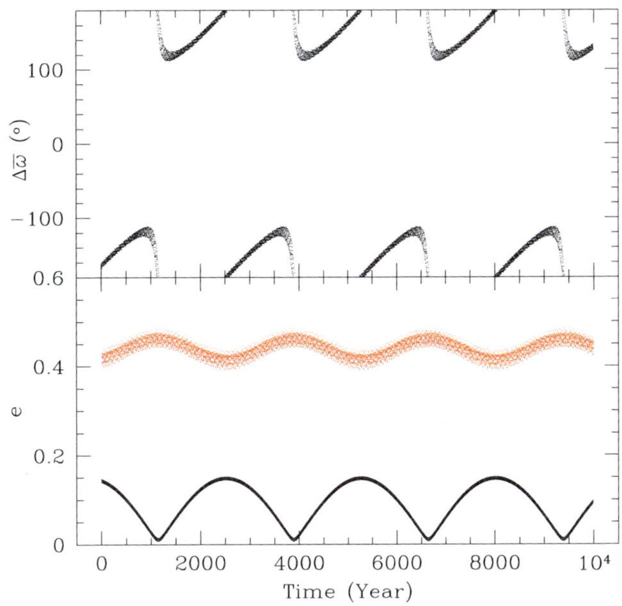

Fig. 7.2. The secular oscillations of the planets in orbit about HIP 14810 (Wright et al., 2007). Note that the oscillations of both e's (inner is black, outer red) and $\Delta\varpi$ have the same period. Also note that when the e's are at their extrema, $\Delta\varpi$ is at its equilibrium value, 180° in this case.

In nearly all planetary system cases, secular theory predicts the e's and $\Delta\varpi$'s (the difference between two ϖ's) oscillate. The movement of ϖ is often called precession. An example of a secular interaction is shown in Fig. 7.2. Secular theory assumes a is constant, and therefore conservation of angular momentum requires that as one planet's eccentricity drops, the other rises (orbital angular momentum is proportional to e). If $\Delta\varpi$ oscillates about 0, the system is experiencing aligned libration. If $\Delta\varpi$ librates about π, then the system is undergoing anti-aligned libration. If $\Delta\varpi$ oscillates through 2π then the "apsides" (the points of closest and furthest approach to the origin of an orbit) are circulating. The type of oscillation depends on initial conditions. Exoplanet examples of these types of behavior are shown in Fig. 7.3. The motion of $\Delta\varpi$ is analogous to that of a swinging pendulum. When the pendulum swings back and forth, the oscillation is libration. If the pendulum swings all the way around, the oscillation is circulation. Note there is a clear boundary between these types of motion: the swing that brings the pendulum up to a perfectly vertical position. This boundary between qualitatively different types of oscillation is known as a "separatrix".

Recently it has been noted that many systems lie near an "apsidal separatrix" (Ford et al., 2005; Barnes & Greenberg, 2006a,c) (the point on the orbit that is closest to the origin is known as the "apse"). The simplest apsidal separatrix is the boundary between libration (either aligned or anti-aligned) and circulation. For

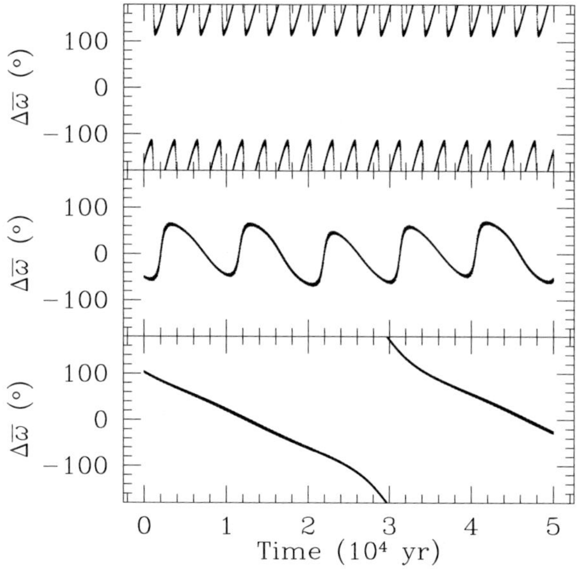

Fig. 7.3. Examples of apsidal oscillations in exoplanetary systems The y-axis is the difference between the longitudes of periastron, $\Delta\varpi$. *Top:* HD 12661 undergoes apsidal libration in an anti-aligned mode. When exactly anti-aligned ($\Delta\varpi = \pi$) the direction of periastron of one planet is the same as apoastron of the other. *Middle:* HD 37124 c-d undergoes apsidal libration in an aligned mode. When aligned ($\Delta\varpi = 0$) the ellipses are oriented such that the two periastra point in the same direction. *Bottom:* The planets in HD 168443 undergo apsidal circulation since $\Delta\varpi$ rotates through 360°.

systems of just two planets, the apsidal separatrix can only separate circulation and libration. This type of separatrix is a "libration-circulation separatrix". An example of the libration-circulation separatrix is shown in the left panels of Fig. 7.4.

In systems of more than two planets, things get more complicated. In addition to the libration-circulation separatrix, the system may interact with different numbers of rotations of $\Delta\varpi$ through 360° during one eccentricity oscillation. The boundary between interactions with different numbers of circulations in one eccentricity cycle is called a "circulation-mode separatrix", and an example is shown in the right panels of Fig. 7.4.

For an interaction to lie near a separatrix, the amplitude of eccentricity oscillations are generally two orders of magnitude or more. Since $0 \leq e < 1$ for bound planets, this means that at least one planet in near-separatrix interactions (both libration-circulation and circulation-mode) periodically is on a nearly circular orbit. The proximity to the separatrix can be parameterized by ϵ, which is approximately equal to the minimum e divided by the average e over a secular cycle (Barnes & Greenberg, 2006c). When $\epsilon = 0$ the pair is on an apsidal separatrix, and one eccentricity periodically reaches zero.

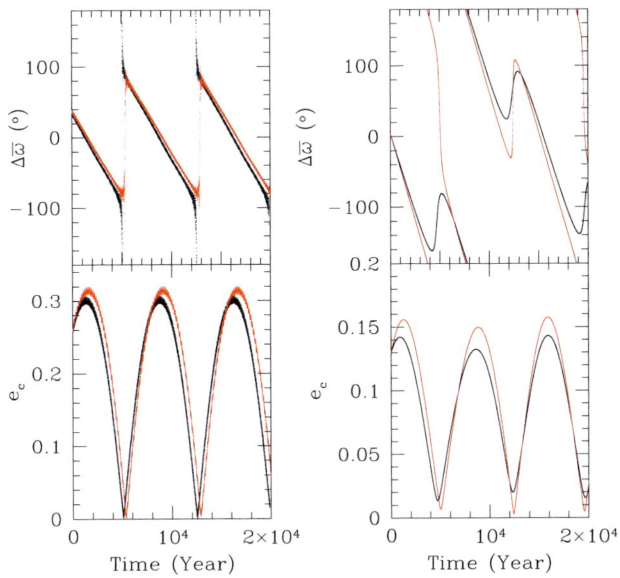

Fig. 7.4. Examples of near-separatrix motion in planetary systems. *Left Panels:* A libration-circulation separatrix. Two possible evolutions of υ And c and d (the middle and outer planet of the system) assuming different estimates of the current orbits. The black points are the system from (Butler et al., 2006), the red from (Ford et al., 2005). Although the best-fit orbits in these two cases are very similar, they result in qualitatively different types of evolution of $\Delta\varpi$. The older data predict aligned libration, whereas the updated data predict circulation (top). Note that the evolution of e_c is similar in both cases, and periodically reach near-zero values (bottom). *Right Panels:* A circulation-mode separatrix. HD 69830 c and d evolve near the circulation-mode separatrix. The black data are from (Lovis et al., 2006), and the red data are for a fictitious system in which the inner planet's, b's, eccentricity was changed from 0.1 to 0.15. In the first 10^4 years, the actual $\Delta\varpi$ undergoes 1 complete rotation through 360°, but in the fictitious system, $\Delta\varpi$ undergoes 2 complete circulations (top). We again see that the middle planet's eccentricity periodically drops to near-zero values in both cases (bottom).

Although secular theory provides a method for identifying components of the dynamics of planetary systems, it must be used with caution on extra-solar planetary systems. Their proximity to the apsidal separatrix, can obscure the true motion of the system (Barnes & Greenberg, 2006a). The inclusion of additional terms may be more useful in these cases (Lee & Peale, 2003; Michtchenko & Malhotra, 2004; Libert & Henrard, 2005; Veras & Armitage, 2007), however N-body methods may be the most practical method for determining the secular behavior of exoplanets.

Finally, it is necessary to digress and discuss the term "secular resonance". Currently two definitions exist in the literature for this term, which often leads to confusion. One definition states that $\Delta\varpi$ is librating. The other definition is that the ratio of two (or more) precessional frequencies is close to a ratio of two small integers. The latter definition is preferable as it is closer to the true meaning of a resonance, a

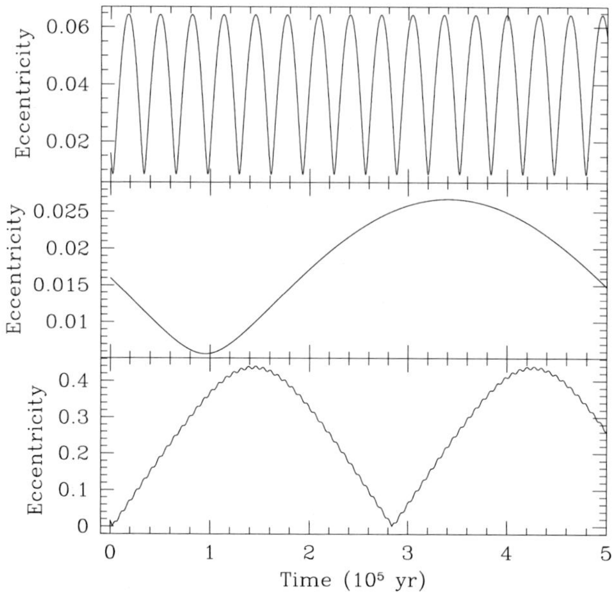

Fig. 7.5. Example of a secular resonance in the υ And system *without* general relativity. General relativity suppresses the resonance by inducing a high frequency apsidal precession in υ And b. *Top:* The eccentricity evolution of υ And b from just planet c. *Middle:* b's evolution from just planet d. *Bottom:* b's evolution from both c and d. The amplitude of eccentricity oscillation is nearly an order of magnitude larger than from either perturber alone, a direct result of the secular resonance.

commensurability of frequencies, so we will use this definition. The former should be referred to as apsidal libration. The difference is illustrated in Figs. 7.3 and 7.5 with exoplanet examples. In Fig. 7.3 the librational behavior of HD 12661 (anti-aligned) and HD 37124 c-d (aligned) are shown (top and middle panels). Fig. 7.5 shows a secular resonance in the υ And system, with orbital elements from (Ford et al., 2005). The top two panels show the eccentricity evolution of υ And b due to only planet c (top) and only d (middle), while the bottom panel is b's evolution affected by both planets. The secular resonance pushes b's eccentricity to values much larger than that of either planet alone. Note that for this example we have neglected the effects of general relativity, which overwhelms the secular resonance in this system. For more on libration and secular resonances consult (Barnes & Greenberg, 2006a), who also show that a secular resonance is impossible in a two-planet system.

Resonant Interactions

Two bodies may be in a mean motion resonance (MMR) when the ratio of their periods is close to a ratio of small integers. When this occurs, the planets periodically line up at the same points in their orbits, which introduces a repetitive force that cannot be assumed to average to zero over long timescales. Resonant effects can be comparable to secular effects, depending on masses and orbits. Eight systems have two planets that are in a resonance.

Resonances can stabilize a system by preventing close approaches that might eject a planet. Stable resonances tend to prevent conjunction from occurring near the minimum distance between two orbits. Consider the 3:2 case of Neptune and Pluto: Although the orbits cross, the resonance is such that conjunction can never occur at this danger zone.

Resonances are often described in terms of the planets' mean longitudes λ. The mean longitude is similar to the true longitude, but it measures the position of a planet assuming its angular velocity is constant (only true for a circular orbit). When resonances occur, the mean longitudes and angles of periastron evolve in certain, regular ways. Resonances also force circular orbits to become non-circular. If conjunction occurs at periastron of the inner planet, then at the following conjunction the apsides will not be perfectly aligned because the change in e will change the apsidal frequency. This non-alignment will introduce a net torque on the orbit that tends to pull the orbits back toward alignment. In this way, resonances maintain themselves, but the alignment will oscillate about an equilibrium position.

From the qualitative description above, it is clear that a resonance occurs if certain combinations of angles librate about fixed values. If we denote the outer planet with a prime, then the resonant dynamics are important if the "resonant argument",

$$\phi = j_1\lambda' + j_2\lambda + j_3\varpi' + j_4\varpi, \tag{7.1}$$

varies slowly relative to the orbital motion. Note the integers j_k obey

$$j_1 + j_2 + j_3 + j_4 = 0 \tag{7.2}$$

in all terms of the disturbing function. For any pair of planets, integers can be identified that solve Eqs. (7.1 – 7.2), but the resonance will only be effective if its order is low enough. The order of a resonance is defined as the difference between $|j_1|$ and $|j_2|$. If the order is ~ 4 or less and the larger number (j_1) is small ($\lesssim 5$) then the resonance is at least as important as secular effects in an exoplanet systems. High order resonances are present and important in the Solar System, including resonances between three planets (Murray & Holman, 1999), but their role is unknown in exoplanets because the observational errors are too large for the effects of these interactions to be unambiguously determined.

In exoplanet systems, some resonances show simple behavior: 1 or more resonant arguments are always librating, see top panel of Fig. 7.6. But some peculiar examples of resonances have been uncovered. The planets around HD 108874, for example, evolve with one resonant argument always librating, but the other arguments alternate between libration and circulation (Barnes & Greenberg, 2006c), as shown in the bottom panel of Fig. 7.6. Note that this system has an ϵ value of 0.2, suggesting it lies far from the apsidal separatrix. However, the resonance alters the apsidal motion, and therefore ϵ is not always a valid description of near-separatrix motion in the case of mean motion resonances.

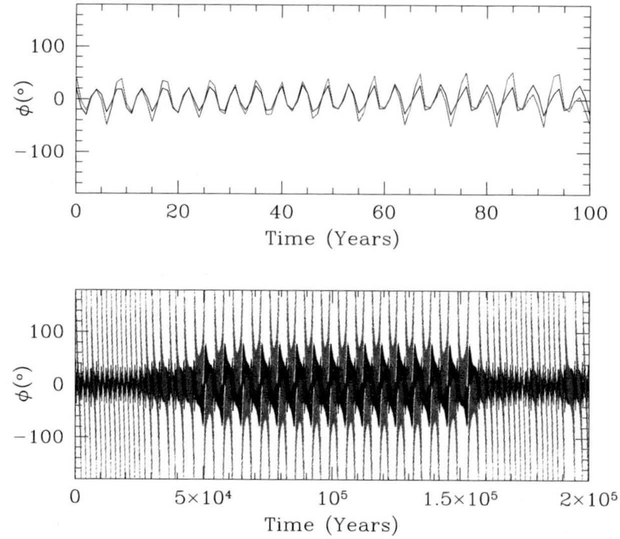

Fig. 7.6. Examples of librating resonance arguments in exoplanet systems. The y-axis, ϕ, represents the resonant arguments. *Top:* The planets GJ 876 c and b are in a 2:1 resonance, and both possible combinations of resonant angles librate. The short period means the oscillation should be observable. *Bottom:* The planets in HD 108874 are in a 4:1 resonance and 4 possible resonant angles exist. In this system one angle, $\phi = 4\lambda' - \lambda - \varpi - 2\varpi'$, (black curve) is always librating, but $\phi = 4\lambda' - \lambda - 3\varpi'$ (red curve) alternates between libration and circulation on 10^5 year timescales.

If ϕ librates for multiple combinations of j's, then the system is in an "apsidal corotation resonance" (Ferraz-Mello et al., 2005; Michtchenko, Beaugé & Ferraz-Mello, 2006), and $\Delta\varpi$ will also librate. For more on the physics of resonances, consult (Peale, 1976; Greenberg, 1977; Beaugé et al., 2003), or (Murray & Dermott, 1999, Chap. 8).

7.2.2 N-body Integrations

The most accurate method to determine the evolution of a system is through an N-body calculation. Although more accurate than analytic methods, it does not provide the researcher with terms that may be interpreted, and may require substantial computational resources.

In general an N-body code solves the second order differential equations of acceleration due to gravity. The accuracy is contingent on two factors: the size of the timestep, Δt, and the order of the integration, which is a measure of the accuracy of the method itself. N-body codes update a coordinate, $r(t)$, to $r_0 + v_r\Delta t$, where r_0 is the position at the start of the timestep, and v_r is the velocity at the start of the timestep, which is determined in an analogous manner with the acceleration. This example is of a first order scheme. Higher order schemes involve calculating positions, velocities and accelerations more frequently through the timestep. They have higher accuracy, but also have more terms and more calculations. As the order increases, the fractional gain in accuracy decreases, which creates an optimal order for algorithms: the order that maximizes speed, but minimizes truncation errors. Most modern codes use second to fourth order schemes.

Modern integration methods are "symplectic", which means that truncation errors grow linearly with time, and are therefore the preferred method for N-body integrations (errors in non-symplectic methods, such as Runge-Kutta, grow faster). For exoplanet systems, symplectic codes need only conserve energy to 1 part in 10^4 to produce reliable results (Barnes & Quinn, 2004). For more on symplectic integrators refer to (Gladman, Duncan & Candy, 1991) or (Yoshida, 1993). Several symplectic N-body codes are publicly available and widely used throughout the planetary dynamics community. These codes are well-tested and reliable. The three most prevalent are SWIFT (Levison & Duncan, 1994)[1], HNBODY[2], and MERCURY (Chambers, 1999)[3].

A code like MERCURY can integrate a few bodies for upwards of 1 Gyr in about 1 month on a 3 GHz processor and therefore permits integrations of planetary systems for the lifetime of the system. Alternatively, these codes may be used to run numerous shorter simulations to explore parameter space of known planetary systems (Barnes & Quinn, 2001, 2004) or model late stage planet formation (Chambers, 1999; Raymond, Quinn & Lunine, 2004; Lissauer, 2007). These codes therefore provide tools to understand both long-term behavior as well as general characteristics of planetary systems.

7.2.3 Dynamical Stability and Chaos

Chaos is a general term that describes a system whose motion is non-repeating over a given timescale, that is, the motion appears random. Stability describes the "boundedness" of a system; a system is stable if changes in its evolution are confined to a certain range. Therefore, one of the most fundamental features of a chaotic system is stability (for a more complete review of chaos theory, consult (Chirikov, 1979)). For example, the Solar System is a chaotic system, but it is stable in the sense that the orbits of the planets do not interchange or become unbounded, and the oscillations of orbital elements, like eccentricity, occur over a finite range. Alternatively the Solar System is unstable in the sense that the minor planets' orbits can evolve in a non-repeating manner, as was spectacularly displayed when comet Shoemaker-Levy 9 impacted Jupiter. So is the Solar System stable? It depends on the bodies in question and the timescale. The orbits of many comets are not stable on timescales comparable to the age of the Solar System, but the orbits of the planets clearly are (they're still here undergoing periodic evolution). But on longer timescales, the planets' orbits are not stable; the most unstable planet in the Solar

[1] http://www.boulder.swri.edu/~hal/swift.html
[2] http://janus.astro.umd.edu/HNBody
[3] http://star.arm.ac.uk/~jec/mercury/mercury6.tar

System, Mercury, may be lost to the Solar System in another 10^{12} years (Lecar et al., 2001). The example of our Solar System elucidates an important dichotomy in chaotic systems: a system may be formally unstable, but, for all practical purposes, is stable. It is irrelevant that Mercury could collide with Venus or the Sun in 10^{12} years because it will be engulfed by the Sun when it enters its red giant phase in 5×10^9 years. So, practically speaking, the planets in the Solar System are on stable orbits. But from a rigorous definition from chaos theory, the planets cannot be said to be stable; the Solar System's lifetime is just less than the timescale for instabilities to arise.

For a system to be chaotic, its motion must be 1) governed by nonlinear equations, and 2) sensitive to initial conditions. These requirements are met for systems with 2 or more planets that are close enough to each other. How close is "close enough" is a subject of intense research. In linear, non-chaotic motion, two nearby trajectories diverge at a constant rate, like two balls thrown together; their random motions increase their separation at a constant rate (their relative velocity times the time). In chaotic systems two nearby trajectories diverge at an exponential rate. Take, for example, two water molecules in a stream that begin right next to each other. Although in general the water flows downhill, the paths of the molecules will eventually become divergent due to rocks, vortices, tributaries, etc. Once one molecule reaches the ocean, the other may be stuck kilometers upstream. A planetary example of chaotic motion is represented in the Kirkwood gaps in the asteroid belt (Kirkwood, 1888; Moons, 1997; Tsiganis, Varvoglis & Hadjidemetriou, 2002). These gaps result from the ejection of asteroids in resonances with Jupiter. Asteroids next to the gaps have evolved regularly (the motion is repeating) for billions of years, but those in the gap were ejected in just millions of years (Lecar et al., 2001).

Most exoplanet systems of 2 or more planets are chaotic, and we would like to know if they are dynamically stable. Several meanings of stability with regard to planetary systems have arisen that complicate discussions. A system in which no planet is ejected and the semi-major axes remain bounded for all time is known as "Lagrange stable". This definition is the preferable definition of stability, as it implies a system will behave the way it does now for all time. Unfortunately, there is no known way to prove Lagrange stability at this time (although numerical simulations may disprove it). A more subtle form of stability exists when the ordering of the planets remain constant. This type is known as "Hill stability" or "hierarchical stability" and it can be proven analytically for non-resonant, two-planet systems (Marchal & Bozis, 1982; Milani & Nobili, 1983; Gladman, 1993; Chambers et al., 1996; Barnes & Greenberg, 2006b). In this type of stability the outermost planet may escape, but not the inner; the ordering of the bodies remains constant for all time.

Unfortunately the term *hierarchical* has two meanings that must be explained here. In stability analyses, a system is hierarchical if it satisfies a simple equation. However the term "hierarchical" is now also employed to describe exoplanet systems for which the ratio of the semi-major axes (a/a') is low (≤ 0.3) (Lee & Peale, 2002; Goździewski & Konacki, 2004). These conflicting definitions naturally lead to the problem that a "hierarchical planetary system" may not be "hierarchically stable", if the eccentricities are large enough.

Recently it has been shown that the Hill and Lagrange boundaries may be quite close to each other (see 7.2.3 or (Barnes & Greenberg, 2006b, 2007b)). The proximity of a system to the Hill boundary may be parameterized by β. If $\beta = 1$, the system is on the Hill boundary, and if $\beta > 1$, the system is Hill stable. (Barnes & Greenberg, 2006b) found that for two systems (47 UMa and HD 12661), the Lagrange stability boundary corresponded to β values of about 1.02 and 1.1, respectively. Although the expression for Hill stability is only valid for systems of 2 planets outside of resonance, many observed systems have only two known companions. Therefore β can be calculated for the majority of observed multiple planet systems to determine their proximity to instability (Barnes & Greenberg, 2007b).

A system's sensitivity to initial conditions is often measured by the Lyapunov time. This time is a measure of the divergence between two initially nearby trajectories. The Lyapunov time does not necessarily predict the onset of irregular (non-repeating) motion. The Earth has a Lyapunov time of about 5 million years (Laskar, 1989; Sussman & Wisdom, 1988, 1992). This value does not mean that in 5 million years the Earth's orbit will begin to change wildly, it just means that 5 million years from now the Earth's position cannot be known with arbitrarily high precision. Furthermore one must not think of the Lyapunov time as a measure of the "degree" or "amount" of chaos. A system is either chaotic or it is not. For more on chaos in planetary systems see (Lecar et al., 2001).

The Lyapunov exponent has been exploited in one code in common use in dynamical analyses of exoplanets: MEGNO (Cincotta & Simó, 2000). This code determines the Lyapunov time in a grid of parameter space, and stability is inferred from this time. Although evolution from a given set of initial conditions was not proven to be stable, if the Lyapunov time is long enough, the configuration is assumed stable (again, "long enough" is not rigorously defined); the Lyapunov time is assumed to be a proxy for stability.

7.3 Dynamics of Individual Systems

As multiple-planet systems are discovered, dynamicists rush to predict and interpret their dynamical properties. In this section we review the dynamics of individual systems in light of the general dynamical properties described above. Not surprisingly researchers have studied the first systems discovered the most extensively. Note that the naming convention for planets is based on the discovery sequence: the first planet detected is "b", the second "c", etc. In the descriptions below, the planets are listed in order of increasing semi-major axis.

47 UMa. Two planets orbit the star 47 UMa, known as b and c (Fischer et al., 2002), however the existence of the second planet remains controversial (Naef et al., 2004). Investigations of this system using the best determination of orbits have shown that its apsidal oscillation is on the boundary between libration and circulation ($\epsilon = 0$) (Laughlin, Chambers & Fischer, 2002; Barnes & Greenberg, 2006a,c) because the uncertainty in planet c's eccentricity is so large that observers have set it to 0 (Butler et al., 2006). For the system to be stable, the eccentricity of planet c must be less than about 0.2 (Goździewski, 2002; Barnes & Quinn, 2004; Barnes & Greenberg, 2006b). Considerable attention has been paid to this system as the companions orbit at relatively large semi-major axes (2.1 and 3.8 AU, respectively) and relatively low eccentricities (0.06 and 0, respectively), making this system a prime candidate for stable Earth mass planets (Thébault, Marzari & Scholl, 2002; Jones, Sleep & Chambers, 2001; Jones & Sleep, 2002; Goździewski, 2002; Noble, Musielak & Cuntz, 2002; Cuntz et al., 2003; Menou & Tabachnik, 2003; Asghari et al., 2004; Ji et al., 2005; Laakso, Rantala & Kaasalainen, 2006; Rivera & Haghighipour, 2007). This system is stable for billions of years (Barnes & Quinn, 2004), and has a β value of 1.025 (Barnes & Greenberg, 2007b).

55 Cnc. Four planets orbit this star, e, b, c, and d (Marcy et al., 2002; McArthur et al., 2004). The inner planet has been tidally circularized and planets b and c are in a 3:1 resonance. The existence of planet c remains controversial (Naef et al., 2004). Planets b and c are the only pair thought to be in a 3:1 resonance leading to considerable research into the interaction's properties and origin (Ji et al., 2003b; Voyatzis & Hadjidemetriou, 2006). The first-determined orbits placed this system in "asymmetric" apsidal libration, in which $\Delta\varpi$ oscillated about 250° instead of 0 or 180° (Ji et al., 2003c; Zhou et al., 2004; Voyatzis & Hadjidemetriou, 2006). The subsequent revision in orbital elements (Butler et al., 2006) has changed that assessment, and the pair now appears to circulate (Barnes & Greenberg, 2006c). All three pairs are undergoing apsidal circulation with ϵ values of 0.067, 0.11 and 0.158 for the inner, middle and outer pairs, respectively. Studies of dynamical stability (Ji et al., 2003c; Marzari, Scholl & Tricarico, 2005) have shown that the resonance stabilizes the system. The gap between planets c and d is quite large, and includes the habitable zone, (a terrestrial planet with an Earth-like atmosphere could support liquid water on the surface (Kasting, Whitmire & Reynolds, 1993), see Ch. 10), leading to several investigations into the possibility of Earth-mass planets in this region (von Bloh et al., 2003; Menou & Tabachnik, 2003; Barnes & Raymond, 2004; Raymond & Barnes, 2005; Raymond, Barnes & Kaib, 2006; Rivera & Haghighipour, 2007). These studies have revealed that, should an hypothetical Earth-mass planet form shortly after the gas giants reach their final masses and orbits, it may form with enough water to be habitable (Raymond, Barnes & Kaib, 2006).

GJ 876. The three planets in this system are d, c, and b. Planets c and b were the first discovered to show evidence for an MMR (Marcy et al., 2001), 2:1 in this case. The innermost planet is tidally evolved, and one of the smallest mass exoplanets known (Rivera et al., 2005). The orbits of all three planets are short enough for the secular effects to be directly observed (Laughlin & Chambers, 2001; Laughlin et al., 2005). Other studies have shown that the system is stabilized by the 2:1 MMR (Kinoshita & Nakai, 2001; Rivera & Lissauer, 2001; Ji, Liu & Li, 2002; Hadjidemetriou, 2002; Goździewski, 2002; Psychoyos & Hadjidemtriou, 2005b; Marzari, Scholl, & Tricarico, 2006; Hadjidemetriou, 2006) and that the resonant pair is in an apsidal corotation resonance with libration about $\Delta\varpi = 0$ ($\epsilon = 0.34$) (Lee & Peale, 2002; Ji et al., 2003c; Beaugé et al., 2003; Lee, 2004; Laughlin et al., 2005;

Barnes & Greenberg, 2006c). The inner pair is on the apsidal separatrix because the inner planet's eccentricity has been set to zero by observers (Butler et al., 2006).

Gliese 581 Three planets orbit this star, b, c, and d (Udry et al., 2007). The masses of the planets in this system are some of the smallest known ($5-15M_\oplus$). The inner planet has probably been tidally evolved, and if the middle, $5M_\oplus$ planet is terrestrial (rocky), then it has probably tidally evolved as well. Both pairs of planets circulate with ϵ values of 0.15 and 0.2 for the inner and outer pair, respectively,

HD 12661. Planets b and c orbit this star at 0.83 and 2.86 AU, respectively (Fischer et al., 2003), orbital distances similar to υ And c and d. Dynamical studies of this system have shown that it lies close to the border between circulation and anti-aligned libration ($\epsilon = 0.003$) (Goździewski & Maciejewski, 2003; Ji et al., 2003c; Lee & Peale, 2003; Zhou & Sun, 2003; Rodríguez & Gallardo, 2005; Libert & Henrard, 2006; Barnes & Greenberg, 2006c), making this interaction one of the few that could be undergoing anti-aligned libration. Stability in this system requires the eccentricities to be less than 0.3 (Kiseleva-Eggleton et al., 2002; Goździewski, 2003; Goździewski & Maciejewski, 2003; Barnes & Greenberg, 2006b), and $\beta = 1.2$ (Barnes & Greenberg, 2007b). These planets may lie in the the 6:1, 11:2, (Goździewski, 2003; Lee & Peale, 2003; Goździewski & Maciejewski, 2003) or 5:1 resonance (Libert & Henrard, 2007).

HD 37124. Three planets, b, c, and d, orbit this star, but the orbits are poorly constrained as two multiplanet fits are nearly equally likely (Vogt et al., 2005). For the fit that is slightly better, the inner pair circulates near the circulation-mode separatrix ($\epsilon = 0.009$), and the outer pair librates in an aligned state ($\epsilon = 0.096$), see Fig. 7.3 (Barnes & Greenberg, 2006c). Both fits appear to be stable (Vogt et al., 2005).

HD 38529. Two planets are known to orbit this star (Fischer et al., 2003), b and c, in well separated orbits. This system is dynamically stable (Kiseleva-Eggleton et al., 2002) ($\beta = 2.07$) (Barnes & Greenberg, 2007b), and can support additional planets in between those that are known (Menou & Tabachnik, 2003; Érdi et al., 2004; Barnes & Raymond, 2004; Raymond & Barnes, 2005; Raymond, Barnes & Kaib, 2006). Although such a planet is unlikely to have a significant water content (Raymond, Barnes & Kaib, 2006). The apsides of this system circulate ($\epsilon = 0.44$) (Libert & Henrard, 2006; Barnes & Greenberg, 2006c).

HD 69830. Three Neptune-mass planets orbit this star (Lovis et al., 2006), including an inner tidally circularized planet. The apsidal motion for these two pairs of planets both circulate (Barnes & Greenberg, 2006c) with ϵ values of 0.095 and 0.04 for the inner and outer pair, respectively. The system could also support asteroid belts between the planets, but terrestrial planets are unlikely (Ji et al., 2007).

HD 73526. The two planets in this system, b and c, are in a 2:1 MMR (Tinney et al., 2006). This system's best-fit is probably unstable, but a slight change in orbital elements will result in regular motion (Sándor, Kley & Klagyivik, 2007). The best-fit (Tinney et al., 2006) librates in an anti-aligned state near the apsidal separatrix ($\epsilon = 0.006$) (Barnes & Greenberg, 2006c), and has a β value of 0.982 (Barnes & Greenberg, 2007b).

HD 74156. Two planets, b and c, orbit this star (Naef et al., 2004). This system is dynamically stable ($\beta = 1.542$) (Dvorak et al., 2003; Barnes & Greenberg, 2007b) and $\Delta\varpi$ circulates ($\epsilon = 0.36$) (Libert & Henrard, 2006; Barnes & Greenberg, 2006c). The semi-major axis of the outer planet was initially thought to be significantly closer to the parent star, leading to some confusion over the possibility of a stable zone between the planets (Dvorak et al., 2003; Barnes & Raymond, 2004; Raymond & Barnes, 2005; Raymond, Barnes & Kaib, 2006). As this book was going to press, a third planet, d, was discovered in between the two that were known (Bean, 2007).

HD 82943. This system contains two planets, b and c, in a 2:1 resonance, with semi-major axes near 1 AU (Mayor et al., 2004). An alternative interpretation of the data posits that the planets in the 2:1 resonance are actually in a 1:1 resonance (Beaugé et al., 2007) (co-orbital "trojans" (Laughlin & Chambers, 2002; Dvorak et al., 2004; Schwarz et al., 2005; Érdi & Sándor, 2005)), and that a third companion is present in the system (Goździewski & Konacki, 2006). For the former system, investigations have shown that the 2:1 resonance stabilizes the system ($\beta = 0.946$) (Goździewski et al., 2001; Ji et al., 2002; Barnes & Quinn, 2004; Ferraz-Mello et al., 2005; Psychoyos & Hadjidemtriou, 2005b; Marzari, Scholl, & Tricarico, 2006; Ji & Liu, 2006; Hadjidemetriou, 2006; Lee et al., 2006; Barnes & Greenberg, 2007b). The best-fit orbits for this system underwent a major revision (Butler et al., 2006) that changed the previously determined apsidal motion of the system (Ji et al., 2002; Hadjidemetriou, 2002; Ji et al., 2003a,b; Ferraz-Mello et al., 2005; Ji & Liu, 2006; Lee et al., 2006). The apsidal motion is now best described as being close to the boundary between aligned libration and circulation ($\epsilon = 0.004$) (Lee et al., 2006; Barnes & Greenberg, 2006c).

HD 83443. This system was announced as a press release by the Geneva planet search group. The outermost planet was thought to be in a 10:1 resonance with the innermost, tidally circularized planet. The outer planet's existence was never confirmed (Butler et al., 2002), but the inner planet's orbit has been well established (Mayor et al., 2004; Butler et al., 2006). Although no additional planets have been detected in this system, the dynamics of a tidally circularizing planet with perturbations due to an external planet were developed in the context of this system (Wu & Goldreich, 2002).

HD 108874. Two planets, b and c, orbit this star in a stable 4:1 MMR (Vogt et al., 2005; Goździewski, Konacki & Maciejewski, 2006). The apsidal motion of this system is peculiar due to the resonance, see Fig. 7.6. The oscillation of $\Delta\varpi$ actually switches between anti-aligned libration and circulation on a $\sim 10^5$ year timescale (Barnes & Greenberg, 2006c), even though $\epsilon = 0.2$. The inner planet in this system is located in the habitable zone, and could support a trojan terrestrial planet (Schwarz et al., 2007). The β value for this system is 1.11, the largest value for any two-planet, resonant system (Barnes & Greenberg, 2007b).

HD 128311. Two planets, b and c, orbit HD 128311 in a 2:1 MMR (Vogt et al., 2005), although the two planets may actually be trojans in a 1:1 MMR (Goździewski & Konacki, 2006). The apsidal mode of these planets is circulation (Barnes & Greenberg, 2006c), which may be problematic for models of the formation of resonant planets (Beaugé, Michtchenko & Ferraz-Mello, 2005; Sándor & Kley, 2006). The configuration for this system is such that $\epsilon = 0.091$ (Barnes & Greenberg, 2006c) and $\beta = 0.968$ (Barnes & Greenberg, 2007b).

HD 155358. Two planets, b and c, orbit this star (Cochran et al., 2007). The current best-fit to this system is stable for at least 10^8 years (Cochran et al., 2007) with a β value of 1.04 (Barnes & Greenberg, 2007b). The apsides oscillate in anti-aligned libration with an ϵ value of 0.21.

HD 168443. Two of the most massive known planets orbit HD 168443 (Marcy et al., 2001), labeled b and c. This system is dynamically stable ($\beta = 1.94$) (Barnes & Quinn, 2004; Barnes & Greenberg, 2007b), but cannot support planets between the two that are known (Érdi et al., 2004; Barnes & Raymond, 2004). The lines of periastron of this system circulate ($\epsilon = 0.22$), see Fig. 7.3 (Lee & Peale, 2003; Libert & Henrard, 2006; Barnes & Greenberg, 2006c).

HD 169830. The two planets in this system, b and c, (Naef et al., 2004) undergo apsidal circulation ($\epsilon = 0.33$) (Libert & Henrard, 2006; Barnes & Greenberg, 2006c). This system is stable, $\beta = 1.28$ (Barnes & Greenberg, 2007b), for at least 1 Gyr (Goździewski & Konacki, 2004), but probably cannot support additional planets in between the two known planets (Érdi et al., 2004).

HD 190360. Two planets, c and b, orbit this star (Vogt et al., 2005). The inner planet in this system, c, has probably been tidally circularized. This system is stable ($\beta = 1.7$) (Barnes & Greenberg, 2007b), and $\Delta\varpi$ circulates ($\epsilon = 0.38$) (Barnes & Greenberg, 2006c).

HD 202206. The two planets in this system, b and c, are dynamically stable only if they are in a 5:1 MMR, the highest order yet discovered (Correia et al., 2005; Goździewski, Konacki & Maciejewski, 2006; Libert & Henrard, 2007). This system has the lowest β value known at 0.88 (Barnes & Greenberg, 2007b). The pair undergoes apsidal circulation, although aligned libration is possible ($\epsilon = 0.096$) (Barnes & Greenberg, 2006c).

HD 217107. Two planets, c and b, orbit this star (Vogt et al., 2005). The inner planet in this system has been tidally circularized (Vogt et al., 2005; Butler et al., 2006). The relative longitudes of periastron circulate with the largest known value of ϵ (0.46) (Barnes & Greenberg, 2006c) and the separation between the two planets gives this system the largest β value, 7.19, by far (Barnes & Greenberg, 2007b).

HIP 14810. Two planets, c and b, orbit this star (Wright et al., 2007). The inner planet in this system has been tidally circularized, and the apsides oscillate in a state of anti-aligned libration ($\epsilon = 0.05$) (Barnes & Greenberg, 2007a). The system is stable, but with the lowest β value for any tidally evolved pair, 1.2 (Barnes & Greenberg, 2007b)

μ **Arae.** Four planets, c, d, b, and e, orbit this star, sometimes called HD 160691 (Pepe et al., 2007; Goździewski, Maciejewski & Migaszewski, 2007). The orbits and number of planets in this system have changed over time, but two groups have now independently confirmed the existence of four planets on relatively circular orbits. Planets d and b (the middle pair) are probably in a 2:1 MMR (Goździewski, Maciejewski & Migaszewski, 2007). The apsides of c and d and d and b circulate near a circulation mode separatrix ($\epsilon = 0.002$ and 0.003, respectively), while the outer pair circulates ($\epsilon = 0.13$) (Barnes & Greenberg, 2007a).

υ **Andromedae.** This system was the first multiple planet discovered (Butler et al., 1999). It contains three planets, b, c and d, at semi-major axes 0.06, 0.83 and 2.54, respectively. The inner planet's orbital evolution is dominated by tides and general relativity. Dynamical analyses of this system have revealed: 1) the limits of stability of the system (the orbits are stable for the age of the system if the eccentricities of c and d are low enough, and the system is not too inclined to the line of sight) (Laughlin & Adams, 1999; Laskar, 2000; Stepinski, Malhotra & Black, 2000; Rivera & Lissauer, 2000; Jiang & Ip, 2001; Barnes & Quinn, 2001; Ito & Miyama, 2001; Lissauer & Rivera, 2001; Goździewski et al., 2001; Barnes & Quinn, 2004) , 2) the apsidal motion of the outer two companions has been revised from aligned libration (Laughlin & Adams, 1999; Rivera & Lissauer, 2000; Chiang, Tabachnik & Tremaine, 2001; Malhotra, 2002; Zhou & Sun, 2003; Michtchenko & Malhotra, 2004; Michtchenko, Ferraz-Mello & Beaugé, 2006) to being very close to the boundary between libration and circulation ($\epsilon = 2.8 \times 10^{-4}$, the smallest known value) (Ford et al., 2005; Barnes & Greenberg, 2006a,c, 2007a), 3) the region between b and c can support and additional planet (Rivera & Lissauer, 2000; Barnes & Raymond, 2004; Rivera & Haghighipour, 2007) 4) the likelihood that c and d are in a MMR (it is close to the 5:1 and 11:2) (Libert & Henrard, 2007), 5) that the current apsidal motion of the system may have resulted from an impulsive event, such as the ejection of an original member of the system (Malhotra, 2002; Ford et al., 2005; Barnes & Greenberg, 2007a), and 6) that general relativistic effects may have played a role in shaping the final architecture of this system (Adams & Laughlin, 2006).

7.4 Distributions of Dynamical Properties

In this section some emerging trends in multiple planet system interactions are highlighted. Since only 23 multiple-planet systems are known, these distributions may not represent the actual population of multiple-planet systems. Nonetheless, the distribution of planetary orbits can validate, as well as inspire, models of planet formation. A naive distribution would be a tabulation of the frequency of individual orbital elements, i.e. the number of planets within a certain range of a or e, etc. This approach leads to the problem that a 2-planet system is described by 12 parameters (5 orbital elements and 1 mass, per planet). Therefore research has focused instead on describing the dynamical interactions. Often these interactions can be characterized by a single parameter, which describe the system as a whole. Moreover, multiple-planet systems are dynamical systems, and research should focus on their dynamical properties, not on the orbital elements the planets happen to have today.

7.4.1 Types of Interactions

Three types of dynamical effects can dominate the interactions between adjacent pairs of planets: secular, resonant and tidal. All three can affect the interaction, but so far, it appears in multiple-planet systems that only one tends to be dominant. A pair is tidally dominated if the inner planet's a is less than 0.1 AU (Rasio et al., 1996), resonantly dominated if one or more resonant argument librates, and is secularly dominated if the former two are not important. These "classes" provide a quick description of the motion (Barnes & Greenberg, 2006c). However, as orbital elements are revised, a system's classification may also change.

Currently the secular class appears to be the most common with 16 of 34 pairs in this type of interaction, including the gas giants in our Solar System. Tidal pairs account for 10 interactions and resonant interactions dominate the remaining 8 pairs. The observational uncertainties associated with the radial velocity surveys are such that resonant interactions are difficult to identify, but identifying tidal pairs is relatively easy. Therefore, we might expect the actual frequency of resonant interactions to be higher, and tidal interactions to be lower.

7.4.2 Frequency of Mean Motion Resonances

The most dramatic (dynamically speaking) aspect of a planetary systems is the presence of MMRs. Of the 31 known pairs of exoplanets, eight appear to be in an MMR: GJ 876 b-c (2:1), HD 82943 (2:1), HD 128311 (2:1), HD 73526 (2:1), μ Ara d-b (2:1), 55 Cnc b-c (3:1), HD 108874 (4:1), and HD 202206 (5:1). Several researchers have suggested that other systems may be in higher order resonances, i.e. HD 12661 in 11:2 or 6:1 (Goździewski, 2003; Lee & Peale, 2003), and 47 UMa in 5:2 (Psychoyos & Hadjidemetriou, 2005a) or 7:3 (Laughlin, Chambers & Fischer, 2002). From the eight systems known to be in resonance, it appears that the 2:1 resonance is most likely. If this trend is real, it may be because this resonance is the strongest, and therefore the most efficient at trapping planets (Kley et al., 2004, 2005; Sándor & Kley, 2006).

Also of interest is a system's proximity to an MMR. Even if a system is not in resonance, the resonance can still affect the secular motion. The classic example of this perturbation is the "Great Inequality"; the orbits of Jupiter and Saturn are close to, but not in, a 5:2 MMR (Varadi, Ghil & Kaula, 1999; Michtchenko & Ferraz-Mello, 2001). Similar phenomena may be present in HD 12661 and υ And (Libert & Henrard, 2007). The HD 38529, HD 168443, HD 74156 and HD 169830 systems are not affected by MMRs, despite the latter's proximity to the 9:1 MMR (Libert & Henrard, 2007).

7.4.3 Apsidal Motion

Considerable attention has been directed toward identifying the apsidal motions, which should only be determined by N-body calculations, see Sect. 7.2.1 (Barnes & Greenberg, 2006a). Nearly half of all planetary interactions, including the gas

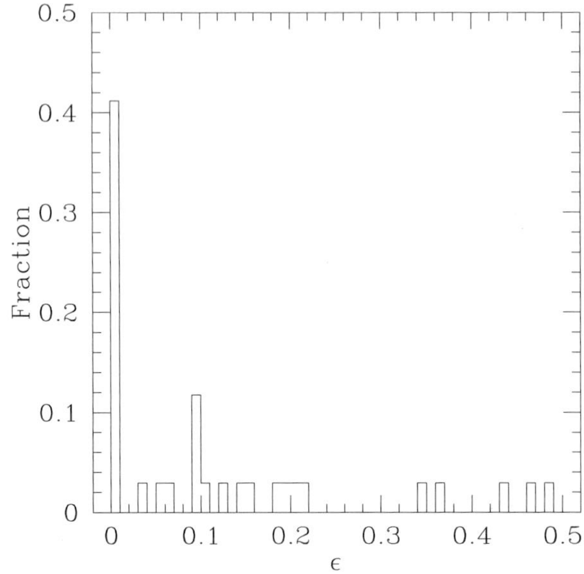

Fig. 7.7. The distribution of proximities to an apsidal separatrix (defined as $\epsilon = 0$). For systems with $\epsilon < 0.01$ eccentricity oscillations of two orders of magnitude are present in the system. Nearly half of all adjacent pairs are on orbits such that they are near a separatrix. If the distribution of eccentricities were uniform and random, only a few per cent of systems should interact such that $\epsilon < 0.01$ (Barnes & Greenberg, 2006a).

giants of the Solar System, interact with $\epsilon < 0.01$, and their apsidal motion is best characterized as near-separatrix, see Fig. 7.7. When this type of interaction was identified (Ford et al., 2005), it was suggested that the ejection of an original Jupiter-mass planet may have produced this near-separatrix behavior. However subsequent analysis has demonstrated that that mechanism is unlikely to result in motion near an apsidal separatrix (Barnes & Greenberg, 2007a), leaving the origin of this type of interaction unclear.

From the best fits, an overwhelming majority of adjacent pairs appear to undergo apsidal circulation (Barnes & Greenberg, 2006c), contrary to initial beliefs that libration was the typical state (Zhou & Sun, 2003). Of librating systems, three undergo anti-aligned libration (HD 73526, HD 155358 and HIP 14810), and two are aligned (GJ 876 and HD 37124 c-d). The statistics of these five systems is too small to draw any reliable conclusions, but at this point it appears that, among librating systems, alignment and anti-alignment occur with approximately the same frequency.

7.4.4 Proximity to Dynamical Instability

The most critical aspect of dynamical systems is their stability, and, therefore, a substantial amount of research has investigated the dynamical stability of planetary systems, e.g. (Goździewski & Maciejewski, 2001; Goździewski et al., 2001; Kiseleva-Eggleton et al., 2002; Barnes & Quinn, 2004; Barnes & Greenberg, 2006b, 2007b). Many known systems appear to lie near Lagrange instability (the type of stability in which at least one planet is ejected from the system within several million years, see Sect. 7.2.3).

Investigations into stability have found that Lagrange unstable regions exist within the 1 standard deviation error ellipses for systems with MMRs, such as 55 Cnc and HD 82943, as well as those without, such as HD 12661 and 47 UMa as shown in Fig. 7.8. This figure shows the results of numerical simulations within observationally permitted parameter space for these four systems (see (Barnes & Quinn, 2004) for more details). The parameter space was sampled as a Gaussian with a peak at the best fit values (at the time of the simulations) with a standard deviation equal to the published error. Therefore the centers of each panel are more highly sampled than the edges. The shading indicates the fraction of initial conditions, in a certain range of orbital element space, that give Lagrange stable behavior (no ejections or exchanges) after $\sim 10^6$ years: White regions contained only stable configurations, black only unstable, and darkening shades of gray correspond to decreasing fractions of simulations which predict stability. In this figure P_c/P_b is the ratio of the orbital periods.

The contours represent values of β, the proximity to the Hill stability boundary (Barnes & Greenberg, 2006b, 2007b). For non-resonant systems, the limit of Lagrange stability corresponds to values of β slightly greater than 1. But in the presence of a resonance, Lagrange stable orbits can be found at β values as small as 0.75. Lagrange stability boundaries are qualitatively different for resonant and non-resonant pairs. The former have a "stability peninsula" located at the resonance, while the latter have a large contiguous "stability plateau".

The apparent correspondence between Lagrange stability and values of β is evidence that the expression for β (Eq. [1] in (Barnes & Greenberg, 2006b)), is a valid representation of the limits of dynamical stability in systems of two planets. In Fig. 7.9, we plot the current distribution of β values for two planet systems. Most systems lie near $\beta = 1$, and resonant systems tend to have $\beta < 1$ (the resonance protects the system from instability) (Barnes & Greenberg, 2007b).

Figs. 7.8 – 7.9 suggest that many planetary systems are dynamically full (no additional companions can survive in between the observed planets), and leads to the Packed Planetary Systems (PPS) hypothesis (Barnes & Quinn, 2004; Barnes & Raymond, 2004; Raymond & Barnes, 2005; Raymond, Barnes & Kaib, 2006): All pairs of planets lie close to dynamical instability. The observation that so many systems (using minimum masses) lie near Lagrange instability, despite incompleteness issues (see Sect. 7.1.2), has led to investigations to identify regions of Lagrange stability between the few pairs that are more separated (Menou & Tabachnik, 2003; Barnes & Raymond, 2004; Dvorak et al., 2003; Funk et al., 2004; Raymond & Barnes, 2005; Raymond, Barnes & Kaib, 2006; Rivera & Haghighipour, 2007). This research has shown that the gaps between HD 74156 b-c, HD 38529 b-c, and 55 Cnc c-d are large enough to support additional Saturn-mass planets. The detection of planets

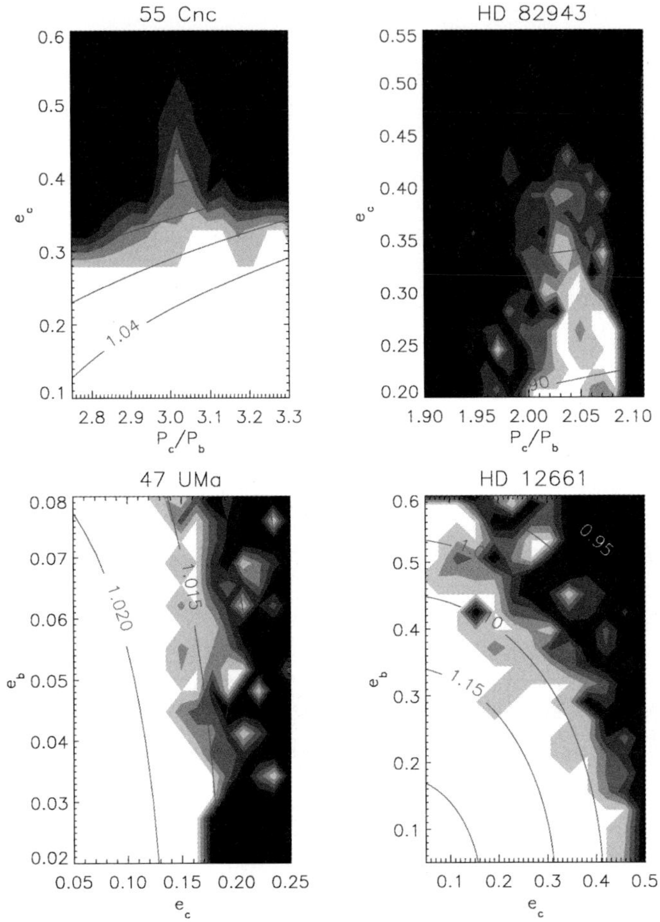

Fig. 7.8. Lagrange stability boundary in relation to the Hill stability boundary for some exoplanetary systems (see text for a discussion of the simulations summarized in these figures). In these plots white regions represent bins in which all configurations were stable, black bins contained no stable configurations, darker shades of gray correspond to regions in which the fraction of stable simulations were smaller (see Barnes & Quinn 2004 for more details). The curves represent contour lines of β. Contour lines follow the shape of the Lagrange stability boundary, except in resonance, where there the Lagrange stability region is larger. *Top left*: Stability of the 55 Cnc system depends on the parameters of the 3:1 resonant pair; the eccentricity of the larger planet, and the ratio of the periods. When $P_c/P_e \neq 3$, the Lagrange stability boundary is located at $\beta \approx 1.03$. *Top right:* HD 82943's stability depends on the eccentricity of the larger planet and the ratio of the planets' periods. The Lagrange stable region shown exists wholly in a region that would be considered unstable from Hill stability theory. *Bottom left:* The stability of 47 UMa depends on the eccentricities of the two planets. The Lagrange stability boundary corresponds to $\beta \approx 1.015$. *Bottom right:* The stability of the HD 12661 system depends on the eccentricities of the two outer planets. The Lagrange stability boundary lies near the $\beta = 1.1$ contour.

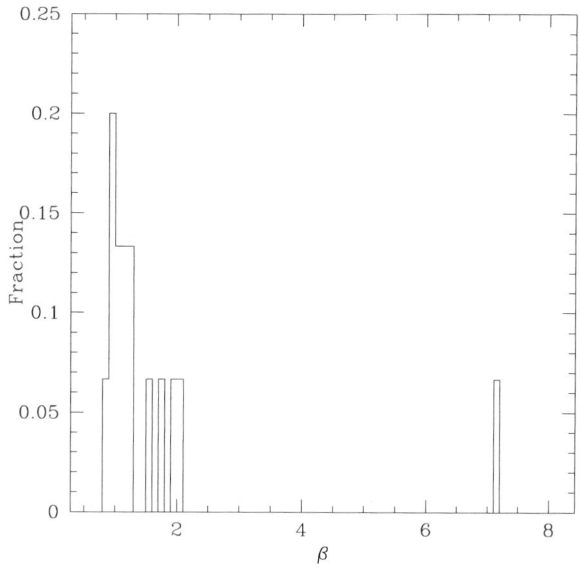

Fig. 7.9. The distribution of proximities to the Hill stability boundary, $\beta = 1$. All systems with $\beta < 1$ are in mean motion resonance, implying that the resonance is protecting the system from instability. Most non-resonant systems have configuration with $\beta < 2$, suggesting these planetary systems are packed (Barnes & Greenberg, 2007b); companions between those that are known would disrupt the system.

in these locations would support the packed nature of planetary systems, and would be an exciting achievement for the nascent field of exoplanet dynamics.

In order to verify or disprove the PPS hypothesis, a quantitative definition of "close" is required. As stated in Sect. 2.3 there exists no quantitative definition of Lagrange stability for any number of planets, and no definition for Hill stability for systems with more than 2 planets. With limited data available, it appears that when $\beta \lesssim 1.5 - 2$ a system is packed (Barnes & Greenberg, 2007b). Future work should reveal how robust this limit is, as well as identify packing limit for systems of more than 2 planets.

The PPS hypothesis received a major boost with the discovery of a Saturn-mass planet in the HD 74156 system between planets b and c (Bean et al., 2007). The revised system is unstable, but the new planet lies close (less than one standard deviation) to the most stable orbit identified by numerical simulations (Raymond & Barnes, 2005). This detection offers strong support for the PPS hypothesis, as it verifies a prediction. At this point, one such detection does not confirm the PPS hypothesis, but it is worth noting that the first 6 multiple planet systems detected now show evidence of dynamical packing.

Table 7.1. Summary of Dynamical Properties of Multiple Planet Systems

System	Pair	MMR	AM	ϵ	β	Class
47 UMa	b-c	-	C[a]	0	1.025	S
55 Cnc	e-b	-	C	0.067	-	T
	b-c	3:1	C	0.11	-	R
	c-d	-	C	0.158	-	S
GJ 876	d-c	-	C[a]	0	-	T
	c-b	2:1	A	0.34	-	R
Gl 581	b-c	-	C	0.15	-	T
	c-d	-	C	0.20	-	T
HD 12661	b-c	-	C	0.003	1.199	S
HD 37124	b-c	-	C	0.009	-	S
	c-d	-	A	0.096	-	S
HD 38529	b-c	-	C	0.44	2.070	S
HD 69830	b-c	-	C	0.095	-	T
	c-d	-	C	0.04	-	S
HD 74156[b]	b-c	-	C	0.36	1.542	S
HD 73526	b-c	2:1	AA	0.006	0.982	R
HD 82943	b-c	2:1	C	0.004	0.946	R
HD 128311	b-c	2:1	C	0.091	0.968	R
HD 108874	b-c	4:1	C/AA[c]	0.2	1.107	R
HD 155358	b-c	-	AA	0.21	1.043	S
HD 168443	b-c	-	C	0.22	1.939	S
HD 169830	b-c	-	C	0.33	1.280	S
HD 190360	c-b	-	C	0.38	1.701	T
HD 202206	b-c	5:1	C	0.096	0.883	R
HD 217107	b-c	-	C	0.46	7.191	T
HIP 14810	b-c	-	AA	0.05	1.202	T
SS	J-S	-	C	0.19	-	S
	S-U	-	C	0.006	-	S
	U-N	-	C	0.004	-	S
μ Ara	c-d	-	C	0.002	-	T
	d-b	2:1	C	0.003	-	R
	b-e	-	C	0.13	-	S
υ And	b-c	-	C	1.8×10^{-4}	-	T
	c-d	-	C	2.8×10^{-4}	-	S

[a] The current eccentricity of one planet is 0, placing the pair on an apsidal separatrix.
[b] These values do not incorporate the new planet (Bean et al., 2007)
[c] This pair alternates between circulation and anti-aligned libration.

7.5 Conclusions

This chapter has laid out analytical methods and preliminary trends in exoplanetary dynamics, using best-determined orbits. These results are summarized in Table 7.1[4]. In this table AM stands for "apsidal motion" and the possibilities are circulation (C), aligned libration (A), or anti-aligned libration (AA). The MMR column lists the resonance, if applicable. The proximities to the apsidal separatrix, ϵ, are the values from the literature (Barnes & Greenberg, 2006c), as are proximities to the Hill stability boundary, β (Barnes & Greenberg, 2007b). The "Class" distinguishes orbits whose evolution are dominated by tidal (T), resonant (R) or secular (S) interactions. Table 1 includes the dynamical properties of the giant planets in our Solar System for comparison.

About half of planetary systems are multiple (Wright et al., 2007), predictions of additional companions are being bourne out (Raymond & Barnes, 2005; Bean et al., 2007), and the current distribuion of planet masses suggest there will be many planets with a mass equal to that of Saturn or less (Marcy et al., 2005). These three observations imply many multiple planet systems will be detected in the future. Hence characterizing planet-planet interactions is becoming a more critical aspect of the study of exoplanets.

The orbits of planets are often a clue to their formation. However the observed orbital elements oscillate due to gravitational interactions. Eccentricities of planets in multiple systems often oscillate by two orders of magnitude (Barnes & Greenberg, 2006c), but proximities to an apsidal separatrix or the Hill stability boundary are fixed quantities. Therefore the consideration of the dynamics of multiple planet system may provide better constraints on models of planet formation than currently observed orbital elements (Barnes & Greenberg, 2007a,b). As the number of multiple systems grows, the distributions of proximities to an apsidal separatrix and instability will be revised, and theorists will need to determine the origins of these distributions.

Although the observations of exoplanetary systems are in general poor, theorists are a fearless lot and have examined these systems both individually and as a whole. With the existence of planets disputed, the behavior of apsidal orientations uncertain, and even the stability of some systems unproven, planetary dynamicists have a considerable amount of research ahead of them. But the field is growing quickly, and motivated by the possibility of detecting life in the universe, observations will improve, models will be refined, and our Solar System's place in the cosmos will be revealed.

Acknowledgments
I would like to thank Richard Greenberg, Sean Raymond, Thomas Quinn, Brian Jackson and Alyssa Sarid for comments and suggestions on the subject matter and presentation of this chapter.

---

[4]see http://www.lpl.arizona.edu/~rory/research/xsp/dynamics for an up-to-date list of these properties

References

Adams, F.C. & Laughlin, G. 2006. Relativistic Effects in Extrasolar Planetary Systems. *Int. J. Mod. Phys. D* **15**, 2133-2140.

Asghari, N. et al. 2004. Stability of terrestrial planets in the habitable zone of Gl 777 A, HD 72659, Gl 614, 47 Uma and HD 4208. *A&A* **426**, 353-365.

Barnes, R. & Greenberg, R. 2006. Extrasolar Planetary Systems Near a Secular Separatrix. *ApJ* **638**, 478-487.

Barnes, R. & Greenberg, R. 2006. Stability Limits in Extrasolar Planetary Systems. *ApJ* **647**, L153-L156.

Barnes, R. & Greenberg, R. 2006. Behavior of Apsidal Orientations in Planetary Systems. *ApJ* **652**, L53-L56.

Barnes, R. & Greenberg, R. 2007. Apsidal Behavior among Planetary Orbits: Testing the Planet-Planet Scattering Model. *ApJ* **659**, L53-L56.

Barnes, R. & Greenberg, R. 2007. Stability Limits in Resonant Planetary Systems. *ApJ*, accepted.

Barnes, R. & Quinn, T.R. 2001. A Statistical Examination of the Short-Term Stability of the υ Andromedae Planetary System. *ApJ* **550**, 884-889.

Barnes, R. & Quinn, T.R. 2004. The (In)stability of Planetary Systems. *ApJ* **611**, 494-516.

Barnes, R. & Raymond, S.N. 2004. Predicting Planets in Known Extrasolar Planetary Systems. I. Test Particle Simulations. *ApJ* **617**, 569-574.

Bean, J.L. et al. 2007. Detection of a Third Planet in the HD 74156 System Using the Hobby-Eberly Telescope. *ApJ*, accepted,

Beaugé, C., Ferraz-Mello, S. & Michtchenko, T.A. 2003. Extrasolar Planets in Mean-Motion Resonance: Apses Alignment and Asymmetric Stationary Solutions. *ApJ* **593**, 1124-1133.

Beaugé, C., Michtchenko, T.A. & Ferraz-Mello, S. 2005. Planetary migration and extrasolar planets in the 2/1 mean-motion resonance. *MNRAS* **365**, 1160-1170.

Beaugé, C. et al. 2007. Co-orbital terrestrial planets in exoplanetary systems: a formation scenario. *A&A* **463**, 359-367.

Bodenheimer, P., Laughlin, G. & Lin, D.N.C. 2003. On the Radii of Extrasolar Giant Planets. *ApJ* **592**, 555-563.

Bois, E. et al. 2003. Conditions of Dynamical Stability for the HD 160691 Planetary System. *ApJ* **598**, 1312-1320.

Butler, R.P. et al. 1999. Evidence for Multiple Companions to υ Andromedae. *ApJ* **526**, 916-927.

Butler, R.P. et al. 2002. On the Double-Planet System around HD 83443. *ApJ* **578**, 565-572.

Butler, R.P. et al. 2006. Catalog of Nearby Exoplanets. *ApJ* **646**, 505-522.

Chambers, J.E. 1999.A hybrid symplectic integrator that permits close encounters between massive bodies. *MNRAS* **304**, 793-799.

Chambers, J.E., Wetherill, G.W. & Boss, A. 1996. The Stability of Multi-Planet Systems. *Icarus* **119**, 261-268.

Chiang, E.I. & Murray, N. 2002. Eccentricity Excitation and Apsidal Resonance Capture in the Planetary System υ Andromedae. *ApJ* **576**, 473-477.

Chiang, E.I., Tabachnik, S. & Tremaine, S. Apsidal Alignment in υ Andromedae. *AJ* **122**, 1607-1615.

Chirikov, B.V. 1979. A universal instability of many-dimensional oscillator systems. *Phys. Rep.* **52**, 263-379.

Cincotta, P. & Simó, C. 2000. Simple tools to study global dynamics in non-axisymmetric galactic potentials - I. *A&AS* **147**, 205-228.

Cochran, W.D. et al. 2007. A Planetary System Around HD 155358: The Lowest Metallicity Planet Host Star. ApJ, in press.

Correia, A.C.M. et al. 2005. The CORALIE survey for southern extra-solar planets. XIII. A pair of planets around HD 202206 or a circumbinary planet? *A&A* **440**, 751-758.

Cuntz, M. et al. 2003. On the possibility of earth-type habitable planets around 47 UMa. *Icarus* **162**, 214-221.

D'Angelo, G., Lubow, S.H. & Bate, M.R. 2006. Evolution of Giant Planets in Eccentric Disks. *ApJ* **652**, 1698-1714.

Dvorak, R. et al. 2003. A study of the stable regions in the planetary system HD 74156 - Can it host earthlike planets in habitable zones? *A&A* **410**, L13-L16.

Dvorak, R. et al. 2004. Extrasolar Trojan planets close to habitable zones. *A&A*, **426**, L37-L40.

Érdi, B. et al. 2004. The dynamical structure of the habitable zone in the HD 38529, HD 168443 and HD 169830 systems. *MNRAS* **351**, 1043-1048.

Érdi, B, & Sándor, Zs. 2005. Stability of Co-Orbital Motion in Exoplanetary Systems. *CeMDA* **92**, 113-121.

Fernandez, J.A. & Ip, W.-H. 1984. Some dynamical aspects of the accretion of Uranus and Neptune - The exchange of orbital angular momentum with planetesimals. *Icarus* **58**, 109-120.

Ferraz-Mello, S., Michtchenko, T.A. & Beaugé, C. 2005. The Orbits of the Extrasolar Planets HD 82943c and b. *ApJ* **621**, 473-481.

Fischer, D.A. et al. 2002. A Second Planet Orbiting 47 Ursae Majoris. *ApJ* **564**, 1028-1034.

Fischer, D.A. et al. 2003. A Planetary Companion to HD 40979 and Additional Planets Orbiting HD 12661 and HD 38529. *ApJ* **586**, 1394-1408.

Ford, E.B., Lystad, V. & Rasio, F.A. 2005. Planet-planet scattering in the upsilon Andromedae system. *Nature* **434**, 873-876.

Funk, B. et al. 2004. Resonances in Multiple Planetary Systems. *CeMDA* **90**, 43-50.

Gladman, B. 1993. Dynamics of systems of two close planets. *Icarus* **106**, 247-263.

Gladman, B., Duncan, M. & Candy, J. 1991. Symplectic integrators for long-term integrations in celestial mechanics. *CeMDA* **52**, 221-240.

Goździewski, K. 2002. Stability of the 47 UMa planetary system. *A&A* **393**, 997-1013.

Goździewski, K. 2003. Stability of the HD 12661 Planetary System. *A&A* **398**, 1151-1161.

Goździewski, K. & Konacki, M. 2004. Dynamical Properties of the Multiplanet System around HD 169830. *ApJ* **610**, 1093-1106.

Goździewski, K. & Konacki, M. 2006. Trojan Pairs in the HD 128311 and HD 82943 Planetary Systems? *ApJ* **647**, 573-586.

Goździewski, K., Konacki, M. & Maciejewski, A. 2006. Orbital Configurations and Dynamical Stability of Multiplanet Systems around Sun-like Stars HD 202206, 14 Herculis, HD 37124, and HD 108874. *ApJ* **645**, 688-703.

Goździewski, K. & Maciejewski, A. 2001. Dynamical Analysis of the Orbital Parameters of the HD 82943 Planetary System. *ApJ* **563**, L81-L85.

Goździewski, K. & Maciejewski, A. 2003. The Janus Head of the HD 12661 Planetary System. *ApJ* **586**, L153-L156.

Goździewski, K., Maciejewski, A. & Migaszewski, C. 2007. On the Extrasolar Multiplanet System around HD 160691. *ApJ* **657**, 546-558.

Goździewski, K. et al. 2001. Global dynamics of planetary systems with the MEGNO criterion. *A&A* **378**, 569-586.

Greenberg, R. 1977. Orbit-orbit resonances in the solar system - Varieties and similarities. *Vistas in Astron.* **21** 209-239.

Hadjidemetriou, J.D. 2002. Resonant Periodic Motion and the Stability of Extrasolar Planetary Systems. *CeMDA* **83**, 141-154.

Hadjidemetriou, J.D. 2006. Symmetric and asymmetric librations in extrasolar planetary systems: a global view. *CeMDA* **95**, 225-244.

Ito, T. & Miyama, S.M. 2001. An Estimation of Upper Limit Masses of ŏ3c5 Andromedae Planets. *ApJ* **552**, 372-379.

Jiang, I.-G. & Ip, W.-H. 2001. The planetary system of upsilon Andromedae. *A&A* **367**, 943-948.

Ji, J-H. & Liu, L. 2006. Stability and 2:1 resonance in the planetary system HD 829431. *ChA&A* **30**, 75-86.

Ji, J., Liu, L. & Li, G.-Y. 2002. The Dynamical Simulations of the Planets Orbiting GJ 876. *ApJ* **572**, 1041-1047.

Ji, J. et al. 2002. The stabilising mechanism of the HD 82943 planetary system. *ChA&A* **26**, 379-385.

Ji, J. et al. 2003. The apsidal motion in multiple planetary systems. *ChA&A* **27**, 127-132.

Ji, J. et al. 2003. Could the 55 Cancri Planetary System Really Be in the 3:1 Mean Motion Resonance? *ApJ* **585**, L139-L142.

Ji, J. et al. 2003. The Librating Companions in HD 37124, HD 12661, HD 82943, 47 Ursa Majoris, and GJ 876: Alignment or Antialignment? *ApJ* **591**, L57-L60.

Ji, J. et al. 2005. Could the 47 Ursae Majoris Planetary System be a Second Solar System? Predicting the Earth-like Planets. *ApJ* **631**, 1191-1197.

Ji, J. et al. 2007. The Secular Evolution and Dynamical Architecture of the Neptunian Triplet Planetary System HD 69830. *ApJ* **657**, 1092-1097.

Jones, B.W. & Sleep, P.N. The stability of the orbits of Earth-mass planets in the habitable zone of 47 Ursae Majoris. *A&A* **393**, 1015-1026.

Jones, B.W., Sleep, P.N. & Chambers, J.E. 2001. The stability of the orbits of terrestrial planets in the habitable zones of known exoplanetary systems. *A&A* **366**, 254-262.

Kasting, J.F., Whitmire, D.P. & Reynolds, R.T. 1993. Habitable Zones around Main Sequence Stars. *Icarus* **101**, 108-128.

Kinoshita, H. & Nakai, H. 2001. Stability of the GJ 876 Planetary System. *PASJ* **53**, L25-L26.

Kley, W. et al. 2005. Modeling the resonant planetary system GJ 876. *A&A* **437**, 727-742.

Kley, W., Peitz, J. & Bryden, G. 2004. Evolution of planetary systems in resonance. *A&A* **414**, 735-747.

Kirkwood, D. 1888. *The Asteroids, or Minor Planets Between Mars and Jupiter*, J. B. Lippencott, Philadelphia.

Kiseleva-Eggleton, L. et al. 2002. Global Dynamics and Stability Limits for Planetary Systems around HD 12661, HD 38529, HD 37124, and HD 160691. *ApJ* **578**, L145-L148.

Laakso, T., Rantala, J. & Kaasalainen, M. 2006. Gravitational scattering by giant planets. *A&A* **456**, 373-378.

Laskar, J. 1989. A numerical experiment on the chaotic behaviour of the solar system. *Nature* **338**, 237-238.

Laskar, J. On the Spacing of Planetary Systems. *Phys. Rev. Lett.* **84**, 3240-3243.

Laughlin, G. & Adams, F.C. 1999. Stability and Chaos in the υ Andromedae Planetary System. *ApJ* **526**, 881-889.

Laughlin, G. & Chambers, J.E. 2001. Short-Term Dynamical Interactions among Extrasolar Planets. *ApJ* **551**, L109-L113.

Laughlin, G. & Chambers, J.E. 2002. Extrasolar Trojans: The Viability and Detectability of Planets in the 1:1 Resonance. *AJ* **124**, 592-600.

Laughlin, G., Chambers, J.E. & Fischer, D.A. 2002. A Dynamical Analysis of the 47 Ursae Majoris Planetary System. *ApJ* **579** 455-467.

Laughlin, G. et al. 2005. The GJ 876 Planetary System: A Progress Report. *ApJ* **622** 1182-1190.

Lecar, M. et al. 2001. Chaos in the Solar System. *ARA&A* **39**, 581-631.

Lee, M.H. 2004. Diversity and Origin of 2:1 Orbital Resonances in Extrasolar Planetary Systems. *ApJ* **611**, 517-527.

Lee, M.H. et al. 2006. On the 2:1 Orbital Resonance in the HD 82943 Planetary System. *ApJ* **641**, 1178-1187.

Lee, M.H. & Peale, S. 2002. Dynamics and Origin of the 2:1 Orbital Resonances of the GJ 876 Planets. *ApJ* **567**, 596-609.

Lee, M.H. & Peale S. 2003. Secular Evolution of Hierarchical Planetary Systems. *ApJ* **592**, 1201-1216.

Levison, H. & Duncan, M. 1994. The long-term dynamical behavior of short-period comets. *Icarus* **108**, 18-36.

Libert, A.-S. & Henrard, J. 2005. Analytical Approach to the Secular Behaviour of Exoplanetary Systems. *CeMDA* **93**, 187-200.

Libert, A.-S. & Henrard, J. 2006. Secular apsidal configuration of non-resonant exoplanetary systems. *Icarus* **183**, 186-192.

Libert, A.-S. & Henrard, J. 2007. Analytical study of the proximity of exoplanetary systems to mean-motion resonances. *A&A* **461**, 759-763.

Lin, D.N.C., Bodenheimer, P. & Richardson, D.C. 1996. Orbital migration of the planetary companion of 51 Pegasi to its present location. *Nature* **380**, 606-607.

Lissauer, J.J. 2007. Planets Formed in Habitable Zones of M Dwarf Stars Probably Are Deficient in Volatiles. *ApJ* **660**, L149-L152.

Lissauer, J.J. & Rivera, E.J. 2001. Stability Analysis of the Planetary System Orbiting υ Andromedae. II. Simulations Using New Lick Observatory Fits. *ApJ* **554**, 1141-1150.

Lovis, C. et al. 2006. An extrasolar planetary system with three Neptune-mass planets. *Nature* **441**, 305-309.

Malhotra, R. 1993. The Origin of Pluto's Peculiar Orbit. *Nature* **365**, 819-821.

Malhotra, R. 2002. A Dynamical Mechanism for Establishing Apsidal Resonance. *ApJ* **575**, L33-L36.

Marchal, C. & Bozis, G. 1982. Hill Stability and Distance Curves for the General Three-Body Problem. *CeMDA* **26**, 311-333.

Marcy, G.W. et al. 2001. A Pair of Resonant Planets Orbiting GJ 876. *ApJ* **556**, 296-301.

Marcy, G.W. et al. 2002. A Planet at 5 AU around 55 Cancri. *ApJ* **581** 1375-1388.

Marcy, G.W. et al. 2005. Five New Extrasolar Planets. *ApJ* 619, 570-584.

Marzari, F. & Weidenschilling, S. 2002. Eccentric Extrasolar Planets: The Jumping Jupiter Model. *Icarus* **156**, 570-579.

Marzari, F., Scholl, H. & Tricarico, P. 2005. Frequency map analysis of the 3/1 resonance between planets b and c in the 55 Cancri system. *A&A* **442**, 359-364.

Marzari, F., Scholl, H. & Tricarico, P. 2006. A numerical study of the 2:1 planetary resonance. *A&A* **453**, 341-348.

Mayor, M. et al. 2004. The CORALIE survey for southern extra-solar planets. XII. Orbital solutions for 16 extra-solar planets discovered with CORALIE. *A&A* **415**, 391-402.

McArthur, B.E. et al. 2004. Detection of a Neptune-Mass Planet in the ρ Cancri System Using the Hobby-Eberly Telescope. *ApJ* **614**, L81-L84.

Menou, K. & Tabachnik, S. 2003. Dynamical Habitability of Known Extrasolar Planetary Systems. *ApJ 583*, 473-488.

Milani, A. & Nobili, A.M. 1983. On topological stability in the general three-body problem. *CeMDA* **31**, 213-240.

Michtchenko, T.A. & Ferraz-Mello, S. 2001. Modeling the 5 : 2 Mean-Motion Resonance in the Jupiter-Saturn Planetary System. *Icarus* **149**, 357-374.

Michtchenko, T.A., Beaugé, C. & Ferraz-Mello, S. 2006. Stationary Orbits in Resonant Extrasolar Planetary Systems. *CeMDA* **94**, 411-432.

Michtchenko, T.A., Ferraz-Mello, S. & Beaugé, S. 2006. Modeling the 3-D secular planetary three-body problem. *Icarus* **181**, 555-571.

Michtchenko, T.A. & Malhotra, R. 2004. Secular dynamics of the three-body problem: application to the υ Andromedae planetary system. *Icarus* **168**, 237-248.

Moons, M. 1997. Review of the dynamics in the Kirkwood gaps. *CeMDA* **65**, 175-204.

Murray, C.D. & Dermott, S.F. 1999. *Solar System Dynamics.* Cambridge UP, Cambridge.

Murray, N. & Holman, M. 1999. The Origin of Chaos in the Outer Solar System. *Science* **283**, 1877-1881.

Naef, D. et al. 2004. The ELODIE survey for northern extra-solar planets. III. Three planetary candidates detected with ELODIE. *A&A* **414**, 351-359.

Namouni, F. 2005. On the Origin of the Eccentricities of Extrasolar Planets. *AJ* **130**, 280-294.

Noble, M., Musielak, Z.E. & Cuntz, M. 2002. Orbital Stability of Terrestrial Planets inside the Habitable Zones of Extrasolar Planetary Systems. *ApJ* **572**, 1024-1030.

Papaloizou, J.C.B. & Terquem, C. 2006. Planet formation and migration. *Rept. Prog. Phys.* **69** 119-.

Peale, S. 1976. Orbital resonances in the solar system. *ARA&A* **14**, 215-246.

Pepe, F. et al. 2007. The HARPS search for southern extra-solar planets. VIII. μ Arae, a system with four planets. *A&A* **462**, 769-776.

Psychoyos, D. & Hadjidemetriou, J.D. 2005. Dynamics of Populations of Planetary Systems, Proceedings of IAU Colloquium #197. Eds Z. Knezevic and A. Milani. Cambridge: Cambridge UP, p.55-62(2005)

Psychoyos, D. & Hadjidemtriou, J.D. 2005. Dynamics Of 2/1 Resonant Extrasolar Systems Application to HD82943 and GLIESE876. *CeMDA* **92**, 135-156.

Rasio, F.A. & Ford, E.B. 1996. Dynamical instabilities and the formation of extrasolar planetary systems. *Science* **274**, 954-956.

Rasio, F.A. et al. 1996. Tidal Decay of Close Planetary Orbits. *ApJ* **470**, 1187-1191.

Raymond, S.N. & Barnes, R. 2005. Predicting Planets in Known Extrasolar Planetary Systems. II. Testing for Saturn Mass Planets. *ApJ* **619**, 549-557.

Raymond, S.N., Barnes, R. & Kaib, N.A. 2006. Predicting Planets in Known Extrasolar Planetary Systems. III. Forming Terrestrial Planets. *ApJ* **644**, 1223-1231.

Raymond, S.N., Quinn, T.R. & Lunine, J.I. 2004. Making other earths: dynamical simulations of terrestrial planet formation and water delivery. *Icarus* **168**, 1-17.(2004)

Rivera, E.J. & Haghighipour, N. 2007. On the stability of test particles in extrasolar multiple planet systems. *MNRAS* **374**, 599-613.

Rivera, E.J. & Lissauer, J.J. 2000. Stability Analysis of the Planetary System Orbiting υ Andromedae. *ApJ* **530**, 454-463.

Rivera, E.J. & Lissauer, J.J. 2001. Dynamical Models of the Resonant Pair of Planets Orbiting the Star GJ 876. *ApJ* **558**, 392-402.

Rivera, E.J. et al. 2005. A $\sim 7.5M_\oplus$ Planet Orbiting the Nearby Star, GJ 876. *ApJ* **634**, 625-640.

Rodríguez, A. & Gallardo, T. 2005. The Dynamics of the HD 12661 Extrasolar Planetary System. *ApJ* **628**, 1006-1013.

Sándor, Zs. & Kley, W. 2006. On the evolution of the resonant planetary system HD 128311. *A&A* **451**, L31-L34.

Sándor, Zs., Kley, W. & Klagyivik, P. 2007. Stability and Formation of the Resonant System HD 73526. *A&A*, accepted.

Schwarz, R. et al. 2005. Trojans in Habitable Zones. *AsBio* **5**, 579-586.

Schwarz, R. et al. 2007. Trojan planets in HD 108874? *A&A* **462**, 1165-1170.

Stepinski, T.F., Malhotra, R. & Black, D.C. 2000. The υ Andromedae System: Models and Stability. *ApJ* **545**, 1044-1057.

Sussman, G.J. & Wisdom, J. 1988. Numerical evidence that the motion of Pluto is chaotic. *Science* **241**, 433-437.

Sussman, G.J. & Wisdom, J. 1992. Chaotic evolution of the solar system. *Science* **257**, 56-62.

Thébault, P., Marzari, F. & Scholl, H. 2002. Terrestrial planet formation in exoplanetary systems with a giant planet on an external orbit. *A&A* **384**, 594-602.

Tinney, C.G. et al. 2006. The 2:1 Resonant Exoplanetary System Orbiting HD 73526. *ApJ* **647**, 594-599.

Tsiganis, K, Varvoglis, H. & Hadjidemetriou, J.D. 2002. Stable Chaos versus Kirkwood Gaps in the Asteroid Belt: A Comparative Study of Mean Motion Resonances. *Icarus* **159**, 284-299.

Udry, S. et al. 2007. The HARPS search for southern extra-solar planets. XI. Super-Earths (5 and 8 $_{\oplus}$M) in a 3-planet system. *A&A*, accepted.

Veras, D. & Armitage, P. 2007. Extrasolar Planetary Dynamics with a Generalized Planar Laplace-Lagrange Secular Theory. *ApJ* **661** 1311-1322.

von Bloh, W. et al. 2003. On the Possibility of Earth-Type Habitable Planets in the 55 Cancri System. *AsBio* **3**, 681-688.

Varadi, F., Ghil, M. & Kaula, W.M. 1999. Jupiter, Saturn, and the Edge of Chaos. *Icarus* **139**, 286-294.

Vogt, S.S. et al. 2005. Five New Multicomponent Planetary Systems. *ApJ* **632**, 638-658.

Voyatzis, G. & Hadjidemetriou, J.D. 2006. Symmetric and asymmetric 3:1 resonant periodic orbits with an application to the 55Cnc extra-solar system. *CeMDA* **95**, 259-271.

Weidenschilling, S. & Marzari, F. 1996. Gravitational scattering as a possible origin for giant planets at small stellar distances. *Nature* **384**, 619-621.

Wright, J.T. et al. 2007. Four New Exoplanets and Hints of Additional Substellar Companions to Exoplanet Host Stars. *ApJ* **657**, 533-545.

Wu, Y. & Goldreich, P. 2002. Tidal Evolution of the Planetary System around HD 83443. *ApJ* **564**, 1024-1027.

Wu, Y. & Murray, N. 2003. Planet Migration and Binary Companions: The Case of HD 80606b. *ApJ* **589**, 605-614.

Yoshida, H. 1993. Recent Progress in the Theory and Application of Symplectic Integrators. *CeMDA* **56**, 27-43.

Zhou, J.-L. & Sun, Y.-S. 2003. Occurrence and Stability of Apsidal Resonance in Multiple Planetary Systems. *ApJ* **598**, 1290-1300.

Zhou, J.-L. et al. 2004. Apsidal corotation in mean motion resonance: the 55 Cancri system as an example. *MNRAS* **350**, 1495-1502.

8 Searching for Exoplanets in the Stellar Graveyard

Steinn Sigurdsson

Summary. There is increasing evidence that planets are ubiquitous, and may form around stars over a wide range of stellar masses. After a star dies, the planets may remain, and in some circumstances there may be a new epoch of planet formation after the main sequence. In this chapter, scenarios for the retention and formation of planets after star death, and the prospects for detection, including current known post-main sequence systems are discussed. Planets in the graveyard are, in many cases, easier observational targets than planets around main sequence stars. Different detection techniques may also be brought to bear, in some cases with much higher sensitivity, allowing the detection of low mass planets. Planets detected in the graveyard reflect the 'live' population of planets, and in some cases provide potentially strong constraints on planet formation processes, and the general planet population.

8.1 The Discovery of Extrasolar Planets

The first extrasolar planets were discovered by Alex Wolszczan and Dale Frail, orbiting a *millisecond pulsar*, PSR B1257+12; two planets were discovered, initially, with masses of $2.8/\sin(i)$ and $3.4/\sin(i)$ Earth masses, where i is the inclination (unknown at the time) to the line of sight of the planets' orbits, and orbital periods of 98.2 and 66.6 days respectively (Wolszczan & Frail, 1992). A third, lower mass planet was discovered subsequently, orbiting interior to the first two discovered, and the additional data also allowed the inclination of the orbital plane of the planets to be determined to be $i = 53°$, implying actual masses of 4.3 ± 0.2 and 3.9 ± 0.2 Earth masses. The third planet has a mass of 0.02 Earth masses, and an orbital period of 25.3 days (Wolszczan, 1994; Konacki & Wolszczan, 2003). All three planets have low orbital eccentricity. These planets are still the lowest mass extrasolar planets known.

8.2 Planets Around Pulsars

Planets were first found around a pulsar, because the observations that can be made of pulsars are the most precise we can make in astronomy, and because they were there to be found.

The essential difficulty in finding planets orbiting around stars is that stars are relatively big and bright, and planets, in comparison, are small and very faint. A number of techniques are available for detecting planets, including direct imaging, but most observations of extrasolar planets to date have relied on indirect detection, observing the effects the planets have on the stars they orbit (or in the case of microlensing, the planets' effect on other intervening stars). To detect the planets, very precise measurements must be made, and pulsars are intrinsically amenable to precision measurements.

8.2.1 Pulsars

Pulsars are the compact remnants of massive stars. High mass stars burn up the hydrogen in their cores, fusing it to higher mass elements. For high enough mass stars, the chain of fusion reactions in the core eventually reaches iron, which is an endpoint for fusion reactions; converting iron to higher mass elements absorbs energy instead of releasing it; the source of power maintaining the core of the star hot and at high pressure is removed, and the star collapses. The core is compressed to a radius of 10-15 km, with the matter compressed to nuclear densities, forming a "neutron star", typically with a mass of about $1.25 - 1.4$ times the mass of the Sun. The gravitational binding energy released in this collapse drives a shock into the outer layers of the star which disassembles explosively, an event we see as a supernova explosion. The details of the explosion depend on the mass, composition and evolutionary history of the star, but generally we think that stars with masses ranging from about 8 times the mass of the Sun, to somewhere between $20 - 40$ times the mass of the Sun will form neutron stars (Heger et al., 2003). The neutron stars formed are hot, rapidly rotating and with high magnetic fields. They are observed to spin after formation with spin periods of less than a second or so. As they spin, narrow beams or fans of coherent radio emission are broadcast from the magnetic field poles that rotate rigidly with the neutron star. We observe this as regular pulses of radio emission, if the beam of radiation intersects the Earth as the neutron star rotates. These neutron stars we observe as radio pulsars. Pulsars are very stable, regular rotators; the arrival of the radio pulses at Earth can be timed, typically to millisecond precision; the energy radiated by the pulsar (only a very small fraction of that which is emitted at radio frequencies) is drawn from the rotational kinetic energy of the neutron star, which "spins-down". The spin down takes place over timescales of thousands to millions of years, and is observed to be very predictable, ramping down the spin of the neutron star almost linearly. For a bright, stable pulsar, the time of arrival of the radio pulses can be measured, and the chain of pulses counted over periods of years. Even with interruptions in observations, the timing of the pulse arrival can be measured "phase coherently", which means that the interval between pulses far apart in time can be consistently linked together, and we know to within a millisecond when to expect a pulse from a given pulsar, years after the initial observation. There are about 100 million seconds in an interval of three years, which means we can measure the arrival of the radio pulses over periods of many years, with a precision of a few parts in a trillion! In particular, if something disturbs the time of arrival of a pulse, by even a little bit, we can observe and measure this.

8.2.2 Searches for Planets

In 1991, a group at the Jodrell Bank Radio Observatory published a letter in *Nature*, claiming that a 10 Earth mass planet was orbiting radio pulsar PSR B1829-10, in a six month orbit (Bailes et al., 1991). The detection relied on the periodic delay and advance of the arrival of the radio pulses from the pulsar at Earth. As the conjectured planet orbited the pulsar, its small but finite mass caused the pulsar to undergo a tiny reflex orbit about the center of mass of the pulsar-planet system. Since the planet orbited at a radius of about a hundred million km, the pulsar orbited the system's center of mass with an orbital radius smaller by a factor equal to the ratio of the mass of the planet to that of the pulsar – a factor of about ten thousand. But the pulsar's motion, over about 10,000 km in radial distance, translated to a periodic delay, and then advance, in radio pulse arrival times, by the time it took for the radio beam to travel the extra distance – a few tens of milliseconds. Since the pulse arrival time was measurable to a precision of a millisecond or so, the periodic change in arrival time was easily measured. Unfortunately, the discovery claim had to be retracted a few months later, in January 1992, at a meeting of the American Astronomical Society. The observed planet was an artefact, caused by an erroneous model for the motion of the Earth within the Solar System. To measure the relative arrival time of the pulses, observers usually subtract the motion of the Earth about the center of mass of the Solar System, but the orbital parameters used in analysing the data on PSR B1829-10 were not precise, and the observations were showing a second harmonic of the Earth's orbit, due to the fact that the Earth's orbit is slightly eccentric. This was not the first claimed discovery of a planet orbiting a pulsar; (Demianski & Proszynski, 1979) had previously found variations in the arrival times of the well-known, bright, 0.714 s pulsar PSR B0329+54, which they suggested were consistent with a planet having a mass less than that of the Earth, but subsequent observations failed to confirm the claim. It seems like that the observed variations in the pulse arrival times for PSR B0329+54 are caused by spin irregularities inherent in this relatively young ($\sim 5 \times 10^6$ year old) neutron star.

Immediately after the retraction on the candidate planet around PSR B1829-10, came the announcment of the planets around PSR B1257+12, with strong confirmation by further observations following within a couple of years (Wolszczan, 1994).

There are three planets in the PSR B1275+12 system: A is only about twice as massive as Earth's Moon and has an orbital period of 25.26 days; B is about four times as massive as the Earth, with an orbital period of 66.54 days; and C is nearly four times as massive as the Earth, with an orbital period of 98.21 days

Fig. 8.1. Artist's impression of the PSR B1257+12 planet system. (adapted from NASA/Caltech-JPL/R. Hurt (Spitzer Science Center), by permission.)

(Konacki & Wolsczcan, 2003). PSR B1257+12 is different from PSR B1829-10 and PSR B0329+54, because it is a *millisecond pulsar*, with a spin period of only 6.2 milliseconds. Millisecond pulsars constitute a few percent of the observed pulsar population. As the name indicates, they have spin periods, and hence pulse intervals, measured in milliseconds, rather than seconds. They typically have weaker magnetic fields than regular pulsars (hundreds of millions of gauss fields, compared with trillions of gauss for regular pulsars and a gauss for the Earth). Because of their weaker magnetic fields, and very high rotational kinetic energy, the millisecond pulsars spin-down over periods of *billions* of years, and therefore can be observed as pulsars for a correspondingly longer time; most of the millisecond pulsars we observe are hundreds of millions to several billion years old. Because of their very high rotational kinetic energy, millisecond pulsars are very stable rotators, and the interval between the arrival of the pulses is stable and measurable to an accuracy of less than a *microsecond* for a bright millisecond pulsar. The measurements of the pulse arrival times can be done phase coherently over intervals of many years, or even decades, leading to a measurement precision of one part in a thousand trillion or so for the best measured pulsars. This compares with the very best laboratory measurement possible for any physical system, and for some time the limiting factor on pulse timing was the long term precision of clocks in the observatories. The clocks have now got better, although over long periods, an ensemble of millisecond pulsars still provides a more stable clock than laboratory clocks.

Such precision permits measurements of the delay in pulse arrival time of a microsecond or less. Since the speed of light is 300,000 km/sec, this corresponds to a displacement in the pulsar of just 300 meters. The pulsar displacement can be measured even if it occurs over a time interval of many years (corresponding to the orbital motion of a planet). This is equivalent to measuring an orbital speed for the pulsar of less than a millimeter per second. This precision enables the detection of planets of much lower masses or longer orbital periods, than can be measured by any other technique.

8.2.3 Origin of the Pulsar Planets

How did PSR B1257+12 come to have planets? The neutron star most likely formed originally in a supernova. Any planets orbiting the star when it exploded as a supernova are highly unlikely to have survived the explosion, so most likely the pulsar acquired the planets after it became a neutron star. A number of scenarios for planet formation around a neutron star have been proposed (Phinney & Hansen, 1993). One scenario is that the supernova explosion had some residual material that was blown back or stalled in the explosion and fell back onto the neutron star, leading to a disk forming around the young neutron star. Intriguingly, infrared radiation has recently been observed around a young neutron star, 4U 0142+61, consistent with just such a "fallback disk" (Wang et al., 2006), and searches for planets around young pulsars continue (Posselt et al., 2006). But, PSR B1257+12 is an old millisecond pulsar, not a young slow spin pulsar. We think millisecond pulsars form when neutron stars accrete gas in compact binary systems, probably so-called "low mass X-ray binaries". In these systems, a low mass star (of solar mass or less) orbits close to a neutron star, transferring mass onto the neutron star, and slowly spinning the neutron star up to millisecond periods over tens of millions years. They are referred to as X-ray binaries, because the energy released during the accretion causes the system to glow brightly in X-rays. We think that only a very small fraction of neutron stars in the Galaxy go through such an accretion phase; maybe one in ten thousand or so of the neutron stars formed. It is difficult to arrange for a close low mass companion to be retained in an orbit around a neutron star when the progenitor star explodes as a supernova.

The process of spin-up affords opportunities for planet formation. It is possible that during accretion some material flows outwards from the star, forming an "excretion disk", which moves outwards and cools, rather than accreting onto the neutron star. Such an excretion disk can provide an environment suitable for the formation of low mass planets in relatively close orbits, such as those which are observed around PSR B1257+12. This scenario requires that when the pulsar "turns on" after it is spun up, that it will then ablate the core of its companion star, destroying it completely, and leaving only the planets, at least in the case of PSR B1257+12. We do see so-called "Black Widow" pulsars, that are in the process of ablating away the last little bit of their stellar companions, although it is not clear that the ablation process will proceed to complete destruction of the star. An alternative version of this scenario short-cuts the accretion disk phase, by postulating that a small fraction of neutron stars physically impact their stellar companions, like old fashioned cannonballs. Supernova explosions can be asymmetric, and the neutron stars when they form could be ejected in some random direction at high speeds. This is observed, but the mechanisms by which the explosion becomes asymmetric are not fully understood. If the neutron star is kicked out in just the right direction and impacts its companion, it may promptly shatter the star, and some of the debris may settle into a disk around the neutron star, serving both to spin up the neutron star to millisecond periods, and to provide a remnant disk from which planets may form in orbit around the newly formed millisecond pulsar (Phinney & Hansen, 1993; Greaves & Holland, 2000; Miller & Hamilton, 2001; Lazio & Fischer, 2004; Bryden et al., 2006; Currie & Hansen, 2007).

The set of three planets orbiting PSR B1257+12 bears an interesting resemblance to a scaled down version of the inner Solar System, and understanding the process of planet formation around pulsars is likely to enhance substantially our understanding of planet formation in general, particularly for terrestrial planets. Currently the pulsar planets are the only example of terrestrial mass planets we have found orbiting stars outside the Solar System.

8.2.4 Planet in Messier 4

Shortly after the original announcement in 1992 of the discovery of the planets around PSR B1257+12, a workshop, "Planets Around Pulsars", was held at the campus of the California Institute of Technology. At that meeting, I proposed that globular cluster pulsars might be a promising target for further planet searches (Sigurdsson, 1992).

There are a little over 150 globular clusters in the Milky Way galaxy, consisting typically of 100,000-1,000,000 stars gravitationally bound in a dense, spherical aggregation. The spatial density of stars in the centers of the denser globular clusters can be a million times higher than in the solar neighbourhood. Globular clusters are generally old, with ages of 11-13 billion years, and the stars within any given cluster are coeval and homogenous in composition. For most globular clusters, the stellar population is metal poor compared with the Sun, by factors of typically 10-100. Globular clusters have long been known to be overabundant in low mass X-ray binaries, and shortly before the discovery in 1990 of PSR B1257+12, the first of many millisecond pulsars had been discovered in the globular cluster Messier 28.

Clearly, the sort of mechanisms by which the planets formed around PSR B1257+12 might also operate for the pulsars formed in globular clusters. However, due to the very high density of stars, there was a potential problem of survival for any planets; the occasional close passage of another star might disrupt the orbits of any planets. But, such passages could also serve to increase the odds of detecting planets around pulsars. If stars in globular clusters had planets around them, then during close passages planets might occasionally be *exchanged*, from their orbits around their parent star, to an orbit around a pulsar. A planet orbiting a typical solar-type star in a globular cluster would be very hard to detect, although attempts have been made to do so (Gilliland et al., 2000), but a planet orbiting a millisecond pulsar in a globular cluster would be comparatively easy to detect. The orbits of exchanged planets would in general be qualitatively different from the orbits of

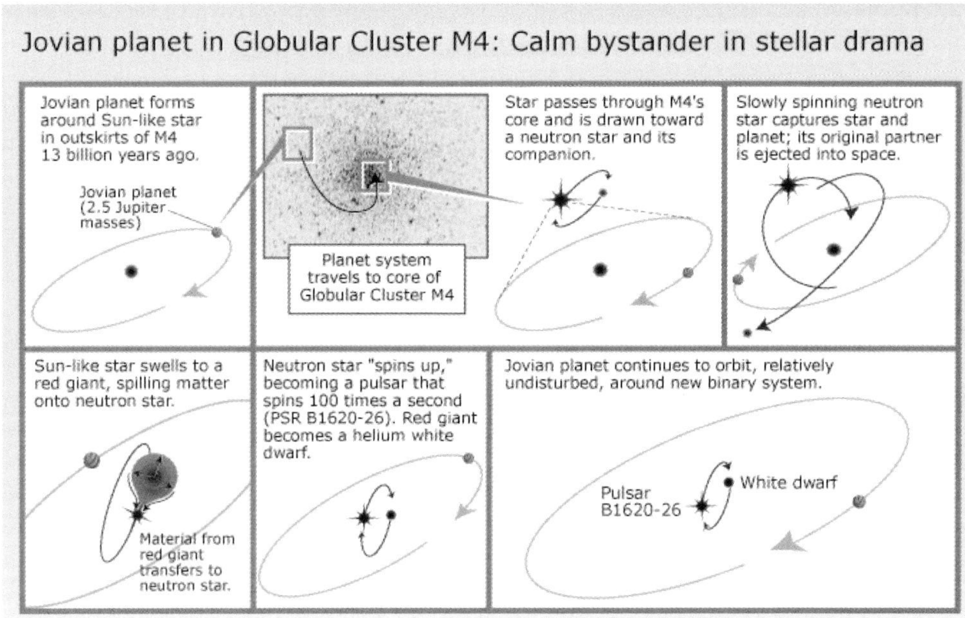

Fig. 8.2. A conceptual schematic of one of the proposed exchange formation mechanisms for the planet around PSR B1620-26 in the globular cluster Messier 4. (Adopted from a Space Telescope Science Institute/NASA press image, http://hubblesite.org/newscenter/archive/releases/2003/2003/19/)

planets formed around the pulsar, such as for PSR B1257+12. The latter are in close circular orbits, and the planets are of relatively low mass; by contrast exchanges tend to slightly favour the most massive planets, and will lead to large, eccentric orbits.

Coincidentally, at the Caltech workshop in 1992, Backer reported anomalous timing residuals for PSR B1620-26 in the metal poor globular cluster Messier 4, the second millisecond pulsar discovered in a globular cluster. Unlike PSR B1257+12, PSR B1620-26 is a binary pulsar, orbited by a low mass white dwarf, the remnant of the star which spun the pulsar up to its current millisecond period. The timing residuals were consistent with a Jovian mass planet in a distant orbit around the pulsar, but did not preclude other explanations (Thorsett et al., 1993). A number of contending models were proposed to explain the system, (e.g., Sigurdsson, 1993; Joshi & Rasio, 1997), and in 1999 additional data strongly constrained the system, requiring the presence of a low mass companion (Thorsett et al., 1999), while new modeling explaind the detailed kinematics of the system more adequately (Ford et al., 2000).

Then, in 2003, at a workshop at the Kavli Institute in Santa Barbara, it was realised that the Hubble Space Telescope had serendipitously imaged the location of the pulsar, and its stellar companion, a low mass white dwarf, was observed. This provided two new pieces of information, a "cooling age" for the system, showing that the white dwarf had formed from its parent star about 500 million years earlier, and an independent constraint on the inclination of the pulsar-white dwarf orbital plane, which in turn allowed independent constraints on the other mystery companion. This is, in fact, a roughly 2 Jupiter mass planet, in a moderately eccentric orbit with an orbital period of about 100 years. Since then, the planet orbit has been observed to cross the periastron point, reversing the sign of its gravitational perturbation on the star, confirming the orbital parameters (Stairs, private communication). The planet is generally thought to have been exchanged into the current system, having originally formed around a $\sim$ 0.85 solar mass star in a fairly wide orbit, some 12.7 billion years ago. Currently there are two competing exchange models, one in which the planet was exchanged into the system when the binary formed (Sigurdsson, 1993), the other conjecturing that the binary pulsar formed first, with the planet exchanged in an independent encounter sometime later (Fregeau et al., 2006). Alternative formation scenarios have been proposed, (e.g. Beer et al., 2004), but they have a hard time accounting for the detailed orbital parameters of the system, in particular the high orbital inclination of the planet relative to the plane of the inner white dwarf orbit.

If the PSR B1620-26 system formed through exchange with a main sequence star, then planet formation started very early in the history of the universe. It is difficult to explain how a giant planet could form in such a metal poor system, and it is possible that the system provides evidence for a second planet formation process.

8.3 Planets Around White Dwarfs

Stars of mass lower than about eight solar masses do not become neutron stars. Rather the fusion in their cores fizzles out before the chain of fusion reactions terminates with formation of an iron care, and the outer envelope of the bloated red giant star is shed, the core contracts and cools passively, forming a so-called white dwarf. White dwarfs do not generate energy internally and are purely pressure supported. Consequently they have high densities, though not as high as neutron stars, and radii comparable with that of the Earth. White dwarfs emerge very hot, and luminous, but cool rapidly and after a hundred million years or so, depending on their exact mass and composition, the luminosity of the white dwarf is much less than its progenitor star.

In the process of shedding its envelope, any inner planets orbiting the red giant star are swallowed up and destroyed. Calculations show that when the Sun ends its life, Mercury and Venus will certainly be destroyed, Mars will probably survive, and the prospects for the survival of the Earth are so marginal that the outcome is dominated by the uncertainties in the modeling. The outer giant planets all survive.

The giant planets are moderately warm, and radiate in the infrared from their intrinsic thermal emission rather than by reflecting the light of their parent star. Since the white dwarf is much fainter than the original star, the contrast between

Fig. 8.3. Hubble Space Telescope Near Infrared Camera image of the immediate surroundings of a nearby DAZ white dwarf. The irregular spread out structure in the upper left is the residual light from the white dwarf after it was masked out by a coronograph, and the image self-subtracted. Two images were taken at different camera roll angles, and then digitally rotated into alignment and subtracted from each other. The light from the white dwarf is suppressed by a factor of several hundred, permitting high contrast searches for close low luminosity objects. The small dot to the left is about ten thousand times fainter than the star and about one arcsecond away, and was consistent in luminousity and colour with a few jupiter mass companion, but subsequent imaging revealed it to be a background object (Debes, private communication).

any outer giant planets and the white dwarf is much lower, typically by a factor of 1,000-10,000, than the contrast between the progenitor star and any planets. Consequently, giant planets that survive the death of their parent star and are still orbiting the remnant white dwarf are comparatively easy targets for direct imaging by high resolution infrared telescopes. A number of projects are currently underway to attempt such imaging of nearby white dwarfs, (e.g. Burleigh et al., 2002; Debes et al., 2005). See Hansen (2004) section 8.5 for a review.

One particularly intriguing aspect of white dwarf astronomy is so-called "DAZ" white dwarfs. These are moderately cool white dwarfs with hydrogen atmospheres, which show evidence for metal absorption lines, In a few instances, there is evidence for warm dust disks around the white dwarfs, containing the equivalent of a few cubic kilometers of metals. Theoretical models suggest the metal absorption lines, and the associated debris disks, came from tidal disruption of a planetesimal, an asteroidal or cometary body (Debes & Sigurdsson, 2002; Jura, 2003). Any surviving planetesimal must come from the outer system, and to get it into a radial orbit which will lead to tidal disruption of the object and a warm disk or dust contamination of the white dwarf atmosphere, requires a planet to perturb the orbit of the planetesimal. It is therefore conjectured that the DAZ white dwarfs are particularly promising targets for imaging any planets which might be in orbit around them. Current telescopes can marginally detect several Jupiter mass planets, if any such are present, and searches to date have not found any. The next generation of telescopes, in particular, the James Webb Space Telescope, will be able to easily image any giant planet around nearby white dwarfs, and in fact map out the orbit of the planet directly from its motion relative to the star over several years. With some additional effort, spectroscopy of giant planet atmospheres would also be possible, testing theories of giant planet atmospheres and carrying out pathfinding science for observation of planets around main sequence stars.

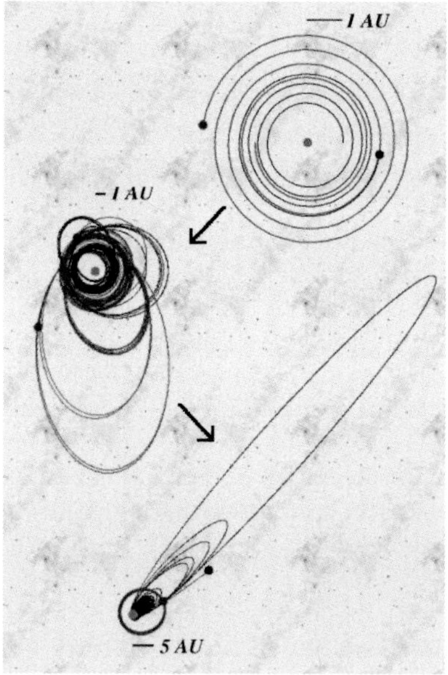

Fig. 8.4. Conceptual image showing the dynamical evolution of some outer planetary systems after the end of the stellar main sequence, as the star becomes a white dwarf. The outer planets spiral outwards in their orbits as the star loses mass during the giant phase. For some orbital configurations, secular evolution drives the orbits of the planets to become chaotic, on time scales of hundred million years or more. The planets' orbits cross and there is a strong dynamical rearrangement, including possible collisions and ejections. The planets settle down into a new orbital configuration, typically this will include an outer planet on a wide eccentric orbit, penetrating the region where Kuiper belt-like objects might be founds, while any surviving inner planets might be scattered into closer orbits, repopulating the orbital region where planets were swallowed during the giant phase (Debes & Sigurdsson, 2002).

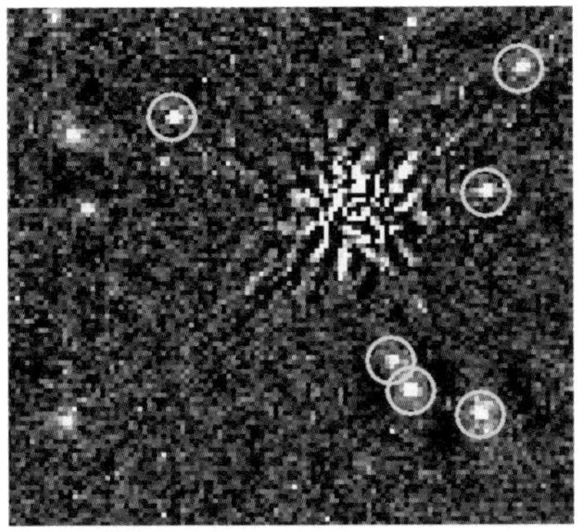

Fig. 8.5. Hubble Space Telescope Near Infrared Camera image of the immediate surroundings of a nearby DAZ white dwarf. This white dwarf is close to the galactic plane where there are a lot of infrared sources, six sources were found which were close in angular separation and had colours and magnitudes consistent with giant planets, none have been confirmed to date as physically associated companions (Debes, private communication).

An additional channel for detecting planets around white dwarfs comes from the possibility of planet formation in excretion disks formed by merging white dwarfs (Livio et al., 1992). Some binary stars evolve to form binary white dwarfs in very tight orbits, which may eventually merge. The process would lead to a very massive white dwarf, or possibly a pulsar, if the combined mass of the white dwarfs is high enough and they can merge without exploding, then in the process an "excretion disk" may form, analogous to the pulsar case. As in the pulsar case, planets may conceivably form in close orbits about the new massive white dwarf, where they may be detected, in an orbital region that would otherwise be scoured clean of planets during the formation of the white dwarfs themselves.

8.3.1 Timing of Pulsating White Dwarfs

During their cooling, white dwarfs may pass through periods of pulsational instability, where the whole star oscillates coherently with a period of minutes. While not as stable or rapid as pulsars, these white dwarfs allow very precise long term observations, including precise measurements of any deviation from stable, steadily slowing oscillations. That is, in the same way as for pulsars, the timing of the white dwarf pulsations allows for high precision detection of any planets orbiting the white dwarfs. Since this requires many years of very high precision measurements, the number of stars which have been observed sufficiently well is still small, but increasing rapidly. Recently, a preliminary detection of a few Jupiter mass candidate planet in a roughly four-year orbit was reported (Mullally et al., 2007). Follow up observations are in progress to confirm the detection and to try to get infrared data on the planet's thermal emission. The white dwarf pulsation timing provides additional prospects for indirect detection of planets in a parameter space of planet mass and orbital period that is otherwise hard to measure.

In addition to white dwarfs, a class of stars known as sdB stars can also undergo stable pulsation, and timing searches are underway to find planets around nearby sdB stars. Preliminary results show that the observations have adequate precision to detect any planets, and there are indications in the data of trends consistent with the presence of massive planets.

At the "Extreme Solar Systems" meeting in Santorini, Greece, in June 2007, Silvotti et al. (2007) announced the possible detection of a giant planet around the extreme horizontal branch star V391 Pegasi. This is a well known pulsating subdwarf, a star that has terminated core hydrogen fusion on the stellar main sequence and evolved through a red giant branch phase. Such stars may spend some time undergoing helium fusion, with only a very thin hydrogen atmosphere, before evolving to the asymptotic giant branch, after which they become white dwarfs. During the subdwarf phase, the star can become unstable and pulse regularly, and the timing of such pulsations provides a good clock that can be used to detect low mass companions in orbit around the star. Silvotti et al. found a few Jupiter mass (3.2 $M_{Jup} \sin i$) companion in an orbit with a radius of about 1.7 astronomical units. The planet must originally have been closer to the star, but moved outwards as the star lost mass, avoiding being swallowed by the red giant envelope as the star expanded. There are only a few pulsating subdwarfs whose pulsations can be timed sufficiently precisely to detect the presence of a planet, but the fact that a planet has already been found around one of these tells us both that planets are indeed ubiquitous, and that they can survive the red giant phase even as close in as an astronomical unit, and may therefore also be seen around white dwarfs.

8.4 Future Prospects

The future for the stellar graveyard looks very promising. As the number of pulsars discovered increases, the odds of discovering additional pulsar planets improves, although high precision observations over long periods are required to confirm any new candidated. The current set of newly discovered pulsars contains one or two pulsars which are suspected to have planetary companions, but additional data are required to confirm their presence. Pulsar planets can provide an extreme perspective on the processes of planet formation, constraining planet formation theories, and in the case of exchanged planets, direct probing of planet formation around normal stars, albeit in extreme environments.

Searches for planets around white dwarfs are intensifying, both with direct imaging of spatially resolved planetary companions, searches for blended infrared excess from unresolved companions, and for indirect signatures such as infrared emission from asteroidal debris disks. Long term observations of white dwarf and sdB star pulsations will continue, with good prospects for planet detection. White dwarf searches are generally restricted to nearby white dwarfs, due to the faintness of the targets, but this permits a broad range of follow up observations to characterise in detail any planets discovered. Future space based missions will increase the prospects for planet detection, and open up new opportunites, such as astrometric detection of giant planets by observing the transverse motion in the position of nearby white dwarfs as they orbit the center of mass of any planetary system around them. Detection of planets around white dwarfs provides perspective on planet formation around stars of different masses, and probes formation of the outer planets with orbital period of decades, which are inaccessible from ground based radial velocity searches for planets.

References

Backer, D.C., 1992, *A pulsar timing tutorial and NRAO Green Bank observations of PSR 1257+12*, In: *Planets Around Pulsars*, PASP Conf Proc. vol **36**, ed. J.A. Phillips, S.E. Thorsett and S.R. Kulkarni, pp. 11

Bailes, M., Lyne, A.G. & Shemar, S.L., 1991, *A planet orbiting the neutron star PSR1829-10*, Nature, **352**, 311

Beer, M.E., King, A.R. & Pringle, J.E., 2004, *The planet in M4: implications for planet formation in globular clusters*, Monthly Notices of the Royal Astronomical Society, **355**, 1244

Blandford, R.D., Romani, R.W. & Applegate, J.H., 1987, *Timing a millisecond pulsar in a globular cluster*, Monthly Notices of the Royal Astronomical Society, **225**, P51

Bryden, G., Beichman, C.A., Rieke, G.H., Stansberry. J.A., Stapelfeldt, K.R., Trilling., D.E., Turner, N.J. & Wolszczan, A., 2006, *Spitzer MIPS Limits on Asteroidal Dust in the Pulsar Planetary System PSR B1257+12*, Astrophysical Journal, **646**, 1038

Burleigh, M.R., Clarke, F.J. & Hodgkin, S.T., 2002, *Imaging planets around nearby white dwarfs*, Monthly Notices of the Royal Astronomical Society, **331**, 41

Currie, T. & Hansen, B.M.S., 2007, *The Evolution of Protoplanetary Disks Around Millisecond Pulsars: The PSR 1257+12 System*, Astrophysical Journal, **666**, 1232

Debes, J.H. & Sigurdsson, S., 2002 *Are There Unstable Planetary Systems around White Dwarfs?*, Astrophysical Journal, Letters, **572**, 556

Debes, J.H., Sigurdsson, S. & Woodgate, B.E., 2005, *Cool Customers in the Stellar Graveyard. II. Limits to Substellar Objects around Nearby DAZ White Dwarfs*, Astronomical Journal, **130**, 1221

Demianski, M. & Proszynski, M., 1979, *Does PSR0329+54 have companions?*, Nature, **282**, 383

Ford, E.B., Joshi, K.J., Rasio, F.A. & Zbarsky, B., 2000, *Theoretical Implications of the PSR B1620-26 Triple System and Its Planet*, Astrophysical Journal, **528**, 336

Fregeau, J.M., Chatterjee, S. & Rasio, F.A., 2006, *Dynamical Interactions of Planetary Systems in Dense Stellar Environments*, Astrophysical Journal, **640**, 1086

Gillilan, R.L., et al., 2000, *A Lack of Planets in 47 Tucanae from a Hubble Space Telescope Search*, Astrophysical Journal, Letters, **545**, 47

Greaves, J.S. & Holland, W.S., 2000, *A search for protoplanetary discs around millisecond pulsars*, Monthly Notices of the Royal Astronomical Society, **316**, 21

Hansen, B.M.S., 2004, *The Astrophysics of Cool White Dwarfs*, Physics Reports, **399**, 1

Hansen, B.M.S., Kulkarni. S. & Wiktorowicz, S., 2006, *A Spitzer Search for Infrared Excesses around Massive Young White Dwarfs*, Astronomical Journal, **131**, 1106

Heger, A., Woosley, S.E., Langer, N. & Hartmann, D.H., 2003, *How Massive Single Stars End Their Life*, Astrophysical Journal, **591**, 288

Joshi, K.J. & Rasio, F.A., 1997, *Distant Companions and Planets around Millisecond Pulsars*, Astrophysical Journal, **479**, 948

Jura, M., 2003, *A Tidally Disrupted Asteroid around the White Dwarf G29-38*, Astrophysical Journal, Letters, **584**, 91

Konacki, M. & Wolszczan, A., 2003 *Masses and Orbital Inclinations of Planets in the PSR B1257+12 System*, Astrophysical Journal, Letters, **591**, 51

Lazio, T.J.W. & Fischer, J., 2004, *Mid- and Far-Infrared Infrared Space Observatory Limits on Dust Disks around Millisecond Pulsars*, Astronomical Journal, **128**, 842

Livio, M., Pringle, J.E. & Saffer, R.A., 1992, *Planets around massive white dwarfs*, Monthly Notices of the Royal Astronomical Society, **257**, P15

Miller, M.C. & Hamilton, D.P., 2001, *Implications of the PSR 1257+12 Planetary System for Isolated Millisecond Pulsars*, Astrophysical Journal, **550**, 863

Mullally, F., Kilic, M., Reach, W.T., Kuchner, M.J., von Hippel, T., Burrows, A. & Winget, D.E., 2007, *A Spitzer White Dwarf Infrared Survey*, Astrophysical Journal, Supplements, **171**, 206

Mullally, F. & Winget, D., 2007, *A Possible Planet Around a White Dwarf*, BAAS, 38, 1129

Phinney, E.S. & Hansen, B.M.S., 1993, *The pulsar planet production process*, In: *Planets Around Pulsars*, PASP Conf Proc. vol **36**, ed. J.A. Phillips, S.E. Thorsett and S.R. Kulkarni, pp. 371

Posselt, B., Neuhäuser, R. & Haberl, F., 2006, *Substellar companions around neutron stars*, In: *On the Present and Future of Pulsar Astronomy, 26th meeting of the IAU, Joint Discussion 2*, IAU, 11

Sigurdsson, S., 1992, *Planets in globular clusters?*, Astrophysical Journal, Letters, **399**, 95

Sigurdsson, S., 1993 *Genesis of a planet in Messier 4*, Astrophysical Journal, Letters, **415**, 43

Sigurdsson, S., Richer, H.B., Hansen, B.M.S., Stairs, I.H. & Thorsett, S.E., 2003, *A Young White Dwarf Companion to Pulsar B1620-26: Evidence for Early Planet Formation*, Science, **301**, 193

Silvotti, R., et al., 2007, *A giant planet orbiting the 'extreme horizontal branch' star V 391 Pegasi*, Nature, **449**, 189–191

Thorsett, S.E., Arzoumanian, Z. & Taylor, J.H., 1993, *PSR B1620-26 - A binary radio pulsar with a planetary companion?*, Astrophysical Journal, Letters, **412**, 33

Thorsett, S.E., Arzoumanian, Z., Camilo, F. & Lyne, A.G., 1999, *The Triple Pulsar System PSR B1620-26 in M4*, Astrophysical Journal, **523**, 763

Wang, Z., Chakrabarty, D. & Kaplan, D.L., 2006, *A debris disk around an isolated young neutron star*, Nature, **440**, 772

Wolszczan, A. & Frail, D.A., 1992 *A planetary system around the millisecond pulsar PSR1257+12*, Nature, **355**, 145

Wolszczan, A., 1994 *Confirmation of Earth Mass Planets Orbiting the Millisecond Pulsar PSR:B1257+12*, Science **264**, 538

9 Formation, Dynamical Evolution, and Habitability of Planets in Binary Star Systems

Nader Haghighipour

Summary. A survey of currently known planet-hosting stars indicates that approximately 25% of extrasolar planetary systems are within dual-star environments. Several of these systems contain stellar companions on moderately close orbits, implying that studies of the formation and dynamical evolution of giant and terrestrial planets, in and around binary star systems have now found realistic grounds. With the recent launch of the space telescope COROT, and the launch of NASA's Kepler satellite in 2009, the number of such dynamically complex systems will soon increase and many more of their diverse and interesting dynamical characteristics will soon be discovered. It is therefore, both timely and necessary, to obtain a deep understanding of the history and current status of research on planets in binary star systems. This chapter will serve this purpose by reviewing the models of the formation of giant and terrestrial planets in dual-star environments, and by presenting results of the studies of their dynamical evolution and habitability, as well as the mechanisms of delivery of water and other volatiles to their terrestrial-class objects. In this chapter, the reader is presented with a comprehensive, yet relatively less technical approach to the study of planets in and around binary stars, and with discussions on the differences between dynamical characteristics of these systems and planetary systems around single stars.

9.1 Introduction

The concept of a "world with two suns" has been of interest to astronomers for many years. Many scientists tried to understand whether planets could form in binary star systems, and whether the notion of habitability, as we know it, could be extended to such environments. Although as a result of their respective works, many dynamical features of *binary-planetary* systems[1] have been discovered, until recently, the subjects of their studies were, in large part, hypothetical. There was no detection of a planet in and/or around a binary system, and planet detection techniques had not advanced enough to successfully detect planets in dual-star environments.

[1]A binary-planetary system is a dual-star system that also hosts planetary bodies.

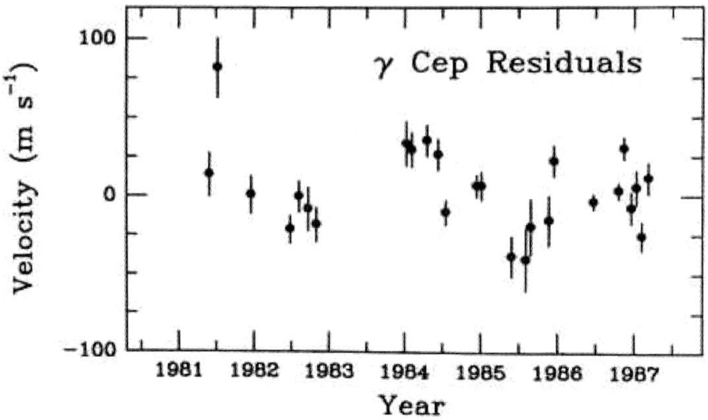

Fig. 9.1. Velocity residuals to γ Cephei after subtracting a second order fit to its original radial velocity data (Campbell, Walker & Yang, 1988). The residuals show periodicity implying the possible existence of a planetary companion.

The discovery of extrasolar planets during the past decade has, however, changed this trend. Although the candidate planet-hosting stars have been routinely chosen to be single, or within wide (>100 AU) binaries [2], the precision radial velocity technique has been successful in detecting planets around the primaries of three moderately close (<40 AU) dual-star systems. As a result, during the past few years, the topic of planets in binaries, once again, found its way to the mainstream research and has now become a real scientific issue that demands theoretical explanations.

The first detection of a planet in a binary system was reported by Campbell, Walker & Yang in 1988. In an attempt to identify planetary objects outside our solar system, these authors measured the variations in the radial velocities of a number of stars, and reported the possibility of the presence of a Jovian-type body around the star γ Cephei (Fig. 9.1, Campbell, Walker & Yang, 1988). This star, that is a K1 IV sub-giant with a mass of 1.59 solar-masses (Fuhrmann, 2004), is the primary of a binary system with a semimajor axis of 18.5 AU and an eccentricity of 0.36 (Griffin, Carquillat, & Ginestet, 2002; Hatzes et al., 2003). The secondary of this system is an M dwarf with a mass of 0.44 solar-masses (Neuhäuser et al., 2007; Torres, 2007). Initial radial velocity measurements of γ Cephei implied that this star may be host to a giant planet with a probable mass of 1.7 Jupiter-masses, in an orbit with a semimajor axis of 1.94 AU (Campbell, Walker & Yang, 1988).

Unfortunately, the discovery of the first binary-planetary system, which could have also marked the detection of the first planet outside our solar system, did not withstand skepticism. In an article in 1992, Walker and his colleagues attributed

[2]As shown by Norwood & Haghighipour (2002), the perturbative effect of the stellar companion on the dynamics of a planetary system around a star becomes important when the binary separation is smaller than 100 AU. At the present, approximately 25% of extrasolar planetary systems detected by radial velocity technique are in binary systems with separations ranging from 250 to 6000 AU.

their measured variations of the radial velocity of γ Cephei to the chromospheric activities of this star, and announced that the possibility of the existence of a giant planet around γ Cephei may be none (Walker et al., 1992). It took observers an additional 12 years to monitor γ Cephei and measure its radial velocity to arrive at the conclusion that the previously observed variations were not due to stellar activities and were in fact representative of a planetary companion (Fig. 9.2). It was the discovery of a giant planet in the binary system of Gl 86 (Queloz et al., 2000), and the (re-)announcement of the detection of a giant planet in γ Cephei system (Hatzes et al., 2003) that opened a new chapter in the theoretical and observational studies of extrasolar planetary systems.

The fact that giant planets exist in moderately close ($<$ 40 AU) binary systems has confronted dynamicists with many new challenges. Questions such as, how are these planets formed, can binary-planetary systems host terrestrial and/or habitable planets, how are habitable planets formed in such dynamically complex environments, and how do such planets acquire the ingredients necessary for life, are among major topics of research in this area. This chapter is devoted to review these issues and present the current status of research on the formation of planets in dual-star systems and habitability of terrestrial bodies in and around binary stars.

The chapter begins with a review of the dynamics of a planet in a binary star system. In general, prior to constructing a theory for the formation of planets, it proves useful to study whether the orbit of a planet around its host star would be stable. In a binary system, such studies are quite important since in these systems the perturbation of the stellar companion may dictate the possibility of the formation of planetary bodies by affecting the stability and dynamics of smaller objects.

The formation of planets in binary star systems is reviewed in the third section. Although the study of the dynamics of planets in and around binary stars dates back to approximately forty year ago, the formation of planets in these systems is an issue that is still unresolved. In spite of the observational evidence that indicates a majority of main and pre-main sequence stars are formed in binaries or clusters (Abt, 1979; Duquennoy & Mayor, 1991; Mathieu, 1994; Mathieu et al., 2000; White & Ghez, 2001), and the detection of potentially planet-forming environments in and around binary stars (Fig. 9.3, also see Mathieu, 1994; Akeson, Koerner & Jensen, 1998; Rodriguez et al., 1998; White et al., 1999; Silbert et al., 2000; Mathieu et al., 2000), planet formation theories are still unclear in explaining how planets may form in multi-star environments. The focus of Sect. 9.3 is on discussing the formation of giant and terrestrial planets in moderately close binary-planetary systems, and reviewing the current status of planet formation theories in this area.

The habitability of a binary system is presented in Sect. 9.4. The notion of habitability is defined based on the habitability of Earth and life as we know it. Such a definition requires a habitable planet to have the capability of retaining liquid water in its atmosphere and on its surface. The latter is determined by the luminosity of the central star, the size of the planet, and also the distribution of water in the protoplanetary disk from which terrestrial-class objects are formed. In Sect. 9.4, a review of the current status of the models of habitable planet formation

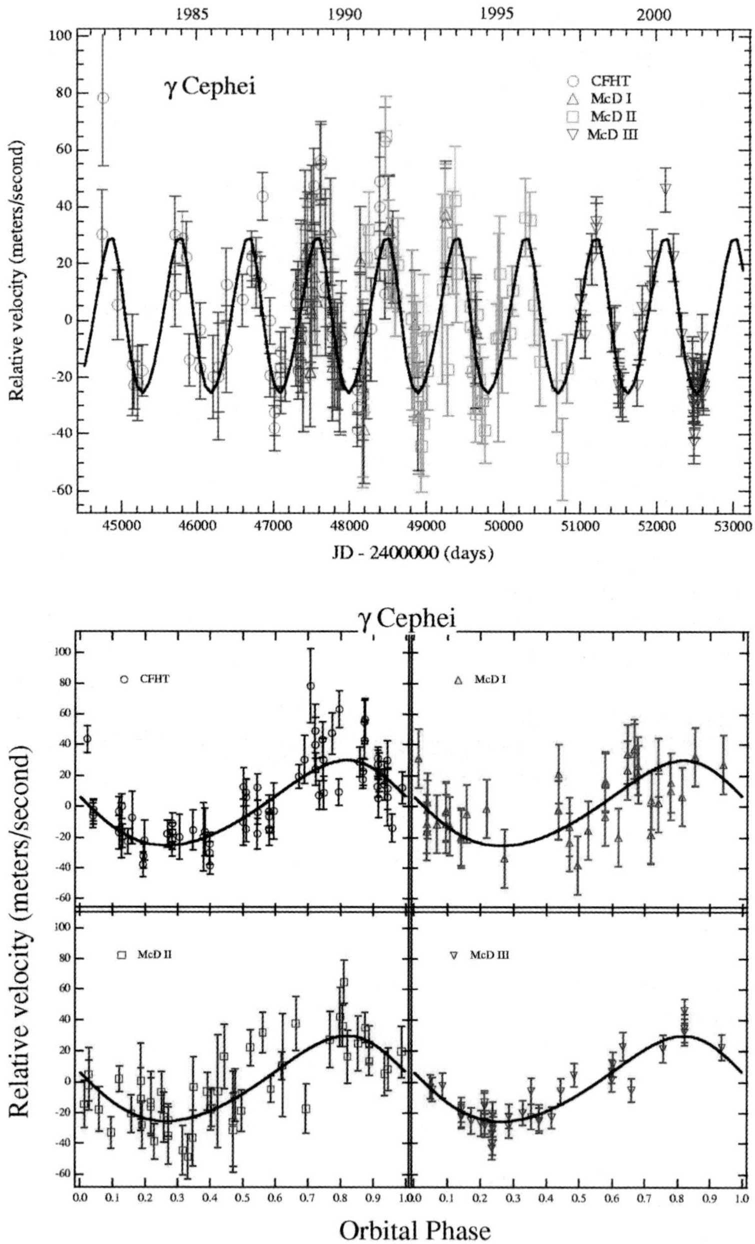

Fig. 9.2. Radial velocity measurements of the primary of γ Cephei binary system. The graph on the top shows the orbital solution for the planet (solid line) and the residual velocity measurements after subtracting the contribution due to the binary companion (data points). The graph on the bottom depicts the phased residual radial velocity measurements (data points) compared to the planet orbital solution (solid line) (Hatzes et al., 2003).

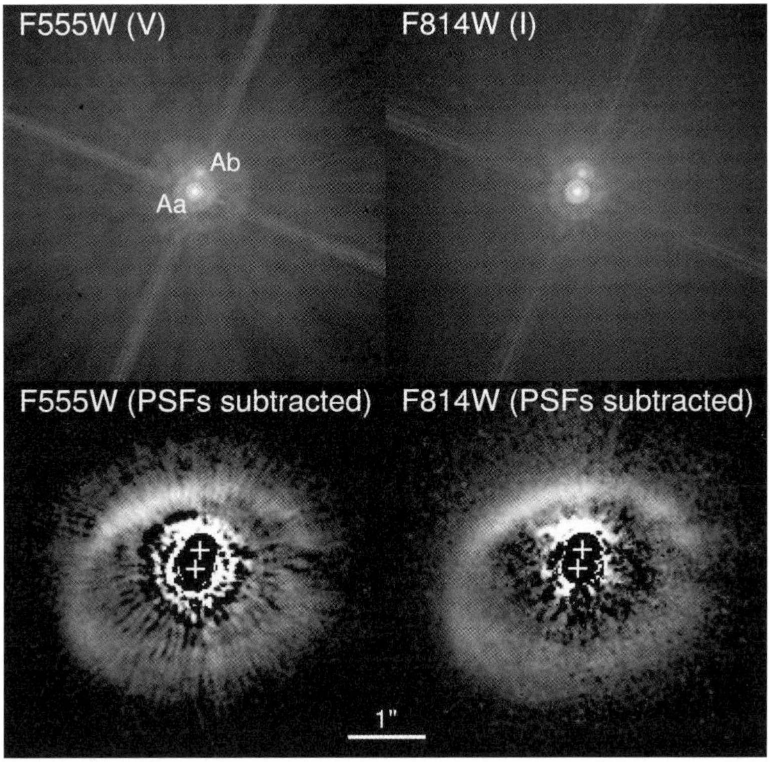

Fig. 9.3. Image of the circumbinary disk of GG Tau (Krist et al., 2005) taken by Advanced Camera for Surveys (ACS) on board of Hubble Space Telescope (HST). The binary separation is 35 AU. The locations of the binary components are marked with crosses.

in and around binary systems are presented, and their connections to models of terrestrial planet formation and water-delivery around single stars are discussed. Finally the chapter ends by discussing the future prospects of research in the field of planets in binaries.

9.2 Dynamical Evolution and Stability

In general, one can consider orbital stability synonymous with the capability of an object in maintaining its orbital parameters (i.e., semimajor axis, eccentricity, and inclination) at all times. In other words, an object is stable if small variations in its orbital parameters do not progress exponentially, but instead vary sinusoidally. Instability occurs when the perturbative forces create drastic changes in the time variations of these parameters and result in either the ejection of the object from the system (i.e., leaving the system's gravitational field), or its collision with other bodies.

The concept of stability, as explained above, although simple, has been defined differently by different authors. A review by Szebehely in 1984 lists 50 different definitions for the stability of an object in a multi-body system (Szebehely, 1984). For instance, Harrington (1977) considered the orbit of an object stable if, while numerically integrating the object's orbit, its semimajor axis and orbital eccentricity would not undergo secular changes. Szebehely (1980), and Szebehely & McKenzie (1981), on the other hand, considered integrals of motion and curves of zero velocity to determine the stability of a planet in and around binary stars. In this chapter, the values of the orbital eccentricity of an object and its semimajor axis are used to evaluate its stability. An object is considered stable if, for the duration of the integration of its orbit, the value of its orbital eccentricity stays below unity, it does not collide with other bodies, and does not leave the gravitational field of its host star.

The study of the stability of a planetary orbit in dual stars requires a detailed analysis of the dynamical evolution of a three-body system. Such an analysis itself is dependent upon the type of the planetary orbit. In general, a planetary-class object may have three types of orbit in and around a binary star system. Szebehely (1980) and Dvorak (1982) have divided these orbits into three different categories. As indicated by Szebehely (1980), a planet may be in an *inner* orbit, where it revolves around the primary star, or it may be in a *satellite* orbit, where it revolves around the secondary star. A planet may also be revolving the entire binary system in which case its orbit is called an *outer* orbit. As classified by Dvorak (1982), on the other hand, a study of the stability of resonant periodic orbits in a restricted, circular, three-body system indicates that a planet may have an S-type orbit, where it revolves around only one of the stars of the binary, or may be in a P-type orbit, where it revolves the entire binary system (Fig. 9.4). A planet may also be in an L-type orbit where it librates in a stable orbit around the L_4 or L_5 Lagrangian points.

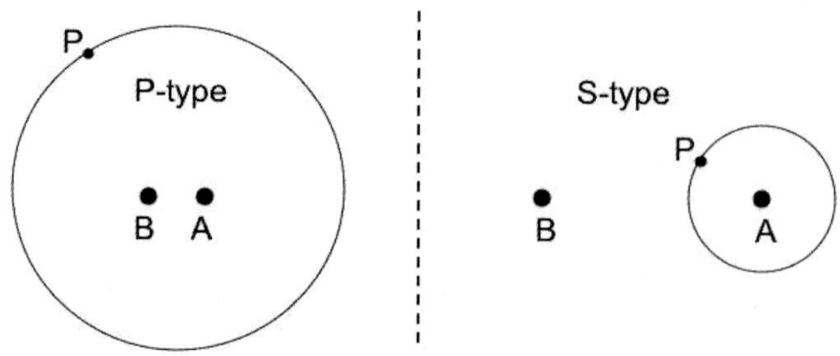

Fig. 9.4. S-type and P-type binary-planetary systems. A and B represent the stars of the binary, and P depicts the planet.

The rest of this section is devoted to a review of the studies of the stability of planets in S-type and P-type orbits. Given that in a binary star system, a planet is subject to the gravitational attraction of two massive bodies (i.e., the stars), it would be important to understand how the process of the formation of planetary objects, and the orbital dynamics of small bodies would be affected by the orbital characteristics of the binary's stellar components. In general, except for some special cases for which analytical solutions may exist, such studies require numerical integrations of the orbits of all the bodies in the system. In the past, prior to the invention of symplectic integrators (Wisdom & Holman, 1991), which enabled dynamicists to extend the studies of the stability of planetary systems to several hundred million years, and before the development of fast computers, the majority of such studies were either limited to those special cases, or were carried out numerically for only a few orbital periods of a binary. Examples of such studies can be found in articles by Graziani & Black (1981), Black (1982), and Pendleton & Black (1983), in which the authors studied the orbital stability of a planet in and around a binary star. By numerically integrating the equations of the motion of the planet, these authors showed that, when the stars of the binary have equal masses, the orbital stability of the planet is independent of its orbital inclination (also see Harrington, 1977). Their integrations also indicated that, when the mass of one of the stellar components is comparable to the mass of Jupiter, planetary orbits with inclinations higher than 50° tend to become unstable.

The invention of symplectic integrators, in particular routines that have been designed specifically for the purpose of integrating orbits of small bodies in dual-star systems[3] (Chambers et al., 2002), have now enabled dynamicists to extend studies of planet formation and stability in dual-star environments to much larger timescales. In the following, the results of such studies are discussed in more detail.

9.2.1 Stability of S-type Orbits

As mentioned above, instability occurs when the perturbative effects cause the semimajor axis and orbital eccentricity of a planet change in such a way that either the object leaves the gravitational field of the system, or it collides with another body. For a planet in an S-type orbit, the gravitational force of the secondary star is the source of these perturbations. That implies, a planet at a large distance from the secondary, i.e., in an orbit closer to its host star, may receive less perturbation from the binary companion and may be able to sustain its dynamical state for a longer time (Harrington, 1977). Since the perturbative effect of the stellar companion varies with its mass, and the eccentricity and semimajor axis of the binary

[3]Symplectic integrators, as they were originally developed by Wisdom & Holman (1991), are not suitable for numerically integrating the orbits of small bodies in the gravitational fields of two massive objects. These integrators have been designed to integrate the orbits of planetary or smaller objects when they revolve around only one massive central body. Recently Chambers et al. (2002) have developed a version of a symplectic integrator that is capable of integrating the motion of a small object in the gravitational fields of two stellar bodies.

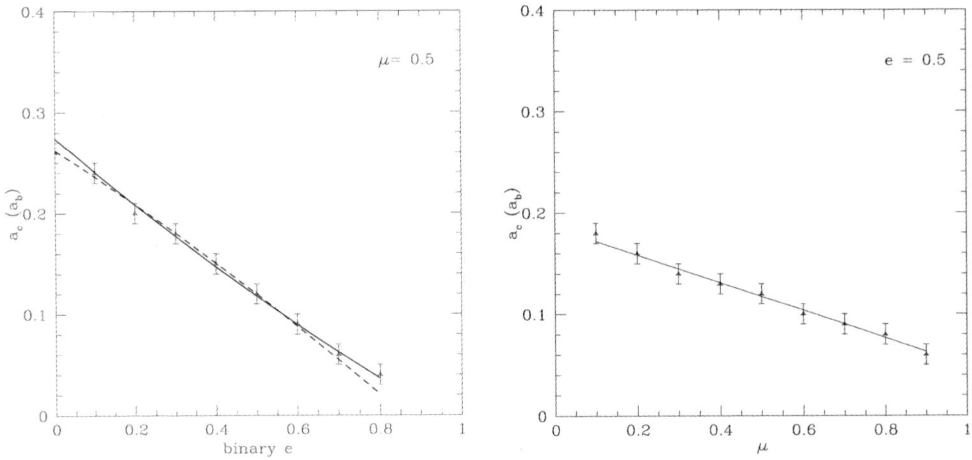

Fig. 9.5. Graphs of the critical semimajor axis (a_c) of an S-type binary-planetary system, in units of the binary semimajor axis (Holman & Wiegert, 1999). The graph on the left shows a_c as a function of the binary eccentricity for an equal-mass binary. The graph on the right corresponds to the variations of the critical semimajor axis of a binary with an eccentricity of 0.5 in term of the binary's mass-ratio. The solid and dashed line on the left panel depict the empirical formulae as reported by Holman & Wiegert (1999) and Rabl & Dvorak (1988), respectively.

(which together determine the closest approach of the secondary to the planet), it is possible to estimate an upper limit for the planet's distance to the star beyond which the orbit of the planet would be unstable. As shown by Rabl & Dvorak (1988) and Holman & Wiegert (1999), the maximum value that the semimajor axis of a planet in an S-type orbit can attain and still maintain its orbital stability is a function of the mass-ratio and orbital elements of the binary, and is given by (Rabl & Dvorak, 1988; Holman & Wiegert, 1999)

$$
\begin{aligned}
a_c/a_b = & (0.464 \pm 0.006) + (-0.380 \pm 0.010)\mu + (-0.631 \pm 0.034)e_b \\
& +(0.586 \pm 0.061)\mu e_b + (0.150 \pm 0.041)e_b^2 + (-0.198 \pm 0.047)\mu e_b^2 \,. \quad (9.1)
\end{aligned}
$$

In this equation, a_c is *critical* semimajor axis, $\mu = M_1/(M_1 + M_2)$, a_b and e_b are the semimajor axis and eccentricity of the binary, and M_1 and M_2 are the masses of the primary and secondary stars, respectively. The $\pm$ signs in Eq. (9.1) define a lower and an upper value for the critical semimajor axis a_c, and set a transitional region that consists of a mix of stable and unstable systems. Such a dynamically *gray* area, in which the state of a system changes from stability to instability, is known to exist in multi-body environments, and is a characteristic of any dynamical system.

Equation (9.1) is an empirical formula that has been obtained by numerically integrating the orbit of a test particle (i.e., a massless object) at different distances from the primary of a binary star (Rabl & Dvorak, 1988; Holman & Wiegert, 1999). Figure 9.5 shows this in more detail. Similar studies have been done by Moriwaki &

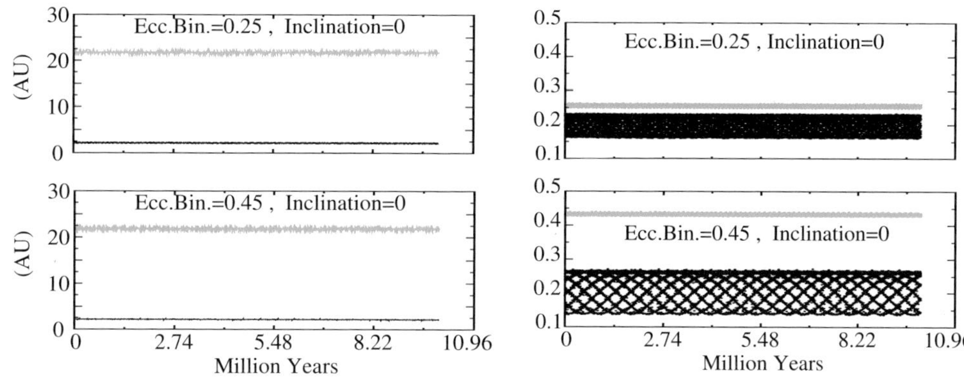

Fig. 9.6. Graphs of the semimajor axes (left) and eccentricities (right) of the giant planet (black) and binary (green) of γ Cephei for different values of the eccentricities of the binary (Haghighipour, 2003). The mass-ratio of the binary is 0.2.

Nakagawa (2004), and Fatuzzo et al. (2006) who obtained critical semimajor axes slightly larger than given by Eq. (9.1).

Since the mass of a Jovian-type planet is approximately three orders of magnitude smaller than the mass of a star, such a test particle approximation yields results that are not only applicable to the stability of giant planets, but can also be used in identifying regions where smaller bodies, such as terrestrial-class objects (Quintana et al., 2002; Quintana & Lissauer, 2006; Quintana et al., 2007) and dust particles (Trilling et al., 2007), can have long term stable orbits[4]. In a recent article, Trilling et al. (2007) utilized Eq. (9.1) and its stability criteria to explain the dynamics of debris disks, and the possibility of the formation and existence of planetesimals in and around 22 binary star systems. By detecting an infrared excess of dust particles, these authors confirmed the presence of stable dust bands, possibly resulting from collision of planetesimals, in S-type orbits in several wide binaries.

The stability of S-type systems has been studied by many authors (Benest, 1988, 1989, 1993, 1996; Wiegert & Holman, 1997; Pilat-Lohinger & Dvorak, 2002; Dvorak et al., 2003, 2004; Pilat-Lohinger et al., 2004; Musielak et al., 2005). In a recent article, Haghighipour (2006) extended such studies to the dynamical stability of the Jupiter-like planet of the γ Cephei planetary system. By numerically integrating the orbit of this object for different values of a_b, e_b and i_p (the orbital inclination of the giant planet relative to the plane of the binary), Haghighipour (2006) has shown that the orbit of this planet is stable for the values of the binary eccentricity within the range $0.2 \le e_b \le 0.45$. Figure 9.6 shows the results of such integrations for a coplanar system with $\mu = 0.2$ and for different values of the binary eccentricity. The initial value of the semimajor axis of the binary was chosen to be

[4]In applying Eq. (9.1) to the stability of dust particles, one has to note that this equation does not take into account the effects of non-gravitational forces such as gas-drag or radiation pressure. The motion of a dust particle can be strongly altered by the effects of these forces.

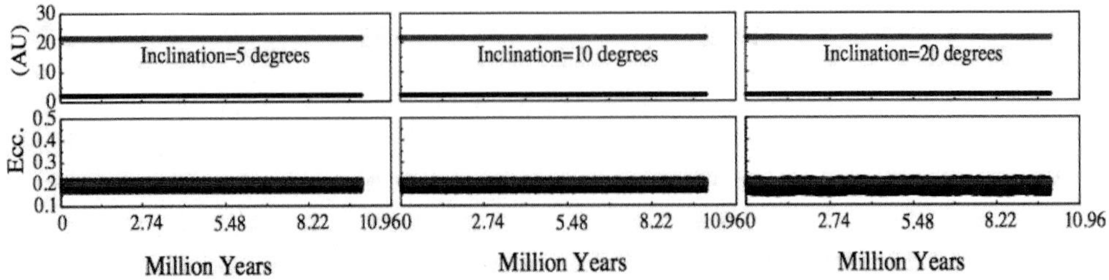

Fig. 9.7. Graphs of the semimajor axes (top) and eccentricities (bottom) of the giant planet (black) and binary (red) of γ Cephei. The initial eccentricity of the binary at the beginning of numerical integration and the value of its mass-ratio were equal to 0.20 (Haghighipour, 2006).

21.5 AU. Integrations also indicated that the binary-planetary system of γ Cephei becomes unstable in less than a few thousand years when the initial value of the binary eccentricity exceeds 0.5.

Interesting results were obtained when the γ Cephei system was integrated for different values of i_p. The results indicated that for the above-mentioned range of orbital eccentricity, the planet maintains its orbit for all values of inclination less than 40°. Figure 9.7 shows the semimajor axes and orbital eccentricities of the system for $e_b = 0.2$ and for i_p=5°, 10°, and 20°. For orbital inclinations larger than 40°, the system becomes unstable in a few thousand years.

Kozai Resonance

An interesting dynamical phenomenon that may occur in an S-type binary, and has also been observed in the numerical simulations of a few of these systems, is the Kozai resonance (Haghighipour, 2003, 2005; Verrier & Evans, 2006; Takeda & Rasio, 2006; Malmberg, Davies & Chambers, 2007). As demonstrated by Kozai (1962), in a three-body system with two massive objects and a small body (e.g., a binary-planetary system), the exchange of angular momentum between the planet and the secondary star, can cause the orbital eccentricity of the planet to reach high values at large inclinations. Averaging the equations of motion of the system over mean anomalies, one can show that in this case, the averaged system is integrable when the ratio of distances are large (the Hill's approximation, Kozai, 1962). The Lagrange equations of motion in this case, indicate that, to the first order of planet's eccentricity, the longitude of the periastron of this object, ω_p, librates around a fixed value. Figure 9.8 shows this for the giant planet of γ Cephei. As shown here, ω_p librates around 90° (Haghighipour, 2003, 2005).

In a Kozai resonance, the longitude of periastron and the orbital eccentricity of the small body (e_p) are related to its orbital inclination as (Innanen et al., 1997)

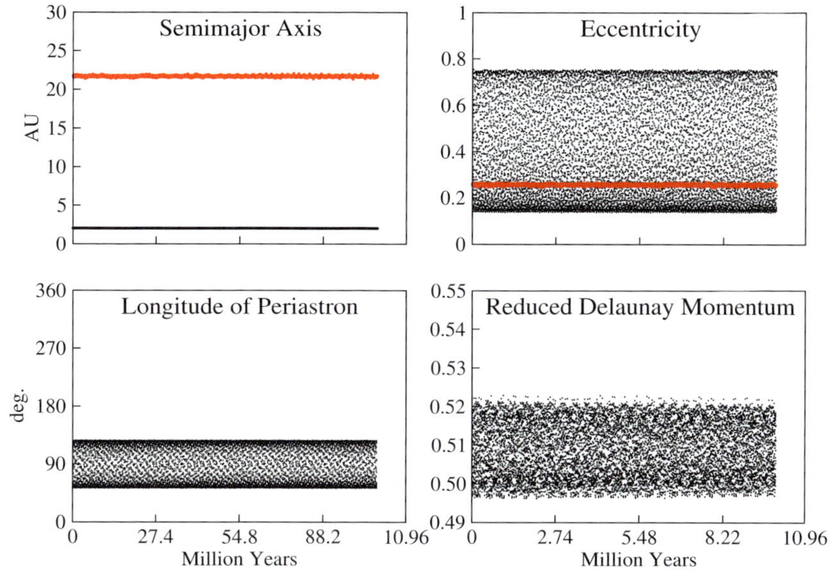

Fig. 9.8. Graphs of the semimajor axis and eccentricity of the giant planet (black) and binary (red) of γ Cephei (top) and its longitude of periastron and reduced Delaunay momentum (bottom) in a Kozai resonance (Haghighipour, 2003, 2005). As expected, the longitude of the periastron of the giant planet oscillates around 90° and its reduced Delaunay momentum is constant.

$$\sin^2\omega_p = 0.4 \csc^2 i_p, \tag{9.2}$$

and

$$(e_p^2)_{\max} = \frac{1}{6}\Big[1 - 5\cos(2i_p)\Big]. \tag{9.3}$$

From Eq. (9.3), one can show that the Kozai resonance may occur if the orbital inclination of the small body is larger than 39.23°. For instance, as shown by Haghighipour (2003, 2005), in the system of γ Cephei, Kozai resonance occurs at $i_p = 60°$. For the minimum value of i_p, the maximum value of the planet's orbital eccentricity, as given by Eq. (9.3), is equal to 0.764. Figure 9.8 also shows that e_p stays below this limiting value at all times.

As shown by Kozai (1962) and Innanen et al. (1997), in a Kozai resonance, the disturbing function of the system, averaged over the mean anomalies, is independent of the longitudes of ascending nodes of the small object (the planet) and the perturbing body (the stellar companion). As a result, the quantity $\sqrt{a(1-e^2)}\,\cos i$ (shown as the "Reduced Delaunay Momentum" in Fig. 9.8) becomes a constant of motion. Since the eccentricity and inclination of the planet vary with time, the fact that the quantity above is a constant of motion implies that the time-variations of these two quantities have similar periods and, at the same time, they vary in such a way that when i_p reaches its maximum, e_p reaches its minimum and vice versa. Figure 9.9 shows this clearly.

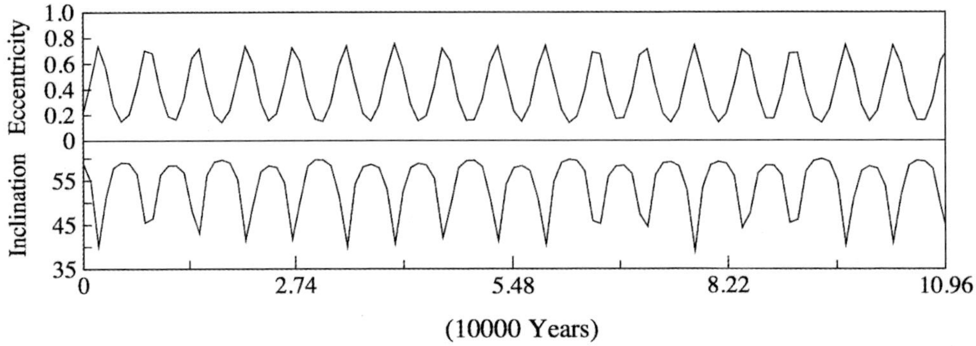

Fig. 9.9. Graphs of the eccentricity and inclination of the giant planet of γ Cephei in a Kozai Resonance (Haghighipour, 2003, 2005). As expected, these quantities have similar periodicity and are 180° out of phase.

9.2.2 Stability of P-type Orbits

Numerical simulations have also been carried out for the stability of P-type orbits in binary-planetary systems (Ziglin, 1975; Szebehely & McKenzie, 1981; Dvorak, 1984, 1986; Dvorak, Froeschlé, & Froeschlé, 1989; Kubala, Black & Szebehely, 1993; Holman & Wiegert, 1999; Broucke, 2001; Pilat-Lohinger, Funk & Dvorak, 2003; Musielak et al., 2005). Similar to S-type orbits, in order for a P-type planet to be stable, it has to be at a safe distance from the two stars so that it would be immune from their perturbative effects. That is, planets at large distances from the center of mass of a binary will have a better chance of being stable. This distance, however, cannot be too large because at very large distances, other astronomical effects, such as galactic perturbation, and perturbations due to passing stars, can render the orbit of a planet unstable.

To determine the critical value of the semimajor axis of a P-type planet in a stable orbit, preliminary attempts were made by Dvorak (1984), who numerically integrated the orbit of a circumbinary planet in a circular orbit around an eccentric binary system and showed that planets at distances 2-3 times the separation of the binary have stable orbits. Subsequent studies by Dvorak (1986), Dvorak, Froeschlé, & Froeschlé (1989), and Holman & Wiegert (1999) complemented Dvorak's results of 1984 and showed that the orbit of a P-type planet will be stable as long as the semimajor axis of the planet stays larger than the critical value given by (Fig. 9.10)

$$\begin{aligned} a_c/a_b &= (1.60 \pm 0.04) + (5.10 \pm 0.05)e_b + (4.12 \pm 0.09)\mu \\ &\quad + (-2.22 \pm 0.11)e_b^2 + (-4.27 \pm 0.17)e_b\mu + (-5.09 \pm 0.11)\mu^2 \\ &\quad + (4.61 \pm 0.36)e_b^2\mu^2 \,. \end{aligned} \tag{9.4}$$

Similar to Eq. (9.1), Eq. (9.4) represents a transitional region with a lower boundary below which the orbit of a P-type planet will be certainly unstable, and an upper

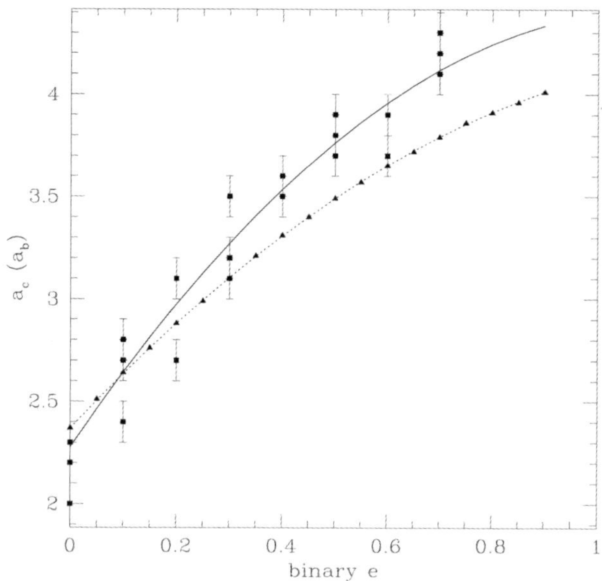

Fig. 9.10. Critical semimajor axis as a function of the binary eccentricity in a P-type system (Holman & Wiegert, 1999). The squares correspond to the result of stability simulations by Holman & Wiegert (1999) and the triangles represent those of Dvorak, Froeschlé, & Froeschlé (1989). The solid line corresponds to Eq. (9.4). As indicated by Holman & Wiegert (1999), the figure shows that at outer regions, the stability of the system fades away.

boundary beyond which the orbit of the planet will be stable. The *mixed zone* between these two boundaries represents a region where a planet, depending on its orbital parameters, and the orbital parameters and the mass-ratio of the binary, may or may not be stable. Recently, by applying the stability criteria of Eq. (9.4) to their observational results, Trilling et al. (2007) have confirmed the presence of stable dust bands, possibly resulting from collision of planetesimals, around close binary star systems.

A dynamically interesting feature of the stable region around the stars of a binary is the appearance of islands of instability. As shown by Holman & Wiegert (1999) islands of instability may develop beyond the inner boundary of the *mixed zone*, which correspond to the locations of (n : 1) mean-motion resonances. The appearance of these unstable regions have been reported by several authors under various circumstances (Hénon & Guyot, 1970; Dvorak, 1984; Rabl & Dvorak, 1988; Dvorak, Froeschlé, & Froeschlé, 1989). Extensive numerical simulations would be necessary to determine whether the overlapping of these resonances would result in stable P-type binary-planetary orbits.

9.3 Planet Formation in Binaries

Despite a wealth of articles on planets in binary star systems, the process of the formation of these objects is still poorly understood. The current theories of planet formation focus only on the formation of planets in a circumstellar disk around a single star, and their extensions to binary environments are limited to either the Sun-Jupiter system, where the focus is on the effect of Jupiter on the formation of inner planets of our solar system (Heppenheimer, 1974, 1978; Drobyshevski, 1978; Diakov & Reznikov, 1980; Whitmire et al., 1998; Kortenkamp, Wetherill & Inaba, 2001), or binaries resembling some of extrasolar planets, in which the secondary star has a mass in the brown dwarf regime (Whitmire et al., 1998). Although attempts have been made to extend such studies to binaries with comparable-mass stellar components (Marzari & Scholl, 2000; Nelson, 2000; Barbieri, Marzari & Scholl, 2002; Quintana et al., 2002; Lissauer et al., 2004), the extent of the applicability of the results of these studies has been only to hypothetical cases since, until recently, there had been no observational evidence on the existence of such binary-planetary systems.

In general, it is believed that planet formation proceeds through the following four stages (Fig. 9.11):

1) coagulation of dust particles and their growth to centimeter-sized objects,

2) collisional growth of centimeter-sized particles to kilometer-sized bodies (planetesimals),

3) formation of Moon- to Mars-sized protoplanets (also known as planetary embryos) through the collision and coalescence of planetesimals, and

4) collisional growth of planetary embryos to terrestrial-sized objects.

The latter is a slow process that may take a few hundred million years. During the first few million years of this process, at larger distances from the star, planetesimals and planetary embryos may form planetary cores several times more massive than Earth, and may proceed to form giant planets.

In a binary star system with a moderate to small separation, the secondary star will have significant effects on the efficiency of each of these processes. As shown by Boss (2006), a binary companion can alter the structure of a planet-forming nebula, and create regions where the densities of the gas and dust are locally enhanced (Fig. 9.12). Also, as shown by Artymowicz & Lubow (1994), and Pichardo, Sparke & Aguilar (2005), a stellar component on an eccentric orbit can truncate the circumprimary disk of embryos to smaller radii and remove material that may be used in the formation of terrestrial planets (Fig. 9.13). As a result, it used to be believed that circumstellar disks around the stars of a binary may not be massive enough to form planets. However, observations by Mathieu (1994), Akeson, Koerner & Jensen (1998), Rodriguez et al. (1998), and Mathieu et al. (2000) have indicated that potentially planet-forming circumstellar disks can indeed exist around the stars of a binary system, implying that planet formation in binaries may be as common as around single stars (Fig. 9.14). Among these circumstaller disks, the two well-separated disks of the system L1551 retain the equivalent of approximately 0.03 to 0.06 solar-masses of their original circumstellar materials in a region

Fig. 9.11. The four stages of planet formation.

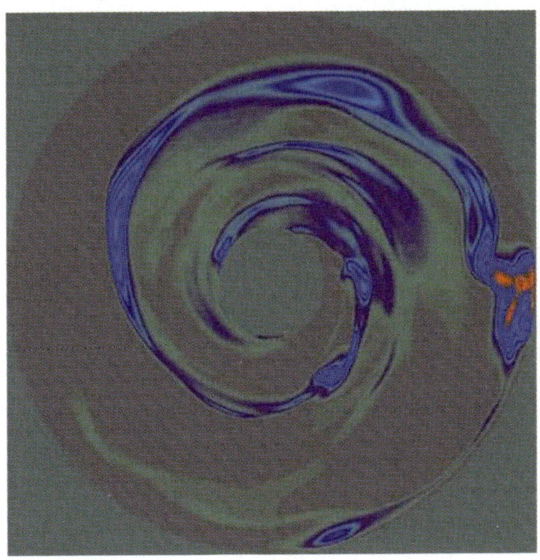

Fig. 9.12. Structure of a circumprimary disk in a double star system. The masses of the primary and secondary stars are 1 and 0.09 solar-masses, respectively. The secondary star is at 50 AU at the top of the figure, and has an eccentricity of 0.5. The figure shows an area of 20 AU around the primary. The structures inside the disk have appeared after 239 years from the beginning of the simulations. The orange structure on the right edge of the graph is an artifact of numerical simulations (Boss, 2006).

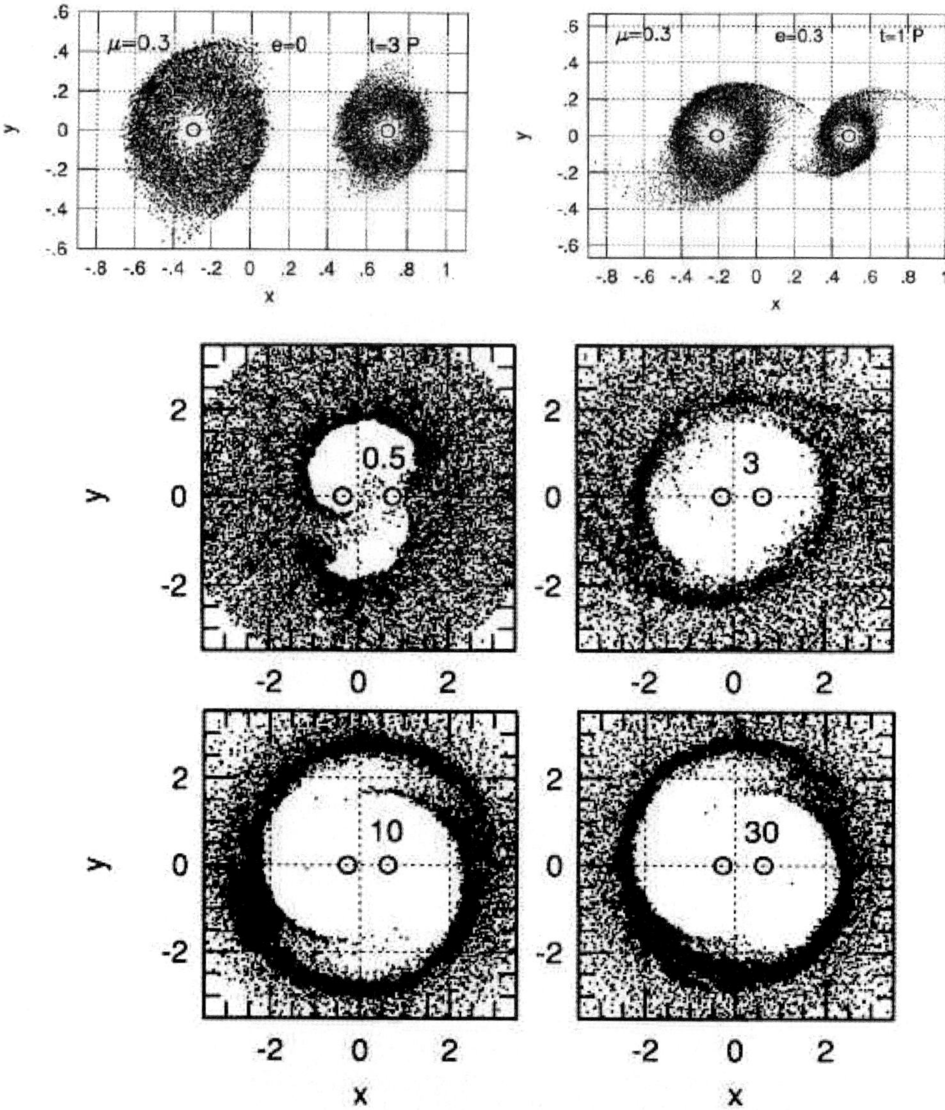

Fig. 9.13. Disk truncation in and around binary systems (Artymowicz & Lubow, 1994). The top graphs show circumstellar disks in a binary with a mass-ratio of 0.3. Note the disk truncation when the eccentricity of the binary is increased from 0 to 0.3. The bottom graphs show a similar effect in a circumbinary disk. The mass-ratio is 0.3 and the binary eccentricity is 0.1. The numbers inside each graph represent the time in units of the binary period. The axes are in units of the binary semimajor axis.

with an outer radius of ∼10 AU (Fig. 9.14, Rodriguez et al., 1998). The masses of these disks are comparable to the minimum solar-mass model of the primordial nebula of our Solar System (Weidenschilling, 1977; Hayashi, 1981), implying that, planet formation in dual-star systems can begin and continue in the same fashion as around our Sun.

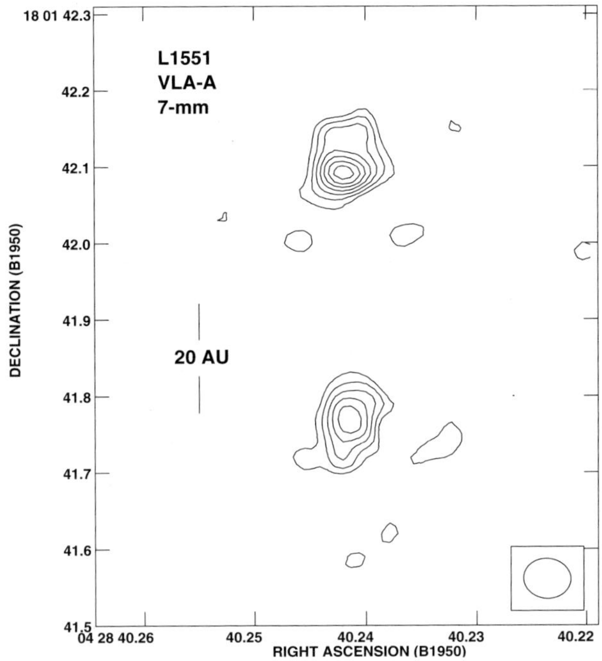

Fig. 9.14. Interferometric observation of the binary system L1551 (Rodriguez et al., 1998). Two compact sources are evident in the map. The separation of the binary is 45 AU and the disk around each core extends to approximately 10 AU.

Despite the observational evidence in support of the existence of planet-forming environments in moderately close binary star systems, the perturbative effect of the binary companion may not always favor planet formation. For instance, as shown by Nelson (2000), giant planet formation cannot proceed through the disk instability mechanism (Boss, 2000) around the primary of a binary star system with separation of $\sim$ 50 AU. Also, when forming planetary embryos, as shown by Heppenheimer (1978), Whitmire et al. (1998), and Thébault et al. (2004), the perturbation of the secondary star may increase the relative velocities of planetesimals and cause their collisions to result in breakage and fragmentation (Fig. 9.15). Results of the studies by these authors suggest that planetesimal accretion will be efficient only in binaries with large separation [50 AU as indicated by Heppenheimer (1978), 26 AU as shown by Whitmire et al. (1998), and 100 AU as reported by Mayer et al. (2005)]. Finally, in a binary star system, the stellar companion may create unstable regions where the building blocks of planets will not maintain their orbits and, as a result, planet formation will be inhibited (Whitmire et al., 1998).

Interestingly, despite all these difficulties, numerical simulations have shown that it may indeed be possible to form giant and/or terrestrial planets in and around a dual-star system. Recent simulations by Boss (2006), and Mayer, Boss & Nelson (2007) indicate that Jupiter-like planets can form around the primary of a binary

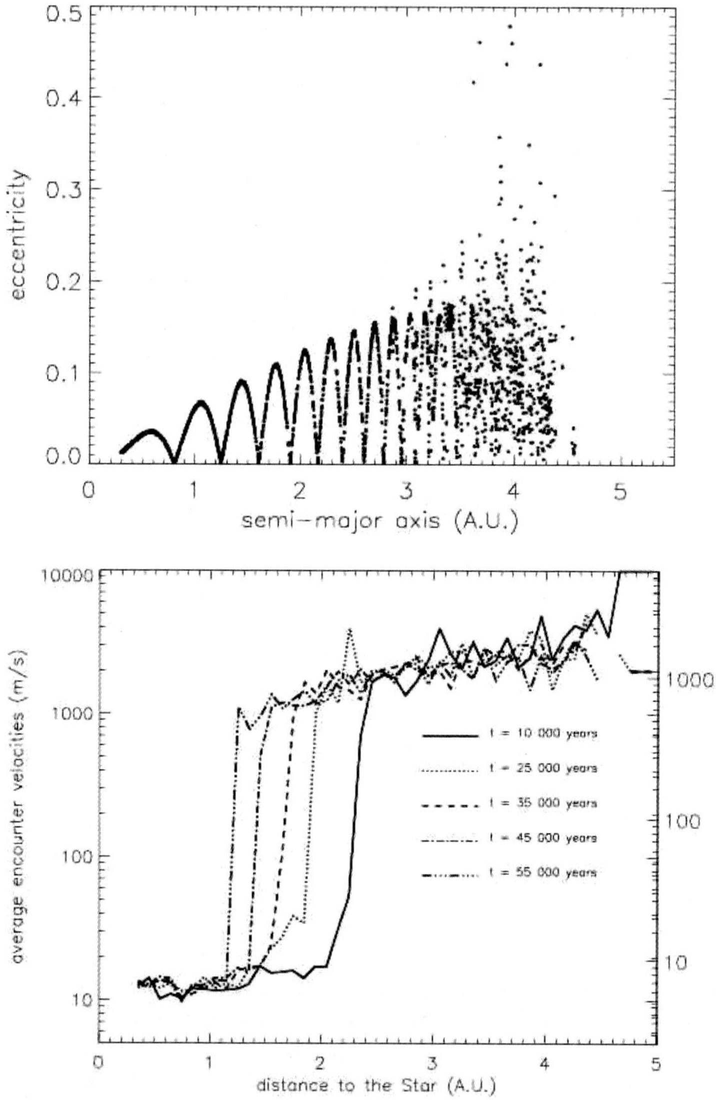

Fig. 9.15. Graphs of the evolution of eccentricity (top) and encounter velocities (bottom) for planetesimals at the region between 0.3 and 5 AU from the primary of γ Cephei (Thébault et al., 2004). The planetesimals disk in the bottom simulation was initially at its truncated radius of 4 AU. As shown here, the perturbative effect of the secondary star increases the eccentricities and relative velocities of these objects.

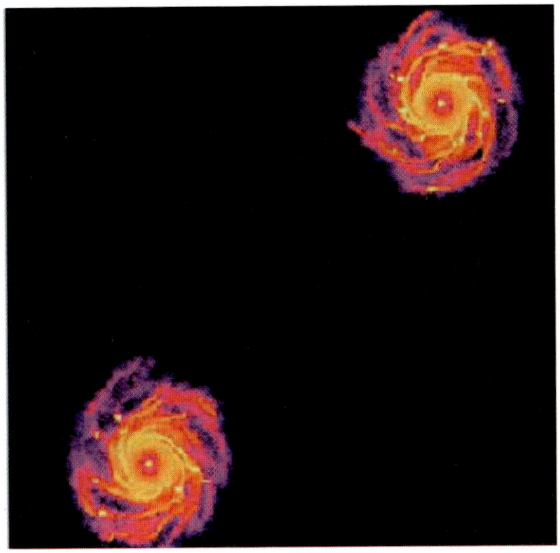

Fig. 9.16. Giant planet formation via disk instability mechanism in a binary system. The separation of the binary is 120 AU and it was initially on a circular orbit. The mass of each disk is 0.1 solar-masses. The snap shot was taken 160 years after the start of the simulations. Figure courtesy of L. Mayer, A. Boss and A. Nelson.

star system via gravitational instability in a marginally unstable circumprimary disk (Fig. 9.16). On the other hand, as shown by Thébault et al. (2004), the core accretion mechanism may also be able to form giant planets around the primary of a binary star. However, as the results of their simulations for planet formation in the γ Cephei system indicate, the semimajor axis of the final gas-giant planet may be smaller than its observed value.

In regard to the formation of terrestrial planets in binary systems, in a series of articles, Quintana and her colleagues integrated the orbits of a few hundred Moon- to Mars-sized objects and showed that terrestrial-class objects can form in and around binaries (Quintana et al., 2002; Lissauer et al., 2004; Quintana & Lissauer, 2006; Quintana et al., 2007). Figure 9.17 shows the results of some of their simulations. As shown here, depending on the mass-ratio of the binary and the initial values of its orbital parameters, in a few hundred million years, terrestrial planets can form around a close (0.01 to 0.1 AU) binary star system.

Quintana and colleagues also studied terrestrial planet formation in binaries with larger separations (Quintana et al., 2007). Figure 9.18 shows the results of their simulations for a binary with a separation of 20 AU. Similar to Fig. 9.17, terrestrial-type objects are formed around the primary of the binary in a few hundred million years. Statistical analysis of their results, as shown in Fig. 9.19, 9.20, and 9.21 indicate that, as expected in binaries with larger perihelia, where disk truncation has been smaller and more planet-forming material is available, terrestrial planet formation is efficient and the number of final terrestrial planets is large. The

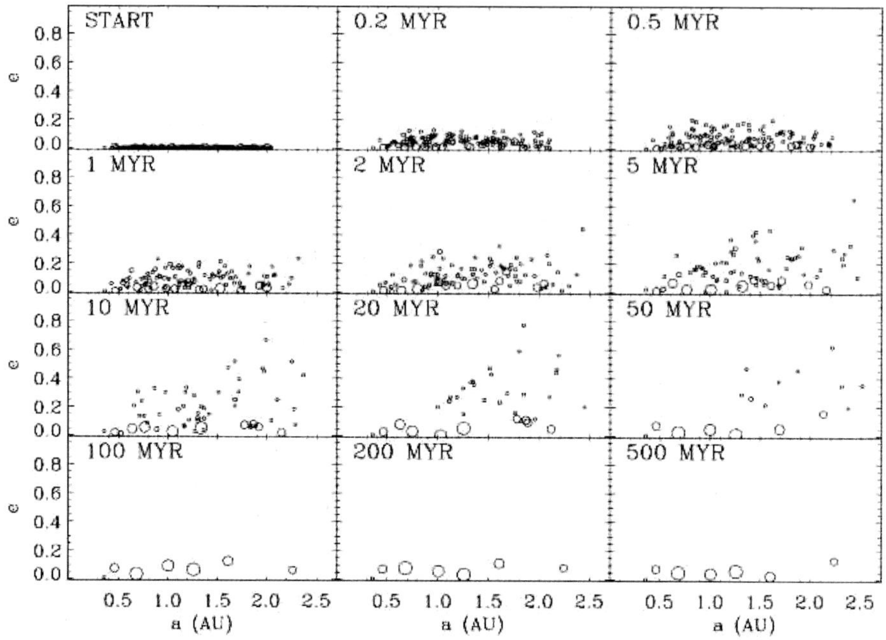

Fig. 9.17. Terrestrial planet formation around a close binary system (Quintana & Lissauer, 2006). The binary is circular and its separation is 0.05 AU. Each star of the binary has a mass of 0.5 solar-masses. A Jupiter-like planet has also been included in the simulation. The circles represent planetary embryos and planetesimals with radii that are proportional to their physical sizes. As shown here, the perturbative effect of the outer giant planet excites the orbits of the bodies at the outer edge of the disk and causes radial mixing as well as truncation. Within the first 100 Myr, several terrestrial-class objects are formed around the binary system.

results of simulations by Quintana et al. (2007) also indicate that in a binary with a periastron distance larger than 10 AU, terrestrial planet formation can proceed efficiently in a region within 2 AU of the primary star. In binaries with periastra smaller than 5 AU, this region may be limited to inside 1 AU (Fig. 9.20).

Despite the destructive role of the binary companion in increasing the relative velocities of planetesimals, which causes their collisions to result in erosion, this efficiency of terrestrial planet formation in binary systems may be attributed to the fact that the effect of the binary companion on increasing the relative velocities of planetesimals can be counterbalanced by dissipative forces such as gas-drag and dynamical friction (Marzari et al., 1997; Marzari & Scholl, 2000). The combination of the drag force of the gas and the gravitational force of the secondary star may result in the alignment of the periastra of planetesimals and increases the efficiency of their accretion by reducing their relative velocities. This is a process that is more effective when the sizes of the two colliding planetesimals are comparable and small. For colliding bodies with different sizes, depending on the size distribution of small

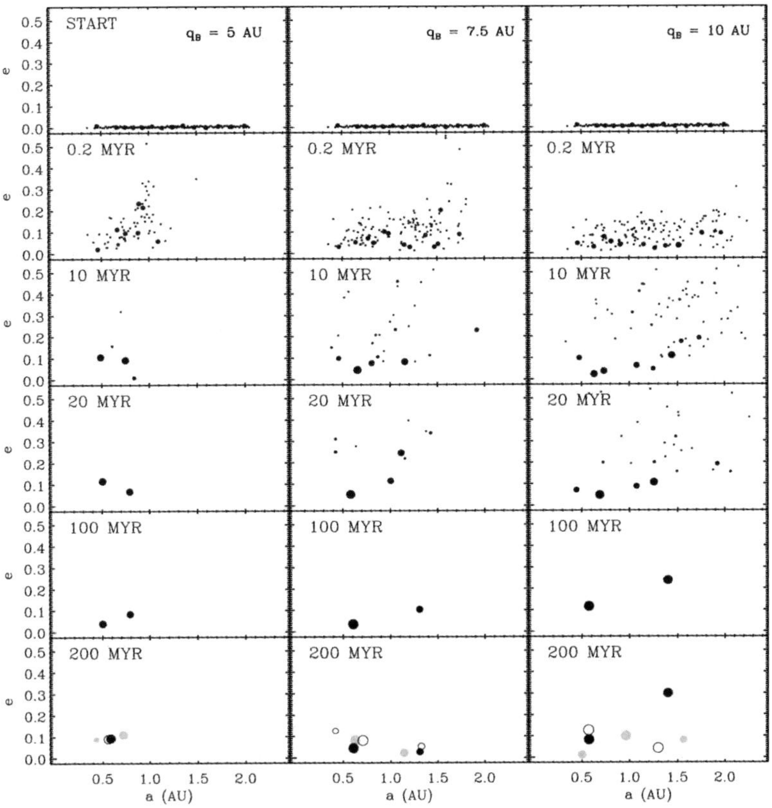

Fig. 9.18. Terrestrial planet formation around the primary of a binary star system (Quintana et al., 2007). The stars of the binary are 0.5 solar-masses with semimajor axis of 20 AU. The eccentricity of the binary is 0.75 (left column), 0.625 (middle column), and 0.5 (right column). As shown here, in each simulation, two terrestrial-type objects are formed after 100 Myr. The last row shows the results of additional simulations of the same systems, with final results showing in black, gray and white, corresponding to different runs. The differences in the final assembly of the planetary system of each simulation are results of the stochasticity of this type of numerical integrations.

objects, and the radius of each individual planetesimal, the process of the alignment of periastra may instead increase the relative velocities of the two objects, and cause their collisions to become eroding (Fig. 9.22, also see Thébault et al., 2006).

Simulations of terrestrial planet formation have also been extended to binaries with larger separations (20-40 AU) that also host a giant planet (Haghighipour & Raymond, 2007). As discussed in the next section, by numerically integrating the orbits of the binary, its giant planet, and a few hundred planetary embryos, these authors have shown that it is possible to form Earth-like objects, with considerable amount of water, in the habitable zone of the primary of a moderately close binary-planetary system.

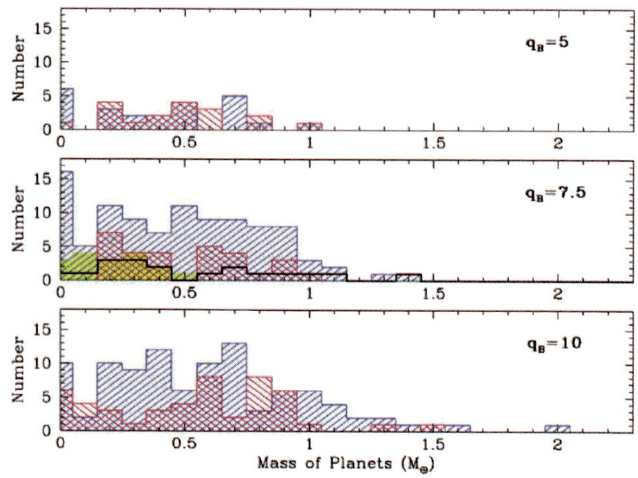

Fig. 9.19. Graphs of the final masses of the terrestrial planets formed in systems of Fig. 9.18. The simulations have been run for three different masses of the binary stars. The red corresponds to simulations in a binary with 0.5 solar-masses stars, the blue represents results in a binary with 1 solar-mass stars, yellow is for a binary with a 1 solar-mass primary and a 0.5 solar-masses secondary, and black represents a binary with a 0.5 solar-masses primary and a 1 solar-mass secondary. These results show that despite the disk truncation in binaries with smaller perihelia, the average masses of the final planets are not significantly altered. (Quintana et al., 2007).

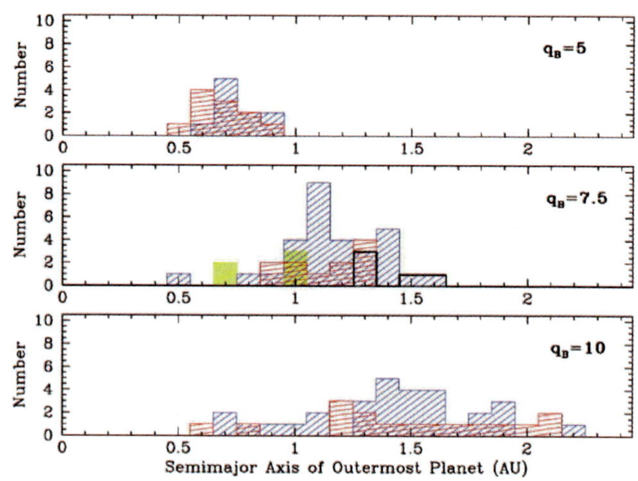

Fig. 9.20. Graphs of the semimajor axis of the outermost planet of the simulations of Fig. 9.18. As shown here, while the outer edge of the disk is affected by the presence of the binary companion, the inner portion of the disk, where terrestrial planets are formed, stays unaffected by this object (Quintana et al., 2007).

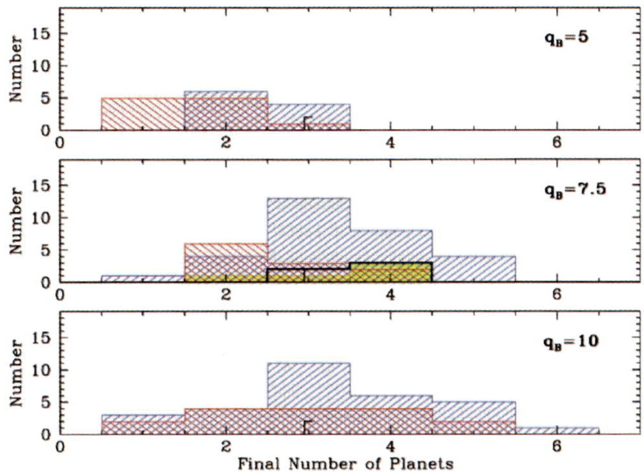

Fig. 9.21. The number of final terrestrial-type planets formed in the binaries of Fig. 9.18. As expected, for a given binary mass-ratio, the number of terrestrial planets increases in systems with larger perihelia. This number also increases when the mass of the binary companion is smaller. (Quintana et al., 2007).

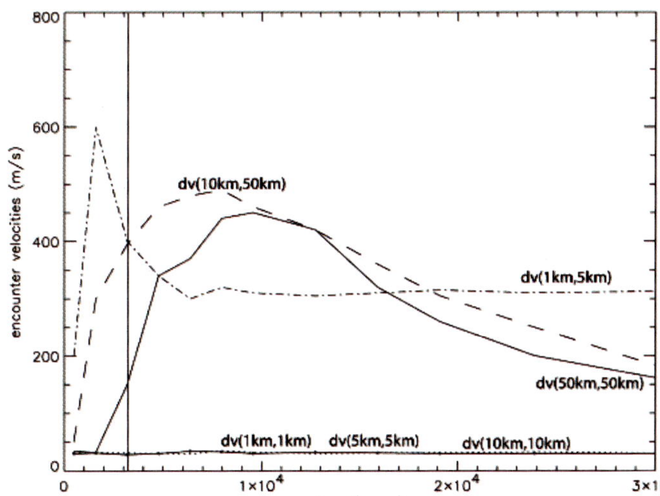

Fig. 9.22. Encounter velocities of planetesimals with different sizes in a binary system with semimajor axis of 10 AU, eccentricity of 0.3, and mass-ratio of 0.5 (Thébault et al., 2006). The simulations include gas-drag. The vertical line shows the time of orbital crossing. As shown here, gas-drag lowers the encounter velocities of smaller equal-size planetesimals through the periastron alignment process.

9.4 Habitability

It is widely accepted that a planet capable of developing life (as we know it), has to be able to continuously maintain liquid water on its surface and in its atmosphere. The capability of a planet in retaining water depends on its size and the processes involving its interior dynamics and atmospheric circulation. It also depends on its orbital parameters (i.e., its semimajor axis and orbital eccentricity) and the brightness of the central star at the location of the planet. While a dynamic interior and atmospheric circulation are necessary for a habitable planet to develop a CO_2 cycle and generate greenhouse effect (which helps the planet to maintain a uniform temperature), a long-term stable orbit, at a right distance from the star, is essential to ensure that the planet will receive the amount of radiation that enables it to maintain liquid water on its surface. These seemingly unrelated characteristics of a potential habitable planet, have strong intrinsic correlations, and combined with the luminosity of the star, determine the system's *habitable zone*.

The inner and outer boundaries of a habitable zone vary with the star's luminosity and the planet's atmospheric circulation models [see Menou & Tabachnik (2003), Jones, Underwood, & Sleep (2005), and Jones, Sleep, & Underwood (2006) for a table of distances of the inner and outer boundaries of the habitable zones of exoplanetary systems]. A conservative estimate of the habitable zone of a star can be made by assuming that its inner edge is located at a distance closer than which water on the surface of the planet evaporates due to a runaway greenhouse effect, and its outer edge is at a distance where, in the absence of CO_2 clouds, runaway glaciation will freeze the water and create permanent ice on the surface of the planet. As shown by Kasting, Whitmire, & Reynolds (1993), such a definition of a habitable zone results in a habitable region between 0.95 AU and 1.15 AU for the Sun (Fig. 9.23). This is a somewhat conservative estimate of the Sun's habitable zone and as noted by Jones, Underwood, & Sleep (2005), the outer edge of this region may, in fact, be farther away (Forget & Pierrehumbert, 1997; Williams & Kasting, 1997; Mischna et al., 2000).

Since the notion of habitability is based on life on Earth, it is possible to calculate the location of the boundaries of the habitable zone of a star by comparing its luminosity with that of the Sun. For a star with the surface temperature T and radius R, the luminosity L is given by

$$L(R,T) = 4\pi\sigma T^4 R^2, \tag{9.5}$$

where σ is Boltzmann constant. Using Eq. (9.5) and the fact that Earth is in the habitable zone of the Sun, the radial distances of the inner and outer edges of the habitable zone of a star can be obtained from (Haghighipour 2006)

$$r = \left(\frac{T}{T_\odot}\right)^2 \left(\frac{R}{R_\odot}\right) r_\odot. \tag{9.6}$$

The quantities $T_\odot$ and $R_\odot$ in Eq. (9.6) are the surface temperature and radius of the Sun, respectively, and $r_\odot$ represent the radial distance of Earth from the Sun.

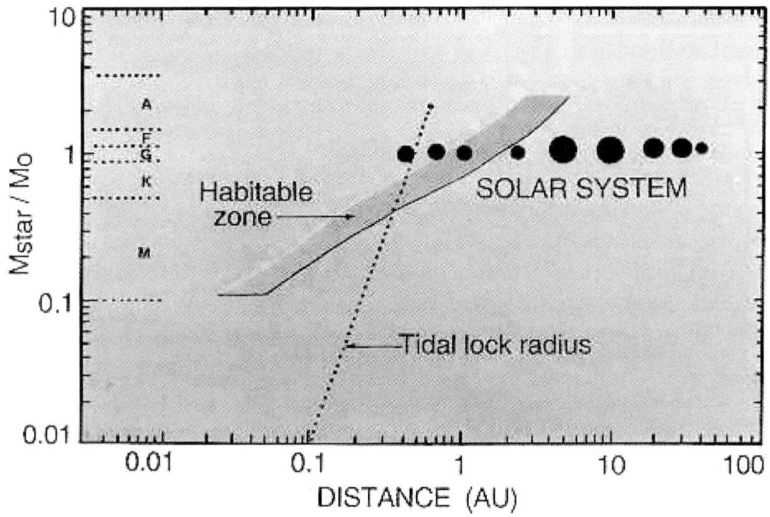

Fig. 9.23. Habitable zone (Kasting, Whitmire, & Reynolds, 1993).

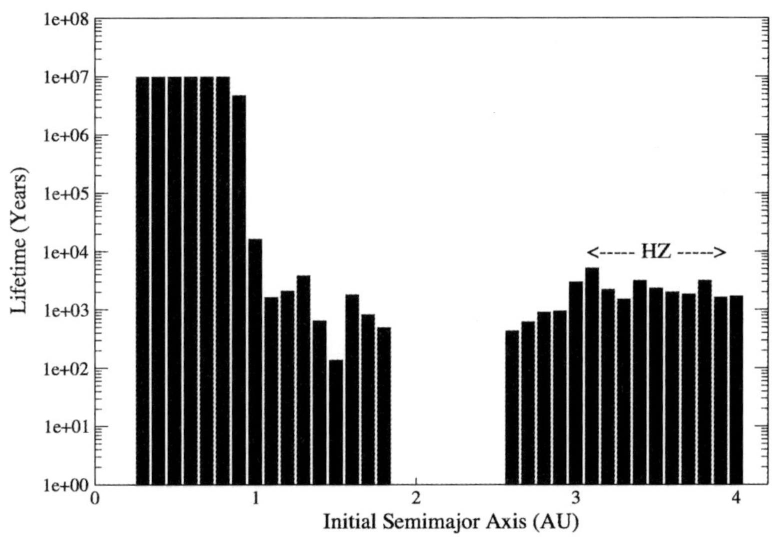

Fig. 9.24. Graph of the lifetime of an Earth-size object in a circular orbit around the primary of γ Cephei. The habitable zone of the primary has been indicated by HZ. No planet was placed in the region between the apastron and periastron distances of the giant planet of the system. As shown here, only Earth-size planets close to the primary star maintain their orbits for long times (Haghighipour, 2006).

Equation (9.6) implies that a habitable zone can be defined as a region around a star where an Earth-like planet can receive the same amount of radiation as Earth receives from the Sun, so that it can develop and maintain similar habitable conditions as those on Earth.

As mentioned above, the orbit of a potential habitable planet in the habitable zone of a star has to be stable over long durations of time. As shown in Sect. 9.2, the stability of the orbit of a planet in a binary system is strongly affected by the orbital motion of the binary companion. In binary systems where the primary hosts other planetary bodies (e.g., giant planets), the dynamics of a habitable planet will also be affected by the gravitational perturbations of these objects. It is therefore important to determine under what conditions a terrestrial-class object will have a long-term stable orbit in the habitable zone of a binary system, prior to constructing a theory for the formation of Earth-like planets in such environments.

Since a terrestrial-class planet is approximately two orders of magnitude less massive than a Jovian-type object, it will not have a significant effect on the motion of the stars and the giant planets of a binary system. Therefore, as explained in Sects. 9.2.1 and 9.2.2, if a binary system does not contain a Jupiter-like planet, any dynamical criterion that is obtained for the stability or instability of a general planetary body, can also be applied to the dynamics of a terrestrial planet. Equation (9.1) and the stability conditions presented by Fig. 9.5 can be used to determine the long-term stability of an Earth-like object in a binary system.

If a binary contains giant planets, however, the situation is different. The gravitational perturbations of the latter objects will have significant effects on the motion and dynamics of terrestrial planets in the system. As shown by Haghighipour (2006), an Earth-size object, in a region between the giant planet and the primary of γ Cephei binary system, can maintain its stability only in orbits close to the primary star and outside the influence zone[5] of the giant body. Integrating the equations of motion of a full four-body system, this author has shown that an Earth-like planet will not be able to sustain a stable orbit in the habitable zone of γ Cephei's primary star (Fig. 9.24). However, it is possible for such an object to have a stable orbit when $0.3 \leq a_T \leq 0.8$ AU, $0° \leq i_T = i_p \leq 10°$, and $e_b \leq 0.4$. Here a_T represents the semimajor axis of the terrestrial planet and i_T is its orbital inclination with respect to the plane of the binary.

As mentioned above, the instability of an Earth-like planet in the habitable zone of γ Cephei can be attributed to the interaction between this object and the giant planet of the system. When the Earth-like planet is outside the giant planet's influence zone (e.g., at closer distances to the primary star) it can maintain its orbit for several hundred million years. Figure 9.24 suggests that, in order for a binary-planetary system to be habitable, its habitable zone has to be outside the influence region of its giant planet. In an S-type binary-planetary system, this implies a primary with a close-in habitable region. In a recent article, Haghighipour

[5] The influence zone of a planetary object with a mass m_p around a star with a mass M is defined as the region between $3R_H - a_p(1-e_p)$ and $3R_H + a_p(1+e_p)$, where a_p is the semimajor axis of the planet, and $R_H = a_p(m_p/3M)^{1/3}$ is its Hill radius

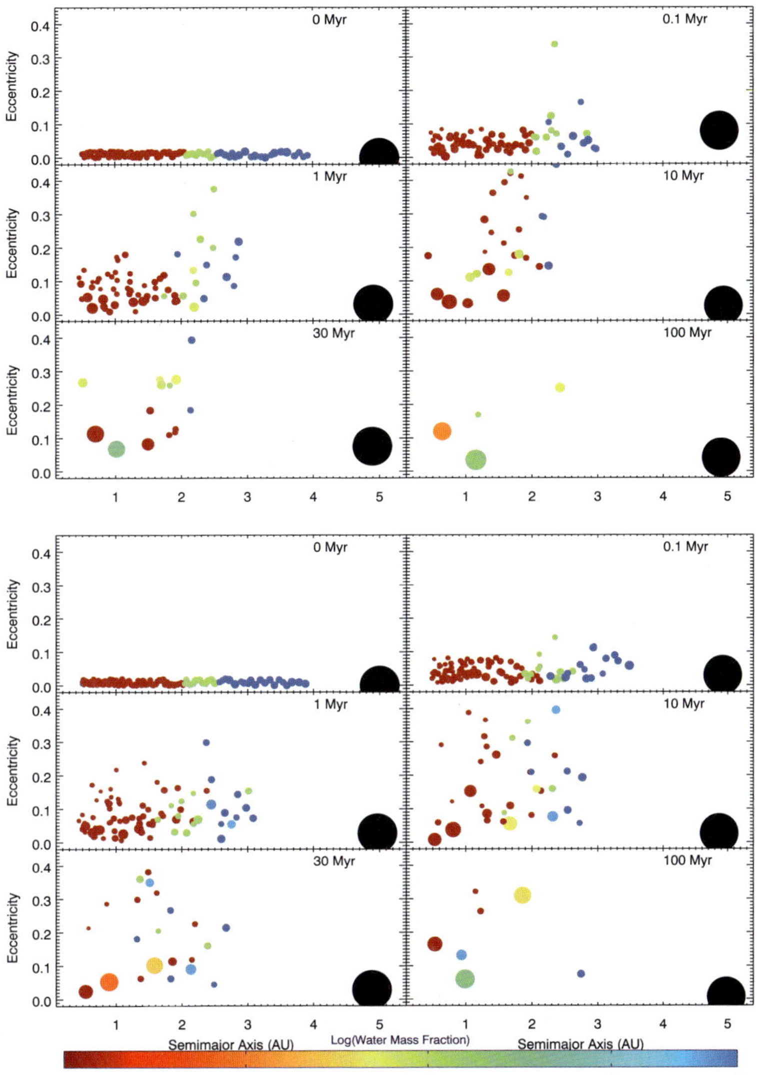

Fig. 9.25. Formation of Earth-like planets in a binary-planetary system. The top panel shows simulations in a binary with a separation of 30 AU, eccentricity of 0.2, and stellar components of 1 solar-mass. As shown here, an Earth-like planet (1.17 Earth-masses) with a water to mass ratio of 0.00164, is formed in the habitable zone of the primary at 1.16 AU, with an eccentricity of 0.02. The bottom panel shows the formation of an Earth-like object in a binary with a solar-mass primary and a 1.5 solar-masses secondary. The separation of the binary in this case is 30 AU, the mass of the Earth-like planet is 0.95 Earth-masses and its water to mass ratio is 0.00226. The semimajor axis of this planet and its orbital eccentricity are equal to 0.99 AU and 0.07, respectively. For the sake of comparison, the Sun's habitable zone is approximately at 0.95-1.15 AU, Earth's orbital eccentricity is 0.017, and Earth's water to mass ratio is $\sim$ 0.001 (Haghighipour & Raymond, 2007).

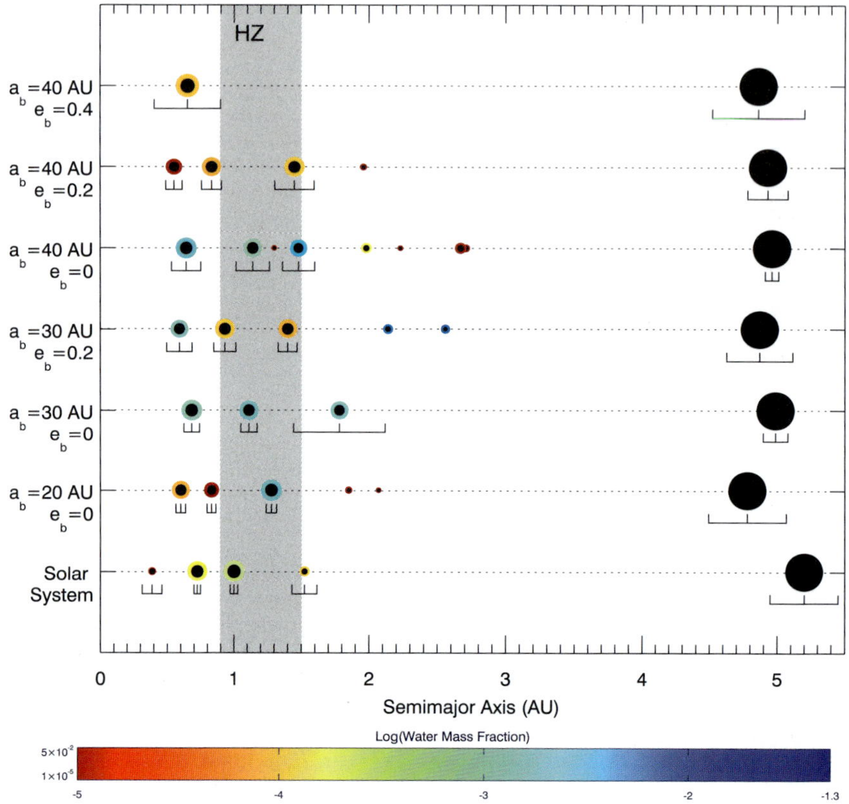

Fig. 9.26. Formation of Earth-like planets in different binary-planetary systems. As shown here, for a given binary mass-ratio, the delivery of water to terrestrial regions becomes less efficient as the periastron distance of the binary becomes smaller (Haghighipour & Raymond, 2007).

& Raymond (2007) have studied the habitability of such a system. By considering a binary with a Sun-like primary star and a Jupiter-sized planet in a circular orbit at 5 AU, and by adopting the model of Morbidelli et al. (2000), which is based on the assumption that water-carrying objects, in the Sun's asteroid belt, were the primary source of the delivery of water to Earth, these authors integrated the orbits of a few hundred protoplanetary (Moon- to Mars-sized) objects, and showed that it is indeed possible to form Earth-sized planets, with substantial amounts of water, in the habitable zone of the primary star (Fig. 9.25). As shown by these authors, the mass and orbital parameters of the secondary star play important roles in the radial mixing of protoplanetary objects and the delivery of water to the habitable zone of the primary star. The giant planet of the system also plays the important role of transferring angular momentum from the secondary star to the disk of protoplanets, and enhancing the radial mixing of these objects. As shown in Fig. 9.26, water delivery is less efficient in binaries with smaller perihelia since

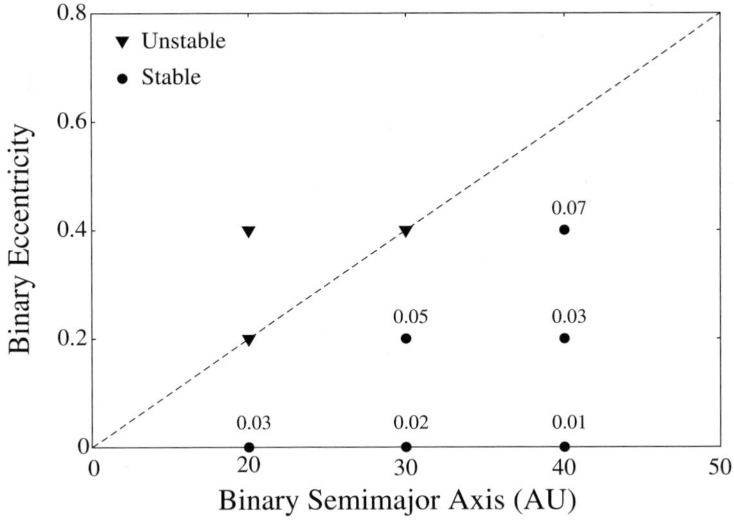

Fig. 9.27. The (e_b, a_b) parameter-space of an equal-mass binary-planetary system. Circles correspond to binaries with initial parameters chosen from Fig. 9.25, in which habitable planets were formed. Triangles represent systems in which the giant planet became unstable. The numbers associated with each circle represent the mean eccentricity of the giant planet of the system at the end of the simulation. As shown here, moderately close binaries with lower eccentricities (larger periastra) are more suitable places for the formation of habitable planets (Haghighipour & Raymond, 2007).

in such systems, the close approach of the binary companion to the giant planet increases its eccentricity, which in turn results in stronger interaction between this object and the disk of protoplanets, causing them to become unstable in a very short time. The results of the simulations by Haghighipour & Raymond (2007) indicate that binary-planetary systems with giant planets at 5-10 AU, and binary perihelion distances of approximately 20 AU to 25 AU, will be more efficient in forming and hosting habitable planets (Fig. 9.27).

9.5 Future Prospects

The discovery of planets in binary star systems is one the interesting surprises of modern astronomy. Despite the long history of the study of planets in such environments, the recent detection of planets in moderately close binaries has confronted astronomers with many new challenges. Many aspects of the formation process of these planets are still unresolved, and questions regarding their frequency and detection techniques demand more detailed investigation.

The habitability of binary systems is also an open question. Although recent simulations of the late stage of terrestrial planet formation in binary systems have

Fig. 9.28. Artistic rendition of the view from the moon of a giant planet in a triple star system. Figure from JPL-Caltech/NASA.

indicated that water-carrying planets can form in the habitable zone of a binary-planetary system, more studies are necessary to understand how protoplanetary objects can develop and evolve in such an environment. Such studies have implications for investigating the habitability of extrasolar planets, and tie directly to several of near-future NASA missions, in particular, the space mission Kepler.

References

Abt, H. A. 1979, The Frequencies of Binaries on the Main Sequence, *Astron. J.*, **84**, 1591

Akeson, R. L., Koerner, D. W. & Jensen, E. L. N. 1998, A Circumstellar Dust Disk around T Tauri N: Subarcsecond Imaging at lambda = 3 Millimeters, *Astrophys. J.*, **505**, 358

Artymowicz, P. & Lubow, S. H. 1994, Dynamics of Binary-Disk Interaction. 1: Resonances and Disk Gap Sizes, *Astrophys. J.*, **421**, 651

Barbieri, M., Marzari, F. & Scholl, H. 2002, Formation of Terrestrial Planets in Close Binary Systems: The Case of Alpha Centauri A, *Astron. Astrophys.*, **396**, 219

Benest, D. 1988, Planetary Orbits in the Elliptic Restricted Problem. I-The Alpha Centauri System, *Astron. Astrophys.*, **206**, 143

Benest, D. 1989, Planetary Orbits in the Elliptic Restricted Problem. II-The Sirius System, *Astron. Astrophys.*, **223**, 361

Benest, D. 1993, Stable Planetary Orbits Around One Component in Nearby Binary Stars. II, *Celest. Mech.*, **56**, 45

Benest, D. 1996, Planetary Orbits in the Elliptic Restricted Problem. III. The η Coronae Borealis System, *Astron. Astrophys.*, **314**, 983

Black, D. C. 1982, A Simple Criterion for Determining the Dynamical Stability of Three-Body Systems, *Astron. J.*, **87**, 1333

Bodenheimer, P., Hubickyj, O. & Lissauer, J. J. 2000, Models of the in Situ Formation of Detected Extrasolar Giant Planets, *Icarus*, **143**, 2

Boss, A. P. 2000, Possible Rapid Gas Giant Planet Formation in the Solar Nebula and Other Protoplanetary Disks, *Astrophys. J.*, **536**, L101

Boss, A. P. 2006, Gas Giant Protoplanets Formed by Disk Instability in Binary Star Systems, *Astrophys. J.*, **641**, 1148

Broucke, R. A. 2001, Stable Orbits of Planets of a Binary Star System in the Three-Dimensional Restricted Problem, *Celest. Mech. Dynamic. Astron.*, **81**, 321

Campbell, B., Walker, G. A. H. & Yang, S. 1988, A Search for Substellar Companions to Solar-Type Stars, *Astrophys. J.*, **331**, 902

Chambers, J. E., Quintana, E. V., Duncan, M. J. & Lissauer, J. J., 2002, Symplectic Integrator Algorithms for Modeling Planetary Accretion in Binary Star Systems, *Astrophys. J.*, **123**, 2884

Diakov, B. B. & Reznikov, B. I. 1980, Computer Simulation of Planet Formation in a Binary Star System Terrestrial Planets, *Moon. Planet.*, **23**, 429

Drobyshevski, E. M. 1978, The Origin of the Solar System - Implications for Transneptunian Planets and the Nature of the Long-Period Comets, *Moon. Planet.*, **18**, 145

Duquennoy, A. & Mayor, M. 1991, Multiplicity Among Solar-Type Stars in the Solar Neighborhood. II - Distribution of the Orbital Elements in an Unbiased Sample, *Astron. Astrophys.*, **248**, 485

Dvorak, R. 1982, *Österr. Akad. d. Wiss., Math-Nat. Klasse* n **191**, 423

Dvorak, R. 1984, Numerical Experiments on Planetary Orbits in Double Stars, *Celest. Mech.*, **34**, 369

Dvorak, R. 1986, Critical Orbits in the Elliptic Restricted Three-Body Problem, *Astron. Astrophys.*, **167**, 379

Dvorak, R., Froeschlé, Ch. & Froeschlé, Ci. 1989, Stability of Outer Planetary Orbits (P-Types) in Binaries, *Astron. Astrophys.*, **226**, 335

Dvorak, R., Pilat-Lohinger, E., Funk, B. & Freistetter, F. 2003, Planets in Habitable Zones:. A Study of the Binary Gamma Cephei, *Astron. Astrophys.*, **398**, L1

Dvorak, R., Pilat-Lohinger, E., Bois, E., Funk, B., Freistetter, F. & Kiseleva-Eggleton, L. 2004, Planets in Double Stars: the γ Cephei System, *RevMexAA (Series de Conferencias)*, **21**, 222

Fatuzzo, M., Adams, F. C., Gauvin, R. & Proszkow, E. M. 2006, A Statistical Stability Analysis of Earth-like Planetary Orbits in Binary Systems, *Pub. Astron. Soc. Pacific.*, **118**, 1510

Forget, F. & Pierrehumbert, R. T. 1997, Warming Early Mars with Carbon Dioxide Clouds That Scatter Infrared Radiation, *Science*, **278**, 1273

Fuhrmann, K. 2004, *Astron. Nachr.*, **325**, 3

Graziani, F. & Black, D. C. 1981, Orbital Stability Constraints on the Nature of Planetary Systems, *Astrophys. J.*, **251**, 337

Griffin, R. F., Carquillat, J. M. & Ginestet, N. 2002, Spectroscopic Binary Orbits from Photoelectric Radial Velocities, Paper 163: HD 213503/4 and HD 220636/7 with a Conjecture Concerning γ Cephei, *The Observatory*, **122**, 90

Haghighipour, N. 2004, On the Dynamical Stability of γ Cephei; An S-type Binary Planetary System, In: *AIP Conference Proceedings 713: The Search for Other Worlds*, ed by S. S. Holt, & D. Deming, Melville, New York, pp 269-272

Haghighipour, N. 2005, Dynamical Stability and Habitability of γ Cephei Binary-Planetary System, astro-ph/0509659

Haghighipour, N. 2006, Dynamical Stability and Habitability of the γ Cephei Binary-Planetary System, *Astrophys. J.*, **644**, 543

Haghighipour, N. & Raymond, S. N. 2007, Habitable Planet Formation in Binary-Planetary Systems, *Astrophys. J.*, **666**, 436

Hagle, J. & Dvorak, R. 1988, An Analytical Study of Stable Planetary Orbits in the Circular Restricted Problem, *Celest. Mech.*, **42**, 355

Harrington, R. S. 1977, Planetary Orbits in Binary Stars, *Astron. J.*, **82**, 753

Hatzes, A. P., Cochran, W. D., Endl, M., McArthur, B., Paulson, D. B., Walker, G. A. H., Campbell, B. & Yang, S. 2003, A Planetary Companion to γ Cephei A, *Astrophys. J.*, **599**, 1383

Hayashi, C. 1981, Structure of the Solar Nebula, Growth and Decay of Magnetic Fields and Effects of Magnetic and Turbulent Viscosities on the Nebula, *Prog. Theor. Phys. Suppl.*, **70**, 35

Hénon, M. & Guyot, M. 1970, Stability of Periodic Orbits in the Restricted Problem, In: *Periodic Orbits, Stability and Resonances*, ed by G. E. O. Giacaglia, D. Reidel Pub. Comp., Netherlands, pp 349

Heppenheimer, T. A. 1974, Outline of a Theory of Planet Formation in Binary Systems, *Icarus*, **22**, 436

Heppenheimer, T. A. 1974, On the Formation of Planets in Binary Star Systems, *Astron. Astrophys.*, **65**, 421

Holman, M. J. & Wiegert, P. A. 1999, Long-Term Stability of Planets in Binary Systems, *Astron. J.*, **117**, 621

Innanen, K. A., Zheng, J. Q., Mikkola, S. & Valtonen, M. J. 1997, The Kozai Mechanism and the Stability of Planetary Orbits in Binary Star Systems, *Astron. J.*, **113**, 1915

Jones, B. W., Underwood, D. R. & Sleep, P. N. 2005, Prospects for Habitable "Earths" in Known Exoplanetary Systems, *Astrophys. J.*, **622**, 1091

Jones, B. W., Sleep, P. N. & Underwood, D. R. 2006, Habitability of Known Exoplanetary Systems Based on Measured Stellar Properties, *Astrophys. J.*, **649**, 1010

Kasting, J. F., Whitmire, D. P. & Reynolds, R. T. 1993, Habitable Zones Around Main Sequence Stars, *Icarus*, **101**, 108

Kortenkamp, S. J., Wetherill, G. W. & Inaba, S. 2001, Runaway Growth of Planetary Embryos Facilitated by Massive Bodies in a Protoplanetary Disk, *Science*, **293**, 1127

Kozai, Y. 1962, Secular Perturbations of Asteroids with High Inclination and Eccentricity, *Astron. J.*, **67**, 591

Krist, J. E., Stapelfeldt, K. R., Golimowski, D. A., Ardila, D. R., Clampin, M., Martel, A. R., Ford, H. C., Illingworth, G. D. & Hartig, G. F. 2005, Hubble Space Telescope ACS Images of the GG Tauri Circumbinary Disk, *Astron. J.*, **130**, 2778

Kubala, A., Black, D. C. & Szebehely, V. 1993, Stability of Outer Planetary Orbits Around Binary Stars - A Comparison of Hill's and Laplace's Stability Criteria, *Celest. Mech. Dynamic. Astron.*, **56**, 51

Lissauer, J. J., Quintana, E. V., Chambers, J. E., Duncan, M. J. & Adams, F. C. 2004, Terrestrial Planet Formation in Binary Star Systems, *RevMexAA (Series de Conferencias)*, **22**, 99

Malmberg, D., Davies, M. B. & Chambers, J. E. 2007, The Instability of Planetary Systems in Binaries: How the Kozai Mechanism Leads to Strong Planet-Planet Interactions, *Mon. Not. Roy. Ast. Soc.*, **377**, L1

Marzari, F., Scholl, H., Tomasella, L. & Vanzani, V. 1997, Gas-Drag Effects on Planetesimals in the 2:1 Resonance With Proto-Jupiter, *Planet. Space. Sci.*, **45**, 337

Marzari, F. & Scholl, H. 2000, Planetesimal Accretion in Binary Star Systems, *Astrophys. J.*, **543**, 328

Mathieu, R. D. 1994, Pre-Main-Sequence Binary Stars, *Ann. Rev. Astron. Astrophys.* **32**, 465

Mathieu, R. D., Ghez, A. M., Jensen, E. L. & Simon, M. 2000, Young Binary Stars and Associated Disks. In: *Protostars and Planets IV*, ed by V. Mannings, A. P. Boss & S. S. Russell, Univ. Arizona Press, Tucson, pp 703

Mayer, L., Boss, A. P. & Nelson, A. F. 2007, Gravitational Instability in Binary Protoplanetary Disks, arXiv0705.3182M

Mayer, L., Wadsley, J., Quinn, T. & Stadel, J. 2005, Gravitational Instability in Binary Protoplanetary Discs: New Constraints on Giant Planet Formation, *Mon. Not. Roy. Ast. Soc.*, **363**, 641

Menou, K. & Tabachnik, S. 2003, Dynamical Habitability of Known Extrasolar Planetary Systems, *Astrophys. J.*, **583**, 473

Mischna, M. A., Kasting, J. F., Pavlov, A. & Freedman, R. 2000, Influence of Carbon Dioxide Clouds on Early Martian Climate, *Icarus*, **145**, 546

Morbidelli, A., Chambers, J., Lunine, J. I., Petit, J. M., Robert, F., Valsecchi, G. B. & Cyr, K. E. 2000, Source Regions and Time Scales for the Delivery of Water to Earth, *Meteorit. Planet. Sci.*, **35**, 1309

Moriwaki, K. & Nakagawa, Y. 2004, A Planetesimal Accretion Zone in a Circumbinary Disk, *Astrophys. J.*, **609**, 1065

Musielak, Z. E., Cuntz, M., Marshall, E. A. & Stuit, T. D. 2005, Stability of Planetary Orbits in Binary Systems, *Astron. Astrophys.*, **434**, 355

Nelson, A. F. 2000, Planet Formation is Unlikely in Equal-Mass Binary Systems with a $\sim$50 AU, *Astrophys. J.*, **537**, L65

Neuhäuser, R., Mugrauer, M., Fukagawa, M., Torres, G. & Schmidt, T. 2007, Direct Detection of Exoplanet Host Star Companion γ Cep B and Revised Masses for Both Stars and the Sub-stellar Object, *Astron. Astrophys.*, **462**, 777

Norwood, J. W. & Haghighipour, N. 2002, On the Stability of υ And. Extrasolar Planetary System: An S-Type Binary-Planetary System with more Than One Planet, *Bul. Am. Ast. Soc.*, **34**, 892

Pendleton, Y. J. & Black, D. C. 1983, Further Studies on Criteria for the Onset of Dynamical Instability in General Three-Body Systems, *Astron. J.*, **88**, 1415

Pichardo, B., Sparke, L. S. & Aguilar, L. A. 2005, Circumstellar and Circumbinary Discs in Eccentric Stellar Binaries, *Mon. Not. Roy. Ast. Soc.*, **359**, 521

Pilat-Lohinger, E. & Dvorak, R. 2002, Stability of S-type Orbits in Binaries, *Celest. Mech. Dynamic. Astron.*, **82**, 143

Pilat-Lohinger, E., Funk, B. & Dvorak, R. 2003, Stability Limits in Double Stars. A Study of Inclined Planetary Orbits, *Astron. Astrophys.*, **400**, 1085

Pilat-Lohinger, E., Dvorak, R., Bois, E. & Funk, B. 2004, Stable Planetary Motion in Double Stars. In: *APS Conference Series 321: Extrasolar Planets: Today and Tomorrow*, ed by J. P. Beaulieu, A. Lecavelier des Etangs, & C. Terquem, pp 410

Queloz, D., Mayor, M., Weber, L., Blécha, A., Burnet, M., Confino, B., Naef, D., Pepe, F., Santos, N., & Udry, S. 2000, The CORALIE Survey for Southern Extrasolar Planets. I. A Planet Orbiting the Star Gliese 86, *Astron. Astrophys.*, **354**, 99

Quintana, E. V., Lissauer, J. J., Chambers, J. E. & Duncan, M. J. 2002, Terrestrial Planet Formation in the Alpha Centauri System, *Astrophys. J.*, **576**, 982

Quintana, E. V. & Lissauer, J. J. 2006, Terrestrial Planet Formation Surrounding Close Binary Stars, *Icarus*, **185**, 1

Quintana, E. V., Adams, F. C., Lissauer, J. J. & Chambers, J. E. 2007, Terrestrial Planet Formation Around Individual Stars Within Binary Star Systems, *Astrophys. J.*, **660**, 807

Rabl, G. & Dvorak, R. 1988, Satellite-Type Planetary Orbits in Double Stars - A Numerical Approach, *Astron. Astrophys.*, **191**, 385

Rodriguez, L. F., D'Alessio, P., Wilner, D. J., Ho, P. T. P., Torrelles, J. M., Curiel, S., Gomez, Y., Lizano, S., Pedlar, A., Canto, J. & Raga, A. C. 1998, Compact Protoplanetary Disks Around the Stars of a Young Binary System, *Nature*, **395**, 355

Silbert, J., Gledhill, T., Duchéne, G. & Ménard, F. 2000, Near-Infrared Imaging Polarimetry of the GG Tauri Circumbinary Ring, *Astrophys. J.*, **536**, L89

Szebehely, V. 1980, Stability of Planetary Orbits in Binary Systems, *Celest. Mech.*, **22**, 7

Szebehely, V. & McKenzie, R. 1981, Stability of Outer Planetary Systems, *Celest. Mech.*, **23**, 3

Szebehely, V. 1984, Review of Concepts of Stability, *Celest. Mech.*, **34**, 49

Takeda, G. & Rasio, F. A. 2006, Eccentricities of Planets in Binary Systems, *Astrophy. Sp. Sci.*, **304**, 239

Thébault, P., Marzari, F., Scholl, H., Turrini, D. & Barbieri, M. Planetary Formation in the γ Cephei System, *Astron. Astrophys.*, **427**, 1097

Thébault, P., Marzari, F. & Scholl, H. 2006, Relative Velocities Among Accreting Planetesimals in Binary Systems: The Circumprimary Case, *Icarus*, **183**, 193

Trilling, D. E., Stansberry, J. A., Stapelfeldt, K. R., Rieke, G. H., Su, K. Y. L., Gray, R. O., Corbally, C. J., Bryden, G., Chen, C. H., Boden, A. & Beichman, C. A. 2007, Debris Disks in Main-Sequence Binary Systems, *Astrophys. J.*, **658**, 1264

Torres, G. 2007, The Planet Host Star γ Cephei: Physical Properties, the Binary Orbit, and the Mass of the Substellar Companion, *Astrophys. J.*, **654**, 1095

Verrier, P. E. & Evans, N. W. 2006, Planets and Asteroids in the γ Cephei System, *Mon. Not. Roy. Ast. Soc.*, **368**, 1599

Walker, G. A. H., Bohlender, D. A., Walker, A. R., Irwin, A. W., Yang, S. L. S. & Larson, A. 1992, Gamma Cephei - Rotation or Planetary Companion? *Astrophys. J.*, **396**, L91

Weidenschilling, S. J. 1977, The Distribution of Mass in the Planetary System and Solar *Nebula, Astrophys. Sp. Sci.*, **51**, 153

White, R. J., Ghez, A. M., Reid, I. N. & Schultz, G. 1999, A Test of Pre-Main-Sequence Evolutionary Models across the Stellar/Substellar Boundary Based on Spectra of the Young Quadruple GG Tauri, *Astrophys. J.*, **520**, 811

White, R. J. & Ghez, A. M. 2001, Observational Constraints on the Formation and Evolution of Binary Stars, *Astrophys. J.*, **556**, 265

Whitmire, D. P., Matese, J. L., Criswell, L. & Mikkola, S. 1998, Habitable Planet Formation in Binary Star Systems, *Icarus*, **132**, 196

Wiegert, P. A. & Holman, M. J. 1997, The Stability of Planets in the Alpha Centauri System, *Astrophys. J.*, **113**, 1445

Williams, D. M. & Kasting, J. F. 1997, Habitable Planets with High Obliquities, *Icarus*, **129**, 254

Wisdom, J. & Holman, M. J. 1991, Symplectic Maps for the N-body Problem, *Astrophys. J.*, **102**, 1528

Ziglin, S. L. 1975, Secular Evolution of the Orbit of a Planet in a Binary-Star System, *Sov. Astron. Let.*, **1**, 194

10 Planetary Environmental Signatures for Habitability and Life

Victoria S. Meadows

Summary. In the vast blackness of space, our home planet is a single sparkling oasis of life. Whether the Universe harbors other worlds that can support even simple life is a question that has been pondered, yet remained unanswered, for over two thousand years. Motivated by the discoveries over the past decade of hundreds of extrasolar planets, NASA and ESA have initiated a series of mission concept studies for space-based astronomical observatories that will find and study extrasolar terrestrial planets, searching for signs of life. This chapter reviews our current understanding of how we will identify planets that might support life around other stars, and incorporates results from the new science of astrobiology that support these searches

10.1 Introduction: Astrobiology and Habitability

The search for life outside our Solar System is just one facet of the field called astrobiology. Astrobiology is the study of life in the Universe, its origins, evolution, distribution and future. Because the research questions addressed by astrobiology are necessarily very broad in scope, astrobiology itself is primarily an interdisciplinary science, where progress on a particular research topic can only be achieved by combining expertise from many different scientific fields. Research topics in astrobiology include but are not limited to: the astronomical and planetary processes and conditions that are conducive to life, the origin of life, the probability of life elsewhere in the Universe, the nature and location of environments beyond the Earth that could harbor life, and the search for life in our own Solar System and beyond. NASA's Astrobiology Roadmap outlines and prioritizes several important goals and tasks for astrobiology (DesMarais et al. (2003); http://astrobiology.arc.nasa.gov/roadmap/)

In searching for life beyond the Solar System we must first find an environment that could harbor life, a so-called habitable world. In its most conservative definition, a habitable world is a solid-surfaced world, either a planet or moon, which can maintain liquid water on its surface. This definition is based on the fact that water is the one common constituent used by an enormous array of life forms on the Earth. It is also possible that life may be present in the atmospheres of planets, or in subsurface water tables or oceans, even in our own Solar System. However, when searching for life beyond our Solar System, we adopt the more conservative definition of the presence of surface water, because this definition also has the advantage of describing worlds that would be more detectable as habitable, even over enormous distances.

The habitability of a planet is in part determined by its formation heritage and early history, which will set its mass, orbit, spin and chemical composition. Habitability will subsequently depend upon the evolution of the planet's interior and its crust, and the evolution of planetary volatiles after they reach the surface of the planet. This is because the temperature and pressure required to maintain liquid water at a planet's surface depends on many factors, including the nature and brightness of the parent star, the planet's orbit (i.e. its distance from the star as a function of time), the planet mass, the reflectivity (or albedo) of the planet's surface, the density and composition of its atmosphere, the presence and nature of clouds, and the availability of internal (geothermal) energy.

10.1.1 Habitable Zones

The Earth has likely been habitable for an extremely long time, and could have become habitable as early as 10-20 Myr after the Moon-forming impact (Zahnle et al., 2007; Martin et al., 2006a). After the late heavy bombardment 3.7 Gya, the Earth was probably habitable continuously, at least in some environments (Catling and Kasting, 2007). Today, our Earth still maintains moderate temperatures and a liquid ocean, in stark contrast to the desert worlds that flank it, Mars and Venus. Earth's habitability, although governed by many factors, is dominated by its distance from the Sun and its near-circular orbit. The regions around a star in which a planet is habitable at a given time, and can be kept habitable over a continuous period of time are known as the 'instantaneous habitable zone' and the 'continuously habitable zone' respectively (Kasting et al., 1993; Kasting & Catling, 2003).

Within this distance range from its parent star, a planet is able to maintain liquid water on its surface. The boundaries of this zone are determined using climate models to calculate the inner star-planet distance at which the Earth's oceans would start to be lost in a process that would lead to a 'runaway greenhouse', and the outer distance at which it would suffer 'runaway glaciation', mediated by the formation of CO_2 ice clouds. For our own Solar System the habitable zone (Fig. 10.1) is conservatively estimated to be between 0.95 and 1.37 AU (Kasting et al., 1993). Venus and Mars, both outside the classical habitable zone limits, have a mean solar distance of 0.7 AU and 1.52 AU respectively. The ellipticity of the planet's orbit can also affect its habitability, as highly eccentric orbits could take the planet in and out of its star's habitable zone. However, the presence of an atmosphere could help to buffer this effect (Williams and Pollard, 2002)

In addition to the instantaneous habitable zone, we are also interested in how long the planet may have been able to maintain habitability, so that life had an opportunity to originate and develop. While this does depend on many planetary environmental factors, a crude estimate can be obtained by estimating the evolution in luminosity of the parent star, and the migration of the habitable zone outward

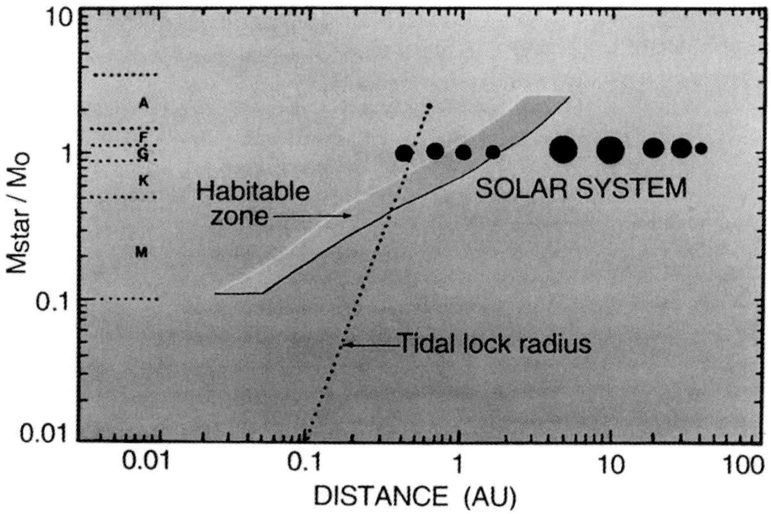

Fig. 10.1. The Habitable Zone. This diagram, based on Fig 16. in (Kasting et al., 1993) shows the calculated habitable zone (diagonal strip) as a function of the distance from the parent star (x-axis) for parent stars of different mass and spectral type (the y-axis). Our Solar System, which orbits a G2V star, is shown for comparision. Note that the Earth sits squarely within the Habitable Zone, and Venus and Mars border it. Also shown is the "tidal lock radius". This is the distance from the parent star at which a planet is likely to become tidally locked to the star. For stars in the K and M classes, the habitable zone is within the tidal lock radius.

over time to calculate a continuously habitable zone (CHZ) over a given time span. It is estimated that our Solar System has had a CHZ spanning 0.95-1.15AU in the past 4.6 Gy. However, in the next 1.1 Gy, the Sun is likely to become 10% brighter and the the Earth's surface temperature may increase to the point where we leave the CHZ in another 500-900 million years (Caldeira and Kasting, 1992).

10.1.2 A Diversity of Habitability

The modern Earth is our only known example of a habitable planet. However, even the conservative definition of habitability still encompasses a vast array of potential worlds that could be considered habitable, without being similar to the modern Earth. For example, the ancient Earth evolved through many different global environments, with different land-mass configurations and atmospheric compositions. Many of these global environments were believed to have still been habitable, and the effects of life on these environments may have been detectable even from space (Meadows, 2006; Kaltenegger et al., 2007). Although abundant oxygen was not found in our atmosphere until after 2.3 Gya (Farquhar et al., 2000), prior to that time there is still ample evidence for early life on our planet, even with an atmosphere that would have had relatively large concentrations of carbon dioxide and methane, rather than oxygen (Martin et al., 2006b; Catling and Kasting, 2007).

It's is also possible to conceive of very different environments to the early or modern Earth. As has been shown by planet formation modeling, there are a wide range of potential planetary system architectures which could contain terrestrial planets whose mass, internal composition, volatile abundance, stellar distance and orbital parameters are completely unlike those in our own Solar System (Raymond et al., 2005, 2006). In particular, the presence of an eccentric Jovian has been shown to reduce the water content of any terrestrial planets formed inward of it, with an extent that is proportional to the Jovian's eccentricity (Raymond et al., 2007). These dessicated planets may in fact never become habitable, even if they form within the habitable zone. At the other end of the spectrum, planetary modelers predict the presence of very water-rich planets formed in the wake of a migrating Jovian (Mandell et al., 2007; Raymond et al., 2006). These waterworlds may well be habitable, but unlike anything seen in our Solar System. The dominance of different life forms with different metabolisms and resultant metabolic products (Pilcher, 2003) could also change the planet's surface and atmosphere to something that is still habitable, but unlike the Earth. Recent discoveries in extrasolar planet detection are also revealing more 'super Earths', planets more massive then our Earth, but below the $10M_{\oplus}$ limit (Beaulieu et al., 2006; Gould et al., 2006; Bonfils et al., 2007). More massive planets may also retain denser atmospheres with different compositions. Depending on subsequent cloud formation, these planets could potentially remain habitable at greater distances from their parent star, or remain habitable even if in slightly non-circular orbits. None of the planet types discussed above would be truly 'Earth-like', but could still potentially be able to support life.

The enormous distances to even the nearest stars preclude sending either robotic probes or human crews to neighboring stellar systems to search for habitable planets and life. Instead, the search will be conducted using powerful astronomical telescopes and remote-sensing characterization techniques that originated in planetary astronomy and the Earth sciences. Using only remotely-sensed information obtained via photometry and spectroscopy we must learn how to recognize worlds that might have habitable conditions, and to discriminate between planets with and without life. The success of these instruments and techniques will depend crucially on the degree to which we can directly detect extrasolar planets, that is, to separate photons from an extrasolar planet from those of the star it orbits, and our ability to recognize the global footprint that life leaves on its home planet.

The new field of extrasolar terrestrial planet characterization is currently theory-based, because existing observing techniques are not yet sensitive enough to directly detect and gather information on Earth-sized planets around other stars. This new field seeks to improve our understanding of the potential range of characteristics for terrestrial planets in our galaxy, and the observational signatures that we are likely to encounter.

10.2 Techniques and Space Missions for Direct Detection of Earth-Sized Worlds

Direct detection of *Earth-sized* planets is currently believed to be infeasible from the ground. Development in this area has concentrated on space-based means to detect terrestrial, Earth-sized worlds (see, for example, http://planetquest.jpl.nasa.gov/TPF/STDT_Report_Final_Ex2FF86A.pdf; http://www.mpia.de/Darwin/CV2007.pdf). These techniques use different means to block the light from a parent star in order to see a companion planet. Currently there are at several candidate mission architectures under consideration, including an infrared nulling interferometer, a visible light coronograph, and more recently, a free-flying occulter.

10.2.1 Infrared Nulling Interferometer

Using the technique of interferometry allows multiple small telescopes on a fixed structure, or on separated spacecraft flying in precision formation, to simulate the angular resolution of a much larger, very powerful telescope (Kaltenegger & Friedlund, 2006). The interferometer would use a 'nulling' technique which relies on the principle of destructive interference to reduce the light from the parent star by more than a factor of a million, thus enabling the detection of the very dim infrared emission from any surrounding planets. To build images, the interferometer must rotate around its line of sight to different relative positions and repeat the exposures. The interferometer could also build up spectra of interesting targets.

10.2.2 Visible Light Coronograph

For this proposed architecture, an optical telescope larger and more precise than the Hubble Space Telescope would collect starlight and the very dim reflected light from the planets, which are up to 10^{10} times fainter than their parent star in the visible. The large telescope would improve angular resolution, allowing the planet to be separated from its parent star, and use the technique of coronography to reduce the starlight by a factor of 10^{11}, thus enabling astronomers to detect and characterize the planets. Coronographic design for planet detection must also have excellent wavefront control. This is best achieved in the space environment, and with a telescope that can actively correct for imperfections in the optics, which can scatter light and degrade the image contrast.

In its simplest form, a coronagraph blocks the light from a bright object so that faint nearby objects and structures can be seen. Coronagraphs have already been used to search for faint, small companions of nearby stars. To study the area around a nearby star, the coronograph must not only minimize the direct light from the nearby bright object, but must also minimize the telescope diffraction, which would otherwise reduce the angular resolution of the image. For a technical review of different coronograph designs and their performance, see (Guyon et al., 2006). A simple round telescope, for example, will produce a diffraction pattern that is dominated by a bright central spot, and a series of concentric Airy rings of decreasing brightness. To see a planet, the first several Airy rings must be suppressed without suppressing the planet. Some coronograph designs use masks to simulate a telescope with a different effective shape, thereby controlling the diffraction pattern so that the starlight is much dimmer closer to the center in some areas, and brighter in others (see, for example (Kasdin et al., 2003)). To enhance detection of a planet, the telescope can also be rotated about its line-of-sight so that the planet image passes in an out of the regions where the starlight is dim. Recent laboratory testbed experiments have come very close to the starlight suppression goal, suppressing scattered and diffracted light near a star-like source to a level of 6×10^{-10} times the peak intensity in a single coronograph image, and down to 1×10^{-11} when a series of images are combined using image processing software (Trauger and Traub, 2007). Coronographs operating at visible wavelengths have the advantage of requiring a smaller telescope to obtain the required angular resolution. Their detectors can also operate at room temperature, and require less thermal control than would be needed for a telescope operating in the thermal infrared.

10.3 Remote Detection of Planetary Characteristics

To understand whether a planet is habitable, or might even already support life, we must learn as much as possible about the planetary environment. However, even with the advanced telescopes planned, it will not be possible to resolve the planet as anything larger than a dimensionless point of light. There will therefore be no explicit constraints on planetary size, nor will we have the ability to image oceans, continents or cloud cover. Instead, all the information available to us will be averaged over the visible disk of the planet, and any signs of life on the planet must be a global phenomenon. Yet even with disk-averaging, a great deal can be learned about the environment of the planet using remote-sensing techniques. Some of these techniques include (i) determining the brightness and color of the planet, and whether there are any changes with time, (ii) analysis of spectra to determine atmospheric and surface composition and physical parameters, and (iii) monitoring spectral features to look for diurnal or seasonal variations in surface reflectivity or atmospheric composition. In all cases however, because we are using remote-sensing techniques, our interpretation of the planetary environment or the presence of life will only be as good as the level of the planet's atmosphere to which we are able to sense. In the case of cloud-covered, or dense or water-rich atmospheres, our view of the surface and lower atmosphere may be significantly obscured, hampering our ability to estimate the planet's surface temperature and likely habitability (Segura et al., 2007).

10.3.1 Planetary System Environmental Characteristics

When an extrasolar terrestrial planet is first detected via direct astronomical observation, the initial characterization will be based on its planetary system environment. The spectral type, luminosity and UV activity of its parent star will ideally have been determined prior to its selection as a target for a planet search. The UV characteristics of the star are particularly important from a habitability point of view, as the star's UV activity affects the surface habitability (Scalo et al., 2007; Kiang et al., 2007), and drives the photochemistry of the planet's atmosphere, including the lifetime and detectability of any atmospheric by-products of life (e.g. Segura et al., 2003, 2005).

Other initial factors in the planetary system that need to be known, and that may be determined in advance, are the type and orbits of other planets in the system, as these too, can affect an inner terrestrial planet's habitability, by depriving it of volatiles during formation (c.f. (Raymond et al., 2007)) or producing dynamical instabilities in the system.

Ultimately though, the most important planetary system environmental characteristic will be the planet's mass and orbital characteristics, specifically its orbital eccentricity and its distance from its parent star. These parameters will allow us to make an initial estimate of the potential habitability of the planet by determining if it is terrestrial and in the habitable zone of its parent star. Planetary mass and orbital characteristics for Earth-mass planets will be best determined using astrometric techniques, such as those to be employed by the planned NASA Space Interferometry Mission (SIM). Otherwise, missions such as TPF and Darwin, which image the planet directly, may also be able to constrain the orbit, although with limited accuracy. In particular, the eccentricity error for TPF/Darwin orbit determination is currently believed to be no better than +/- 0.1 (Traub et al., 2006), which will compromise our ability to determine whether or not the planet is located within, or remains in, the habitable zone.

10.3.2 Photometry and Photometric Variability

Once the Solar System environment is understood, additional information about an extrasolar planet can be extracted by looking at the planet's photometric brightness and variability. Photometric observations have provided a wealth of information about planets in our Solar System, and similar techniques could be applied to extrasolar planets. The planet's apparent brightness in one or more mid-infrared wavelength ranges can be used to estimate its effective temperature, and its visible photometric color could provide clues to the composition of its surface and atmosphere (Traub, 2003), although the potential diversity of extrasolar planet atmospheres and surfaces will make a definitive determination of planetary characteristics from color alone extremely problematic. Instead, color, when combined with additional subsequent data from spectroscopy, may serve as a useful initial characterization once more is known about extrasolar terrestrial planet characteristics and classes. Time-resolved photometric observations acquired as the planet rotates may reveal variability in brightness or color that could indicate the presence of clouds or surface variations (Ford et al., 2001). Photometric observations acquired as the planet moves around its parent star may also provide a way to discriminate between daily and seasonal variations in its environment (Tinetti et al., 2006a,b). If the plane of the planetary system and the planet's rotation axis are fortuitously aligned, it may also be possible to detect the presence of planetary rings or moons (c. f. (Barnes and Fortney, 2004)).

10.3.3 Remote Sensing Spectroscopy

Photometric measurements could yield valuable information about an extrasolar terrestrial planet, and may be the only practical characterization technique for faint planets near the detection limit of a particular mission. However, for brighter planets, by far the most powerful technique available for retrieving the characteristics of planetary surfaces and atmospheres is spectroscopy, which samples radiation reflected, scattered or emitted by the planet at different wavelengths. Ultraviolet, visible and near-infrared radiation from the parent star interacts with the planetary surface and atmosphere as it is scattered and reflected from the planet. Similarly, at typical planetary temperatures, significant thermal infrared (typically wavelengths between 5-50μm) is emitted by the planet and interacts with the planetary atmosphere as it escapes to space.

These interactions produce characteristic absorption and emission bands (Fig- 10.2 that can reveal the composition and physical properties of the atmosphere and surface (c.f. Hanel et al., 2003). In some cases, the wavelengths at which spectral features are observed can be used to straightforwardly reveal the presence of different types of molecules in the atmosphere via comparison with the known spectral features for gases taken from a spectral line list (Rothman et al., 2005, http://vpl.ipac.caltech.edu/spectra/frontpage.htm). However, most planetary atmospheres contain more than just one or two gases, and if the spectral resolution of the instrument is also low, then the presence of a particular trace gas species may have to be disentangled from overlapping spectral features from another species. This may require either higher performance instrumentation, or the use of an atmospheric radiative transfer model to simulate the observed spectrum for a multi-gas atmosphere.

Additionally, with sufficient instrumental precision, the temperature and pressure environment of the molecule can be gleaned from the shape and width of the spectral band. This shape can be affected by thermal Doppler broadening of the constituent molecular lines (an effect that dominates at lower pressures, typically higher in an atmosphere), pressure-induced Lorentz broadening (a function of the frequency of collisions, and so dominant at the higher pressures closer to the planetary surface), and the temperature distribution with altitude in the planet's atmosphere. Clues to the atmospheric temperature distribution are most readily obtained in the MIR, where band shape, i.e. whether a feature is in absorption, emission, or a combination of the two, is a function of the temperature structure and constituent distribution along the line of sight.

For the Earth, water vapour, confined predominantly between the warm surface and cold tropopause, absorbs thermal radiation from the surface, producing absorption features in a thermal spectrum recorded from space. Ozone in the cold lower stratosphere also mainly absorbs thermal radiation from the warm surface and lower atmosphere. However, ozone in the warm upper stratosphere emits enough thermal

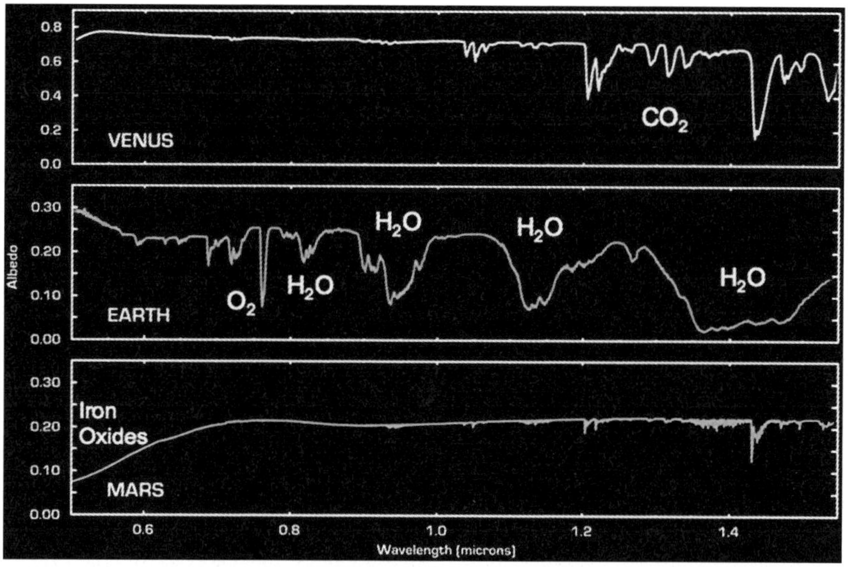

Fig. 10.2. Solar System spectra. Synthetic disk-averaged spectra of Venus, Earth and Mars showing an array of spectral shapes and both atmospheric and surface absorption features. Spectra from the VPL, graphics by Tim Pyle, Spitzer Science Center.

radiation to create an emission feature in the center of the band, and reduce the depth of the observed absorption feature. Consequently, to disentangle the two effects of temperature structure and the vertical distribution of a gas in a planetary atmosphere, temperature structure determinations from line shapes are best done with a gas that is likely to be evenly mixed in the atmosphere, such as CO_2, and not with a vertically inhomogeneously mixed gas such as ozone, water, or methane.

The mid-IR and the optical/near-IR spectral regions provide strongly complementary information on planetary characteristics. Observations in both spectral regions can help to reduce confusion due to overlapping spectral features from different planetary constituents, and provide more robust observations of a single constituent by observing multiple spectral bands. It is highly desirable to remotely explore extrasolar planets over as broad a wavelength range as possible.

Habitability and the Search for an Atmosphere

Once disk-averaged spectra are obtained for an extrasolar terrestrial planet by missions such as TPF and Darwin, we will initially search those spectra for the signs of a planetary atmosphere, including the bulk (principal) and trace (smaller component) gases. The detection of these gases may indicate several environmental characteristics for habitability, the presence of an atmosphere, abundant liquid water, a shield against surface UV radiation, and the presence and amount of greenhouse gases.

Earth's main atmospheric gas, nitrogen (N_2), has no strong spectral signatures at wavelengths other than the far UV, although its presence might be inferred from its scattering properties in the UV to the visible, or its pressure broadening effect on the spectral features of other atmospheric gases. Oxygen (O_2), which constitutes 21% of the Earth's atmosphere is more readily detectable in the spectrum, with a triplet of features in the 0.6-0.76μm region. However, in our Solar System, it is only present in large quantities on the Earth, and so is not a good universal indicator of the presence of an atmosphere.

In contrast, carbon dioxide, which is a trace gas in the Earth's atmosphere and a bulk gas on Venus and Mars, produces strong spectral features in the MIR in all three cases (Fig. 10.3). For Venus, and for Earth-like planets with at least 0.2 bars of CO_2, it also produces strong features in the near-IR between 1.0 and 1.7μm (DesMarais et al., 2002; Segura et al., 2007). This gas is likely to be rare in strongly reducing atmospheres such as those seen on the giant planets. Consequently CO_2 is currently considered to be the most detectable indicator of a terrestrial planet atmosphere.

Another extremely high priority search target will be the presence of water vapor. This gas absorbs readily throughout the visible, near-IR and MIR, and its presence in the spectrum may indicate the existence of a surface ocean. Other indicators of a surface ocean are also being explored, including observations of the glint spot, rainbows and polarization signatures (Ford et al., 2001; Bailey, 2007; Schmid et al., 2006; Stam et al., 2006)

The spectrum of an extrasolar planet may also reveal the presence of an atmospheric shield against surface UV. This UV shield could be a UV-absorbing gas, such

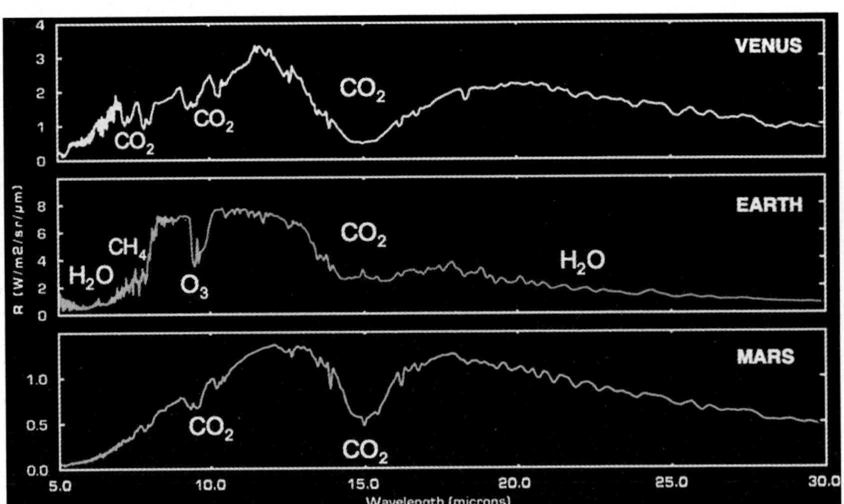

Fig. 10.3. Solar System spectra. Synthetic disk-averaged spectra of Venus, Earth and Mars in the MIR. CO_2 at 15μm can be seen prominently in all three planetary spectra. Spectra from the VPL, graphics by Tim Pyle, Spitzer Science Center.

as ozone, or an aerosol. The presence of a UV shield could be inferred from either a direct detection of the shield material itself, such as the abundant ozone observed at 9.6μm in the Earth's spectrum, or an indirect measurement via the temperature profile inferred from the band shape of an unrelated, but evenly-mixed gas. In the case of the Earth, carbon dioxide is evenly mixed throughout the atmosphere, and can be used to infer the atmospheric temperature structure. In particular, the carbon dioxide feature at 15μm exhibits a central emission peak that indicates a hot stratosphere. The stratosphere is hot in this case due to strong absorption of incoming solar UV radiation by ozone. Consequently, the shape of the CO_2 band serves as a secondary indicator of a UV shield, and could be used to corroborate an otherwise marginal detection of ozone in an extrasolar planetary atmosphere.

Finally, the spectrum of an extrasolar planet atmosphere may reveal the presence of other gases, including those that can enhance a planetary greenhouse effect. The nature, and ideally the abundance, of these gases should be determined to provide the best possible estimate of the planet's surface temperature. Typical terrestrial planet gases, such as O_3, N_2O, NO_2, NO, CO, CH_4, O_2, SO_2 and OCS etc., can produce distinctive spectral signatures at UV, visible, and/or infrared wavelengths, which could potentially be directly detected. Quantification of these gases however, is more problematic. In the visible, an estimate of the atmospheric mass would be required for accurate abundance retrieval, and in the MIR, knowledge of the atmospheric temperature structure would also be required.

Atmospheric Mass

Modeling the effects of broadening seen in infrared bands of absorbing gases can be used to infer the atmospheric pressure at the emitting level in an atmosphere. For relatively transparent atmospheres (e.g. Mars, Earth) radiation from the surface or near-surface atmosphere escapes upward through the atmosphere, allowing the total or near-total mass of the atmosphere to be derived from the planet's spectrum. For opaque atmospheres, (e.g. Venus or the Jovian planets), only the atmospheric mass above a cloud deck or other opaque surface can be determined, because that is the limiting depth from which we can see atmospheric radiation escape. Observations at different wavelengths probe different levels of an atmosphere, and can indicate different effective pressures, thereby providing additional constraints on the vertical variation of pressure in an atmosphere. This information is important both for studies of the planet's climate (which can be used to infer surface temperature) and for quantifying the trace gas mixing ratios for studies of the atmospheric chemistry.

Surface and Atmospheric Temperature

The 'holy grail' of habitability is the ability to determine if a planet has a surface pressure and temperature that is conducive to the existence of water in its liquid phase. Thermal infrared or microwave spectra may return atmospheric, and possibly surface temperatures, but this will depend on whether the wavelength chosen to make the temperature measurement is relatively free of atmospheric absorption, and capable of sensing radiation from the planetary surface. For an Earth-like terrestrial planet whose atmospheric absorption is dominated by H_2O, CO_2, and O_3 (e.g. Earth, Mars), surface temperature data is best sought in the spectral window at wavelengths between 8 and 12μm. At these wavelengths, the ubiquitous water vapor continuum absorption and strong, narrow O_3 band near 9.6μm are the only significant sources of atmospheric absorption. At shorter and longer thermal wavelengths, strong absorption by H_2O and CO_2 largely preclude surface observations. However water vapor does still produce weak continuum absorption throughout the 8-12μm window regions, and in high water vapor atmospheres (as seen in the Earth's tropics) can return a cooler brightness temperature for the atmosphere a kilometer or so above the surface, rather than the surface temperature itself. In addition, planets with atmospheres of different composition may produce an atmospheric window over a different wavelength range, as is seen in the case of an atmosphere that contains 2-bars of CO_2, where the atmospheric window is an extremely narrow region between 8.5 and 9.0μm (Segura et al., 2007). In this case, absorption from the far wings of CO_2 also preclude detection of the surface temperature within the atmospheric window. The accuracy of the surface temperatures derived from these measurements also relies on correctly estimating the planet's emissivity, which is in turn a function of the planet's surface composition, and could be anything from ice to molten rock. If the surface emissivity is not known, then it is very difficult to retrieve an accurate surface temperature.

Because infrared spectral features depend both on the amount of constituent in the atmosphere and the temperature structure, atmospheric temperatures are required to quantify trace gas amounts from thermal radiances. For example, on Earth, accurate estimates of the atmospheric temperature distribution are essential for quantifying the atmospheric concentrations of gases such as water vapour and ozone. Global scale constraints on the atmospheric thermal structure are also valuable for studies of the planet's climate and chemical equilibrium, and to determine whether or not liquid water is stable on the surface. The atmospheric temperature structure can be retrieved from the spectrum of a well-mixed gas with well-characterized absorption features that is free of absorption from other atmospheric gases or particulates. For the terrestrial planets with atmospheres in our Solar System (Venus, Earth, and Mars), observations of the CO_2 15μm band provide the best available constraints on the atmospheric thermal structure. For planets with H_2 dominated atmospheres (Jupiter, Saturn, Uranus, Neptune), the atmospheric temperature structure can be retrieved from the hydrogen continuum absorption at wavelengths longer than 20μm.

Surface Composition

One advantage the optical wavelength regime has over the mid-IR is that although the mid-IR region is sensitive to both surface and atmospheric temperatures and trace gas abundances, it is generally insensitive to underlying surface composition. This is due in part to the relatively small "atmospheric windows" to the surface in the MIR range, from 8-12μm for the Earth, and even smaller in a water vapor- or CO_2-rich atmosphere. This problem is compounded by the typical lack of spectral features, and the uniformly high emissivity of most Earth surface types (ocean/land/vegetation) at these wavelengths. Exceptions exist, however, and quartz sand displays low emissivity between 8-9μm, revealing its presence in MIR spectra taken over the Sahara (Hanel et al., 2003). Another exception is carbon dioxide ice, which has an abrupt change in emissivity across the 10-12μm wavelength range, and might be identified in a disk-average spectrum at mid-infrared wavelengths if present on a planet with a relatively thin atmosphere, as is the case for Mars (Tinetti et al., 2005).

The optical and near-infrared spectral regions contain a rich array of features associated with surface composition, and optical observations may be able to distinguish between a world dominated by oceans, sand, rock, ice, or clouds. In addition to searching for habitable environments, searching for ice is also important. An ice covered planet may be a clear indication that the planet is not habitable, and if the composition of the ice can be determined, provides a further constraint on the surface temperature.

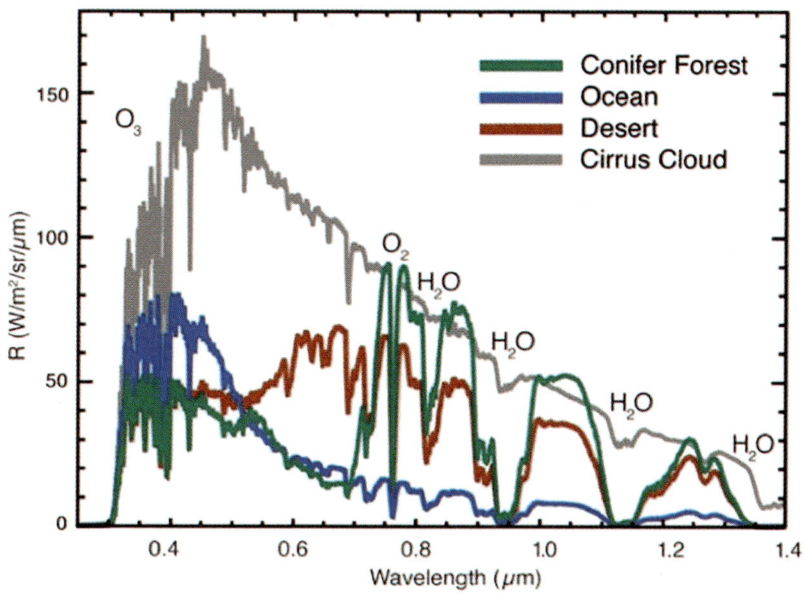

Fig. 10.4. Surface Spectra. This diagram shows simulated spectra of radiance along lines of sight through the Earth's atmosphere over surfaces of different types (vegetation, ocean, desert and clouds). In all cases, the incoming spectrum of sunlight was the same, but the radiation reflected back to the observer is quite diverse. This behavior allows us to understand surface composition in remotely-sensed data.

Phase and Seasonal Variations in Spectral Features

As with time-variable photometry, understanding how different spectral features change with time may help us infer spatial information in disk-averaged observations (e.g the presence of continents or oceans), and may also reveal variations in surface and atmospheric composition that are linked to day-night or seasonal variations e.g seasonal ice caps or dust storms (Tinetti et al., 2005). Variations in the retrieved atmospheric temperature or pressure variations could also yield information about global scale weather systems, or seasonal variations in the total atmospheric mass, like those that characterize the Martian atmosphere. Information on the surface type, temperature variations and climate over the course of an orbit are all important for understanding the potential habitability of a planet.

10.4 Biosignatures: The Global Footprints of Life

Although characterizing a planet for the ability to support life is an exciting first step, it is the precursor to the arguably more exciting search for any indications that the planet already harbors life. In astrobiology parlance, signs of life, either past or present, are called 'biosignatures' or 'biomarkers'. Signs of life that can be inferred from very distant measurements are called "astronomical", or "remote-sensing' biosignatures.

The search for remote-sensing biosignatures is based on the premise that widespread life will modify the atmosphere and surface of its planet, and that these modifications will be detectable on a global scale (DesMarais et al., 2002). On the one planet we know to support life, our Earth, we know this to be true. The Earth exhibits global life-induced changes to our planet's surface and atmospheric spectral properties that are detectable from space, many in a disk-averaged observation (e.g. (Sagan et al., 1993; Arnold et al., 2002; Woolf et al., 2002; Turnbull et al., 2006)). Remote-sensing biosignatures can be divided into three main classes: changes to a planet's atmosphere, changes to a planet's surface, and changes in planetary characteristics over time. These changes due to the presence of life are most robustly identified in the context of the planetary environment. It is therefore extremely important to first characterize the planet's atmosphere and surface and the way these change over time. This will allow us to better identify anomalous changes that may be life-induced, and to guard against the possibility of 'false positives', abiotic planetary phenomena that may mimic a biosignature. For example, an ice-covered planet may be more prone to the buildup of atmospheric oxygen from photolysis, rather than photosynthesis, because the normal sink for oxygen into the planet's rocky surface has been blocked (Schindler et al., 2000). The presence of widespread surface ice on a planet under study would serve as a warning that any oxygen detected may not be of biological origin, and suggest a more thorough study of the planet to determine the most likely source.

10.4.1 Atmospheric Biosignatures

An atmospheric signature of life could either be the detection of a single atmospheric constituent in a sufficiently large quantity that it cannot be explained as a product of a non-biological process, or a combination of atmospheric constituents whose simultaneous presence, or inferred abundance ratio, are again unlikely via non-biological processes. For example, in our own atmosphere, the abundant oxygen (20.95% of the total atmosphere) is largely produced by photosynthetic organisms such as bacteria and vegetation and so is considered to be a global biosignature (Leger et al., 1994). On a planet with liquid water on its surface, this amount would be very difficult to produce via geological or photochemical processes, even in the presence of high stellar UV and a lack of volcanic activity (Segura et al., 2007).

Oxygen can be detected most readily in the Earth's spectrum at visible wavelengths near 0.76 μm, the so-called oxygen 'A-band'. This spectral feature is routinely seen from space, and via disk-averaged observations of the Earth obtained from Earthlight scattered from the Moon (e.g. (Montañés-Rodriguez, 2006)). In the mid-infrared, molecular oxygen has no prominent spectral features, and the presence of O_2 must be inferred from detection of ozone at 9.6μm (Fig. 10.5). Significant concentrations of ozone can be formed even at relatively low oxygen levels (Kasting & Donahue, 1980) making ozone a more sensitive indicator of oxygen then oxygen itself. Once O_3 is detected, O_2 concentration can, in principle, be inferred from atmospheric chemistry models that combine information on the parent star's spectrum and the amount of O_3 observed. However, inferring O_2 abundance this way may not be straightforward in atmospheres with different chemical composition and incoming solar flux to our own. Like O_2, O_3 is considered a stronger biosignature when seen in the presence of water (Leger et al., 1999; Selsis et al., 2002).

It is also important to remember that oxygen as a large fraction of the total atmosphere has only been characteristic of the Earth's spectrum since the Proterozoic, i.e. for about half the time that the Earth has supported life (Holland, 1994; Farquhar et al., 2000). Prior to that time, the dominant microbial life-forms may have produced different gas products, such as methane, or sulfur compounds, such as methanethiol (Pilcher, 2003). Remote-sensing biosignatures detectable before oxygen became a major constituent of our atmosphere are currently not well understood, and it should not be assumed that all planets with life have evolved down the same path taken by our Earth.

Another type of atmospheric biosignature is the simultaneous presence of strongly oxidized and reduced gases that are not in chemical equilibrium e.g. O_2 and CH_4 in the Earth's atmosphere (Margulis and Lovelock, 1974). This type of biosignature is thought to be robust for many different kinds of planetary atmospheres. However, to be interpreted correctly, some understanding of the planetary environment is required to set the different components in context. This is especially important because on another planet, or on Earth earlier in its history, the biosignature pair may not necessarily be O_2 and CH_4, but other combinations of gases that must be assessed relative to the rest of the atmosphere to determine the equilibrium state. Also, these robust two-component indicators are generally

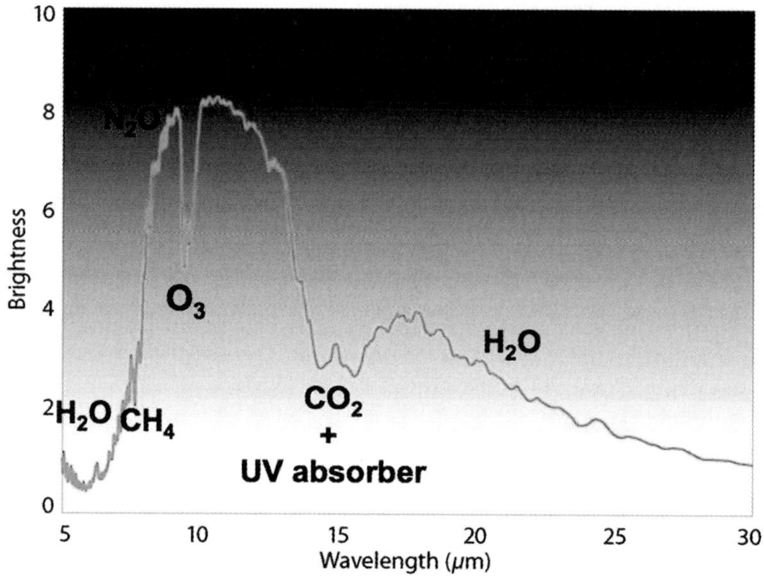

Fig. 10.5. Atmospheric Biosignatures. Synthetic spectrum of a the Earth's atmosphere taken in the mid-infrared. The biosignatures, O_3, CH_4 and N_2O are indicated. CO_2 and H_2O are not biosignatures, but instead serve as "habitability markers" indicating the presence of an atmosphere, the possibility of liquid water on the surface, and in the temperature inversion seen in the CO_2 band profile (the central emission seen in the absorption band), the likelihood that there is a UV absorber present high in the atmosphere that shields surface life.

much harder to detect via remote-sensing techniques. In the case of the Earth, the oxygen is relatively readily detectable, at least in the visible, but the methane is at much lower concentration and is spectrally most active at near-infrared wavelengths near 2.2μm (a wavelength region that is not currently being considered for the first generation planet finding instruments) and at thermal wavelengths near 7-8μm, a spectral range that includes strong water vapor bands. Consequently the methane is much harder to detect, and requires higher sensitivity and spectral resolution to disentangle its contribution to the spectrum from that of water vapor.

Other potential biosignatures, especially when seen in the presence of oxygen, include the products of biomass burning, such as CH_3Cl, and N_2O, which have low abundances and are difficult to detect in the Earth's spectrum, but which may build up to detectable levels on Earth-like planets around cooler stars (Segura et al., 2005) or on planets near the outer edges of the habitable zone (Grenfell et al., 2007) Even ammonia, when seen on a terrestrial planet in the presence of oxygen, may be considered a biosignature.

10.4.2 Surface Signatures

In addition to the rich array of spectral features associated with surface composition, such as the optical and near-infrared signatures of rock, ice or water, the Earth also displays spectral signatures due to the plant life that covers much of its surface. Chlorophyll absorbs strongly in the UV and blue ($< 0.5\mu$m) and in the red (0.6-0.7μm), and has slightly less absorption in the green (0.55μm). This characteristic absorption in the red end of the visible spectrum was seen by the Galileo spacecraft when it took spectra of Earth on its way to Jupiter (Sagan et al., 1993). However, the most detectable plant spectral feature is a large increase in reflectivity just beyond the visible range ($> 0.7\mu$m) due to the change in refractive index between air and the internal leaf structure. This feature, combined with the chlorophyll absorption just shortward of 0.7μm, results in a strong discontinuity in plant reflectance at $\sim 0.7\mu$m, which is known as "the red edge" (c.f. Seager et al., 2005) (Fig. 10.6). This property of plants is widely used for remote-sensing studies of the Earth via satellite, and can be used to monitor vegetation coverage over particular portions of the Earth. However, it has also been shown that it is only weakly visible in the Earth's global spectrum, by observing spectra of Earth light reflected from the dark side of the Moon (e.g. Montañés-Rodriguez, 2006; Hamdani et al., 2006). For the

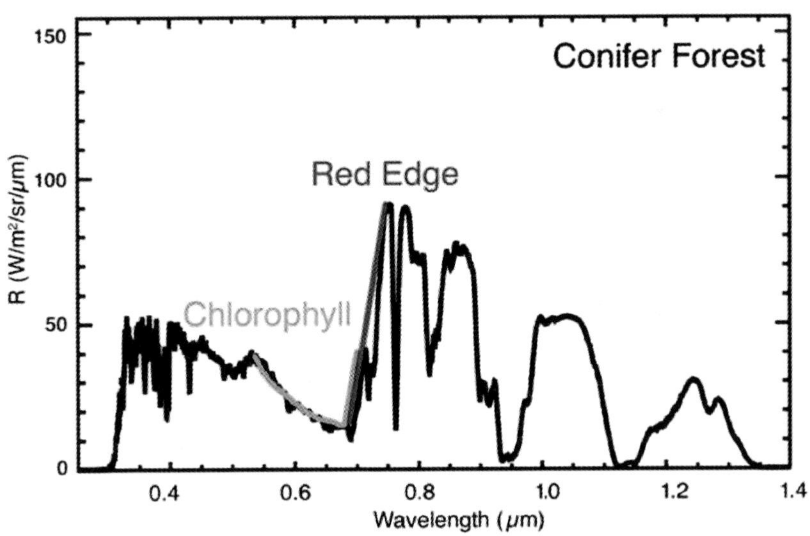

Fig. 10.6. The Red Edge. Synthetic spectrum of a line of sight through the Earth's atmosphere over a conifer forest, with chlorophyll absorption and the red-edge reflectivity marked. Chlorophyll, a potentially important biosignature, has strong absorption in the UV and blue ($< 0.5\mu$m) and in the red (0.6–0.7μm marked in green), and slightly less absorption in the green (0.55μm). Due to changes in the refractive index between air and the internal leaf structure, plants are also highly reflective just beyond the visible range ($> 0.7\mu$m), resulting in a prominent discontinuity (marked in red) known as "the red edge".

Fig. 10.7. Photosynthetic Pigments. Since photosynthesis evolves under the influence of both the parent star's available spectrum and the planet's atmospheric composition, the pigments developed to harvest incoming radiation may be quite different on other worlds (Kiang et al., 2007). This whimsical artist's impression of an alien Earth shows what it might be like to live on a planet where alternative photosynthetic pigments dominate and plants aren't green.

Earth, this signature is potentially much more difficult to detect than the abundant oxygen, but it may be stronger on an extrasolar terrestrial planet with a larger fraction of visible vegetation. Indeed, the photosynthetic pigments evolved, and the spectral position of the red edge itself may be quite different for vegetation on a planet around a star of different spectral type (Kiang et al., 2007; Tinetti et al., 2006b). The important feature to look for will be a sharp, otherwise unexplained, rise in the planet's reflectivity at longer wavelengths.

10.4.3 Temporal Signatures

The Earth's biomass is largely supported either directly or indirectly by photosynthesis, and cycles in the life processes on our planet are tied to the diurnal or seasonal cycles of sunlight. Consequently, a third type of biosignature is a temporal signature, the time-varying behavior in photometric brightness or spectral features. For example, a 'snapshot' spectrum of the Earth would show the presence of CO_2 and CH_4. If not seen in the presence of oxygen or ozone, it would be hard to conclude that these gases are biologically produced, since photochemistry and geological processes also generate them. However, sensitive spectroscopic observations of the Earth taken over a period of time would reveal periodic variations in the atmospheric CO_2 and CH_4 abundance. This behavior could be shown to be correlated with season, which would be unlikely for a geological process. However a photochemical model, and an understanding of the planet's environment, including the parent star's spectrum and the atmospheric composition, would be required to preclude the possibility that these variations were simply photochemically produced. On Earth the observed seasonal cycling of CO_2 and CH_4 (Fig. 10.8) is known to be linked to seasonal variations in the amount and photosynthesis of land plants (Tucker et al., 1986), and can be traced to a surface source, rather than a photochemical product. However, these seasonal variations are very small, and would require a very sensitive instrument to detect them, making temporal variability of atmospheric constituents potentially the hardest type of biosignature to detect for a truly Earth-like planet. This is perhaps beyond the ability of the first generation of planet detection and characterization missions. Another time-variable sign of life might be vegetation coverage as a function of season, which might be detected spectrally or photometrically. One must be cautious, however. Not all time-variable surface signatures are due to life. Numerous astronomers from the late 19th and early 20th century attributed seasonal albedo variations on Mars to variations in vegetation, when the true cause was the seasonal cycle of dust storm activity.

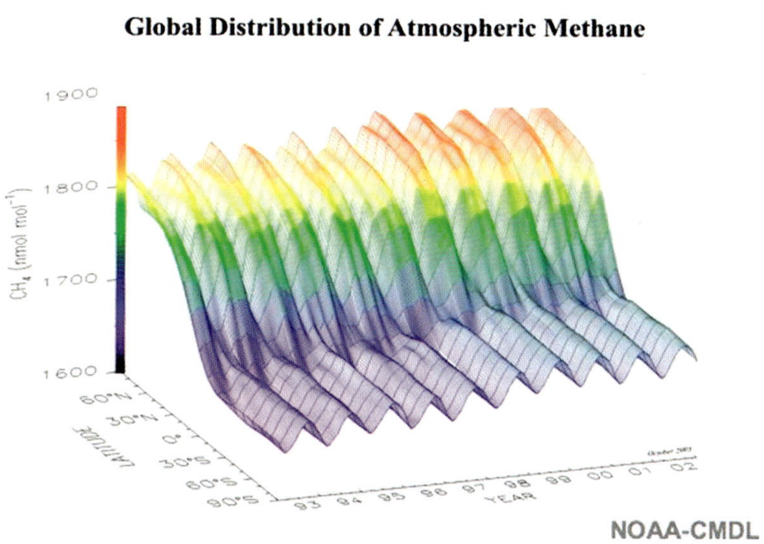

Fig. 10.8. Temporal Biosignatures. This 3-D plot generated by the NOAA-CMDL, represents the measured temporal and spatial variability in the concentration of methane in the Earth's atmosphere over a span of almost 10 yrs. The periodic ripples in the concentration are due to seasonal cycles in the methane, and the sharp step function in the plot from northern to southern latitudes shows the disparity in CH_4 concentration in the land-dominated Northern Hemisphere, and the ocean-dominated Southern Hemisphere.

10.4.4 Sensitivity to Cloud Cover

Another factor that must be considered when attempting to characterize a planet is the potential loss of information due to persistent cloud cover. Typically, clouds are associated with convection and condensation of a volatile species, like the water ice clouds seen in the atmosphere of the Earth, or the CO_2 clouds seen on Mars. Hazes can also be formed via photochemistry, with the planetwide haze layers that dominate the atmospheres of Venus and Titan being two examples in our own Solar System. Cirrus clouds high in the Earth's atmosphere can obliterate even a strong signal due to ozone at mid-IR wavelengths, although clouds at lower levels still allow the detection of O_3 (DesMarais et al., 2002) On the other hand, in the visible, the strong A-band of O_2 is visible in oxygen rich atmospheres, even in the presence of high cloud, although its contrast is somewhat reduced (Tinetti et al., 2006b). Theoretical models also predict that since the optical behavior of clouds is phase dependent, there may in fact be optimal phases at which to observe an extrasolar terrestrial planet to minimize the effect of cloud scattering and detect surface biosignatures (Tinetti et al., 2006b).

Although the photochemical hazes that shroud Venus and Titan are opaque at visible wavelengths, they display "atmospheric windows" at near-IR wavelengths (Meadows and Crisp, 1996; Smith et al., 2006), which allow penetration and remote-sensing of the underlying planetary surface. For Venus, thermal radiation from the hot surface and lower atmosphere escapes through the clouds and can be detected only on the night side of the planet. In the case of Titan, the haze is sufficiently transparent at near-IR wavelengths the surface can be detected even when the satellite is fully illuminated.

10.5 Biosignature Detection

Determining if an extrasolar planet exhibits signs of life will ultimately be extremely challenging. Given the known diversity of extrasolar giant planets and the even higher diversity anticipated for extrasolar terrestrial planets, it is likely that we will see planetary environments quite unlike those in our Solar System. Each search for biosignatures will be best done on a case-by-case basis, without preconceived biases as to what biosignatures will be found. It will be prudent to first assume that all observed characteristics are planetary processes, rather than signs of life, and so the method by which a biosignature is discovered will be one of systematically eliminating all other abiotic planetary processes as its source, within the constraints of existing observations and models. Unless an exact replica of the modern Earth is found, biosignatures will probably not be definitive detections, but rather probabilistic ones.

The first spectra of extrasolar terrestrial planets are likely to have relatively low spectral resolution and signal to noise, and it will be important to study the trade-offs in these two parameters, as lowered spectral resolution could result in a higher S/N required to observe a given molecular band. It will also be important

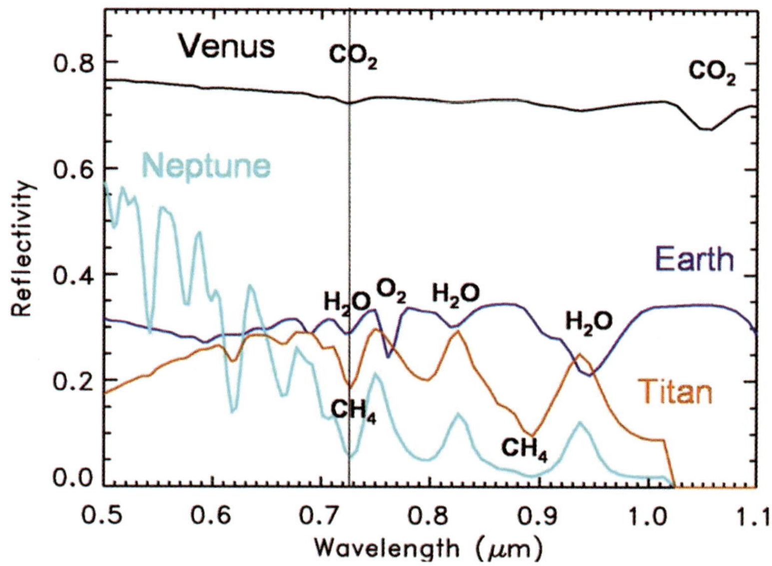

Fig. 10.9. Low Resolution Planetary Spectra. Synthetic and observed spectra for planets in our own Solar System at low spectral resolution. Note that the spectral feature seen near 0.725μm is variously due to CO_2, H_2O or CH_4, depending on the planetary environment. When trying to characterize a planetary atmosphere of unknown composition, it is important to have as large a wavelength range as possible to search for other features that will help distinguish between multiple possible species.

to think of a biosignature as not a single molecular band, for example, but as a suite of observations that corroborate and strengthen each other. In the simplest form, providing a sufficiently large spectral wavelength coverage that two or more bands of the same molecule can be observed, will improve our ability to confirm a detection, or discriminate between multiple candidate molecules for a single feature (see Fig. 10.9). A more advanced example would be not considering O_2 alone to be a biosignature detection, but instead searching for O_2 in the visible, O_3 in the MIR, coupled with observations of water vapour, a planetary albedo that precludes widespread surface ice, and an orbital semi-major access and eccentricity that maintains the planet within the habitable zone of its parent star.

The search for life beyond the Earth is indeed a challenge, and yet the questions we hope to answer are profound. It is truly exciting to know that in the coming decades, humanity will finally gain the technical and scientific capability to undertake that search. We can look forward to being the generation that through interdisciplinary research, and the development of advanced technology and theoretical frameworks, was able to make an enormous advance in our understanding of our place in the Universe.

References

Arnold, L., Gillet, S., Lardire, O., Riaud, P., Schneider, J., 2002, A test for the search for life on extrasolar planets. Looking for the terrestrial vegetation signature in the Earthshine spectrum, *A & A.*, **392**, 231-237.

Bailey, J., 2007, Rainbows, Polarization, and the Search for Habitable Planets, *Astrobiology*, **7(2)**, 320-332.

Barnes, J. W., Fortney, J. J., 2004, Transit Detectability of Ring Systems around Extrasolar Giant Planets, *Ap. J.*, **616(2)**, 1193-1203.

Bekker, A., Holland, H. D., Wang, P.-L., Rumble, D., Stein, H. J., Hannah, J. L., Coetzee, L. L., Beukes, N. J., 2004, Dating the rise of atmospheric oxygen, *Nature*, **427(6970)**, 117-120.

Beaulieu, J.-P.; Bennett, D. P.; Fouqu, P.; Williams, A.; Dominik, M.; Jorgensen, U. G.; Kubas, D.; Cassan, A.; Coutures, C.; Greenhill, J., and 63 additional co-authors., 2006, Discovery of a cool planet of 5.5 Earth masses through gravitational microlensing, *Nature*, **439(7075)**, 437-440.

Bonfils, X., Delfosse, X., Forveille, T., Mayor, M., Udry, S., 2007, Gliese 581: A System with 3 Very Low-mass Planets In: *Proceedings of the conference In the Spirit of Bernard Lyot: The Direct Detection of Planets and Circumstellar Disks in the 21st Century June 04-08, 2007*, ed by P. Kalas, University of California, Berkeley, CA, USA.

Caldeira, K., and Kasting, J.F., 1992, The Life Span of the Biosphere Revisited, *Nature* **360**, 721-723.

Catling, D. , Kasting, J. F., 2007, Planetary atmospheres and life, In: *Planets and Life: The Emerging Science of Astrobiology*, eds Sullivan W, Baross J, Cambridge, MA, Cambridge Univ, 91-116.

DesMarais, D. J., Allamandola, L. J., Benner, S. A., Boss, A. P., Deamer, D., Falkowski, P. G., Farmer, J. D., Hedges, S. B., Jakosky, B. M., Knoll, A. H., Liskowsky, D. R., Meadows, V. S., Meyer, M. A., Pilcher, C. B., Nealson, K. H., Spormann, A. M., Trent, J. D., Turner, W. W., Woolf, N. J., Yorke, H. W., 2003, The Astrobiology Roadmap, *Astrobiology*, **3(2)**, 219-235

DesMarais, D. J., Harwit, M. O., Jucks, K. W., Kasting, J. F., Lin, D. N. C., Lunine, J. I., Schneider, J., Seager, S., Traub, W. A., Woolf, N. J., 2002, Remote Sensing of Planetary Properties and Biosignatures on Extrasolar Terrestrial Planets, *Astrobiology*, **2(2)**, pp153-181.

Farquhar, J., Bao, H.M., Thiemens, M., 2000, Atmospheric influence of Earth's earliest sulfur cycle, *Science*, **289(5480)**, 756-758.

Ford, E. B., Seager, S., Turner, E. L., 2001, Characterization of extrasolar terrestrial planets from diurnal photometric variability, *Nature*, **412**, 885-887.

Gould, A., Udalski, A., An, D., Bennett, D. P., Zhou, A.-Y., Dong, S., Rattenbury, N. J., Gaudi, B. S., Yock, P. C. M., Bond, I. A., Christie, G. W., and 26 additional co-authors, 2006, Microlens OGLE-2005-BLG-169 Implies That Cool Neptune-like Planets Are Common,*Ap. J.*, **644(1)**, L37-L40.

Grenfell, J. L., Stracke, B., von Paris, P., Patzer, B., Titz, R., Segura, A., Rauer, H., 2007, The response of atmospheric chemistry on earthlike planets around

F, G and K Stars to small variations in orbital distance, *Planetary and Space Science*, **55(5)**, 661-671.

Guyon, O., Pluzhnik, E. A., Kuchner, M. J., Collins, B., Ridgway, S. T., 2006, Theoretical Limits on Extrasolar Terrestrial Planet Detection with Coronagraphs, *Ap. J. Supp.*, **167(1)**, 81-99.

Hamdani, S., Arnold, L., Foellmi, C., Berthier, J., Billeres, M., Briot, D., François, P., Riaud, P., Schneider, J., 2006, Biomarkers in disk-averaged near-UV to near-IR Earth spectra using Earthshine observations, *A. & A.*, **460(2)**,617-624.

Hanel, R. A., Conrath, B. J., Jennings, D. E., Samuelson, R. E., 2003, *Exploration of the Solar System by Infrared Remote Sensing*: Second Edition, pp. 534. ISBN 0521818974. Cambridge, UK: Cambridge University Press.

Holland HD. 1994. In: *Early Life on Earth*, ed. S Bengtson, New York: Columbia Univ. Press, 237-244.

Kaltenegger, L.; Fridlund, M., 2006, Characteristics of proposed 3 and 4 telescope configurations for Darwin and TPF-I In: *Direct Imaging of Exoplanets: Science and Techniques, Proceedings of IAU Coll. 200*, ed by C. Aime, F. Vakili, Cambridge University Press, Cambridge, 255-258

Kaltenegger, L., Traub, W. A., Jucks, K. W., 2007, Spectral Evolution of an Earth-like Planet, *Ap. J.*, **658**,598-616.

Kasdin, N. Jeremy, Vanderbei, Robert J., Spergel, David N., Littman, Michael G., 2003, Extrasolar Planet Finding via Optimal Apodized-Pupil and Shaped-Pupil Coronagraphs, *Ap. J.*, **582(2)**, 1147-1161.

Kasting, J. F., Donahue, T. M., 1980, The evolution of atmospheric ozone, *J. Geophys. Res.*, **85**, 3255 - 3263.

Kasting, J. F., Whitmire, D. P., Reynolds, R. T., 1993, Habitable Zones around main sequence stars, *Icarus*, **101**, 108-128.

Kasting, J. F., 1997, Habitable Zones around low mass stars and the search for extraterrestrial life, In: *Origins of Life and Evolution of the Biosphere*, Kluwer Academic Publishers, 27: 291-307.

Kasting, J.F., Catling, D., 2003, Evolution of a Habitable Planet, *Ann. Rev. Atron. Astrophys.* **41**, 429-463.

Kiang, N. Y., Segura, A., Tinetti, G., Govindjee, Blankenship, R. E., Cohen, M., Siefert, J., Crisp, D., Meadows, V. S., 2007, Spectral signatures of photosynthesis. II. Coevolution with other stars and the atmosphere on extrasolar worlds, *Astrobiology*, **7(1)**, pp

Leger, A., Ollivier, M., Altwegg, K., Woolf, N. J., 1999, Is the presence of H_2O and O_3 in an exoplanet a reliable signature of a biological activity?, *A. & A.*, **341**, 304-311.

Leger, A., Pirre, M., Marceau, F. J., 1994, Relevance of oxygen and ozone detections in the search for primitive life in extra solar planets, *Adv. Sp. Res.*, **14(6)**, 117-122.

Mandell, A. M., Raymond, S. N., Sigurdsson, S., 2007, Formation of Earth-like Planets During and After Giant Planet Migration, *Ap. J.* **660(1)**, 823-844.

Meadows, V. S., Crisp, D., 1996, Ground-based near-infrared observations of the Venus nightside: The thermal structure and water abundance near the surface *JGR - Planets*, **101(E2)**, 4595-4622.

Meadows, V.S., 2006. Modeling the diversity of extrasolar terrestrial planets. In: *Direct Imaging of Exoplanets: Science and Techniques, Proceedings of IAU Coll. 200*, ed by C. Aime, F. Vakili, Cambridge University Press, Cambridge, 25-34.

Margulis, L. Lovelock, J. E., 1974, Biological Modulation of the Earth's Atmosphere, *Icarus*, **21**, 471.

Martin, H. ; Albarde, F., Claeys, P., Gargaud, M., Marty, B., Morbidelli, A., Pinti, D., 2006, From Suns to Life: A Chronological Approach to the History of Life on Earth 4. Building a habitable planet, *Earth, Moon, and Planets*, **98(1-4)**, 97-151

Martin, H., Claeys, P., Gargaud, M., Pinti, D., Selsis, F., 2006, From Suns to Life: A Chronological Approach to the History of Life on Earth 6. Environmental Context, *Earth, Moon, and Planets*, **98(1-4)**, 205-245.

Montañés-Rodriguez, P, Palle, E., Goode, P. R., Martin-Torres, F. J., 2006, Vegetation Signature in the Observed Globally Integrated Spectrum of Earth Considering Simultaneous Cloud Data: Applications for Extrasolar Planets, *Ap. J.*, **651(1)**, 544-552.

Pilcher, C. B., 2003, Biosignatures of Early Earths, *Astrobiology*, **3(3)**, 471-486.

Raymond, S. N., Quinn, T., Lunine, J. I., 2007, High-Resolution Simulations of The Final Assembly of Earth-Like Planets. 2. Water Delivery And Planetary Habitability, *Astrobiology*, **7(1)** 66-84.

Raymond, S. N., Mandell, A. M., Sigurdsson, S., 2006, Exotic Earths: Forming Habitable Worlds with Giant Planet Migration, Science, **313**(5792), pp. 1413-1416.

Raymond, Sean N., Quinn, Thomas, Lunine, Jonathan I., 2005, The formation and habitability of terrestrial planets in the presence of close-in giant planets, *Icarus*, **177(1)**, 256-263.

Rothman, L. S., Jacquemart, D., Barbe, A., Benner, D. Chris, Birk, M., Brown, L. R., Carleer, M. R., Chackerian, C., Chance, K., Coudert, L. H., Dana, V., Devi, V. M., Flaud, J. M., Gamache, R. R., Goldman, A., Hartmann, J. M., Jucks, K. W., Maki, A. G., Mandin, J. Y., Massie, S. T., Orphal, J., Perrin, A., Rinsland, C. P., Smith, M. A. H., Tennyson, J., Tolchenov, R. N., Toth, R. A., Vander Auwera, J., Varanasi, P., Wagner, G., 2005, The HITRAN 2004 molecular spectroscopic database, *J. Quant. Spect. & Rad. Trans.*, **96**, 139–204

Seager, S., Turner, E. L., Schafer, J., Ford, E. B., 2005, Vegetation's Red Edge: A Possible Spectroscopic Biosignature of Extraterrestrial Plants, *Astrobiology*, **5(3)**, 372-390.

Sagan, C., Thompson, W. R., Carlson, R., Gurnett, D., Hord, C., 1993, A Search for Life on Earth from the Galileo Spacecraft, *Nature*, **365**, 715-717.

Scalo, J., Kaltenegger, L., Segura, A., Fridlund, M., Ribas, I., Kulikov, Y. N., Grenfell, J. L., Rauer, H., Odert, P., Leitzinger, M., Selsis, F., Khodachenko, M. L., Eiroa, C., Kasting, J., Lammer, H., 2007, M Stars as Targets for Terrestrial Exoplanet Searches And Biosignature Detection, *Astrobiology*, **7(1)**, 85-166.

Schindler, T. L., Kasting, J. F., 2000, Synthetic spectra of simulated terrestrial atmospheres containing possible biomarker gases, *Icarus*, **145**, 262-271.

Schmid, H. M., Beuzit, J.-L., Feldt, M., Gisler, D., Gratton, R., Henning, Th., Joos, F., Kasper, M., Lenzen, R., Mouillet, D., Moutou, C., Quirrenbach, A., Stam, D. M., Thalmann, C., Tinbergen, J., Verinaud, C., Waters, R., Wolstencroft, R., 2006, Search and investigation of extra-solar planets with polarimetry, In: *Direct Imaging of Exoplanets: Science and Techniques. Proceedings of the IAU Colloquium 200*, Edited by C. Aime and F. Vakili. Cambridge, UK: Cambridge University Press,165-170

Segura, A., K. Krelove, J. F. Kasting, D. Sommerlatt, D., V. Meadows, D. Crisp, M. Cohen, E. Mlawer, 2003, Ozone concentrations and ultraviolet fluxes on Earth-like planets around other stars, *Astrobiology*, **3(4)**, 689-708.

Segura, A., Kasting, J. F., Meadows, V., Cohen, M., Scalo, J., Crisp, D., Butler, R. A. H., Tinetti, G., 2005, Biosignatures from Earth-like planets around M dwarfs, *Astrobiology*, **5(6)**, 706-725.

Segura, A., Meadows, V. S., Kasting, J. F., Crisp, D., and Cohen, M., 2007, Abiotic Formation of O_2 and O_3 in High-CO_2 Terrestrial Atmospheres, *Astron. and Astroph.*, **472(2)**, 665–679.

Selsis, F., Despois, D. and Parisot, J.-P., Signature of life on exoplanets: Can Darwin produce false positive detections? *Astr. & Astrophys.* **388**, 985-1003

Smith, P. H.; Lemmon, M. T.; Lorenz, R. D.; Sromovsky, L. A.; Caldwell, J. J.; Allison, M. D., 1996, Titan's Surface, Revealed by HST Imaging, *Icarus*, **119(2)**, 336-349.

Stam, D. M., de Rooij, W. A., Cornet, G., Hovenier, J. W., 2006, Integrating polarized light over a planetary disk applied to starlight reflected by extrasolar planets, *A & A*, **452(2)**, 669-683.

Tucker, C. J., Fung, I. Y., Keeling, C. D., Gammon, R. H., 1986, Relationship Between Atmospheric CO_2 Variations and a Satellite Derived Vegetation Index, *Nature*, **319(6050)**, 195-199.

Tinetti, G., V. Meadows, D. Crisp, W. Fong, H. Snively, 2005, Disk-averaged Synthetic Spectra Of Mars, *Astrobiology*, **5(4)**, 461-482.

Tinetti, G., V. Meadows, D. Crisp, W. Fong, E. Fishbein, T. Velusamy, M. Turnbull and J.-P. Bibring, 2006, Detectability of Planetary Characteristics in Disk-Averaged Spectra I: The Earth Model, *Astrobiology*, **6(1)**, 34-47

Tinetti, G., V. Meadows, D. Crisp, N. Kiang, B. Kahn, E. Fishbein, T. Velusamy, M. Turnbull, 2006, Detectability of Planetary Characteristics in Disk-Averaged Spectra II: Synthetic Spectra and Lightcurves of Earth, *Astrobiology*, **6(6)**, 881-900.

Traub, W. A., 2003, The Colors of Extrasolar Planets, In: *Scientific Frontiers in Research on Extrasolar Planets, ASP Conference Series*, **294**, Eds Drake Deming and Sara Seager, San Francisco, ASP, 595-602.

Traub, W. A., Levine, M., Shaklan, S., Kasting, J., Angel, J. R., Brown, M. E., Brown, R. A., Burrows, C., Clampin, M. Dressler, A., Ferguson, H. C., Hammel, H. B., Heap, S. R., Horner, S. D., Illingworth, G. D., Kasdin, N. J., Kuchner, M. J., Lin, D., Marley, M. S., Meadows, V., Noecker, C., Oppenheimer, B. R.,

Seager, S., Shao, M., Stapelfeldt, K. R., Trauger, J. T., 2006, TPF-C: status and recent progress, In: *Advances in Stellar Interferometry,* Edited by Monnier, John D., Schüller, Markus Danchi, William C., Proceedings of the SPIE, **6268**, 62680T-1-14.

Trauger, J. T., Traub, W. A., 2007, A laboratory demonstration of the capability to image an Earth-like extrasolar planet, *Nature*, **446(7137)**, 771-773.

Turnbull, M. C., Traub, W. A., Jucks, K. W., Woolf, N. J., Meyer, M., Gorlova, N., Skrutskie, M. F., Wilson, J. C., 2006, Spectrum of a Habitable World: Earthshine in the Near-Infrared, *Ap. J.*, **644(1)**, 551-559.

Williams, D M., Pollard, D, 2002, Earth-like worlds on eccentric orbits: excursions beyond the habitable zone, *Int. J. Astrobiol.*, **1(1)**61-69

Woolf, N. J., Smith, P.S., Traub, W. A., Jucks, K. W., 2002, The Spectrum of Earthshine: A Pale Blue Dot Observed From the Ground, *Ap. J.*, **574**, 430-433.

Zahnle, K., Arndt, N., Cockell, C., Halliday, A., Nisbet, E., Selsis, F., Sleep, N. H., 2007, Emergence of a habitable planet, *Space Sci Rev* , **129**, 35-78.

11 Moons of Exoplanets: Habitats for Life?

Caleb A. Scharf

Summary. Moon systems exhibit diverse characteristics, and present unique environments. In our own Solar System the majority of giant planet moons harbor substantial water ice mantles. The inferred internal structure, and observed activity, of many suggests the potential for extensive subsurface liquid water, both currently and in the past. Liquid water is vital for all forms of terrestrial life, through its integrated roles in biochemistry and geophysics. By contrast, the thick atmosphere and rich, low-temperature, hydrocarbon chemistry of Titan points towards a highly complex surface environment paralleling some of the conditions on the early Earth, and conceivably offering alternative pathways for complex phenomena such as life.

There is good reason to hypothesize that giant exoplanets will harbor significant moon systems. These may share many characteristics with those in our own Solar System, as well as represent alternatives - possibly including temperate Mars or Earth sized bodies. Detecting the presence of moons in exoplanetary systems is rapidly approaching feasibility, and will open a new window on such objects and their potential habitability.

11.1 Introduction

"What a wonderful and amazing Scheme have we here of the magnificent Vastness of the Universe! So many Suns, so many Earths, and every one of them stock'd with so many Herbs, Trees and Animals, and adorn'd with so many Seas and Mountains!" "... even the little Gentlemen round Jupiter and Saturn..."

Christiaan Huygens (1695, Cosmotheoros)

Huygens clearly felt strongly that life exists on any suitable body, including the "little Gentlemen". These are, of course, the major satellites of Jupiter and Saturn, and it would seem that he had no difficulty in placing them in a category whereby they too would be well "stock'd" with organisms. It was also true, however, that Huygens was aware that at this distance from the Sun, conditions on these bodies would be cold. A simple blackbody thermodynamic model for an illuminated spherical body yields an equilibrium effective temperature at its surface (Sect. 11.3). For the Galilean moons this implies surface temperatures of some ~ 100 K. At Saturn this drops to ~ 85 K. Over the three hundred years following Huygens' statements there was relatively little thought given to the possibility of the Galilean (or Saturnian) moons being at all suitable for life. A notable exception to this was Proctor (1870), who speculated that Jupiter might be capable of heating its moons to temperate levels. However, the physics behind this assertion was incomplete, and he was motivated by a belief in the plurality of worlds. Interest in the moons of the giant planets began to increase again when Gerard Kuiper and others (e.g. Kuiper, 1957) were able to show, using infrared spectroscopy, that Europa's surface was composed primarily of water ice. Then, during the 1970's, flyby data from the Pioneer – and to a much greater extent the Voyager missions (Smith et al., 1979a,b) – not only confirmed Kuiper's observations but discovered some extraordinary properties for the icy crusts on many moons. In Fig. 11.1 Voyager data partially covering Europa is shown, together with an inset at higher resolution from the later Galileo mission.

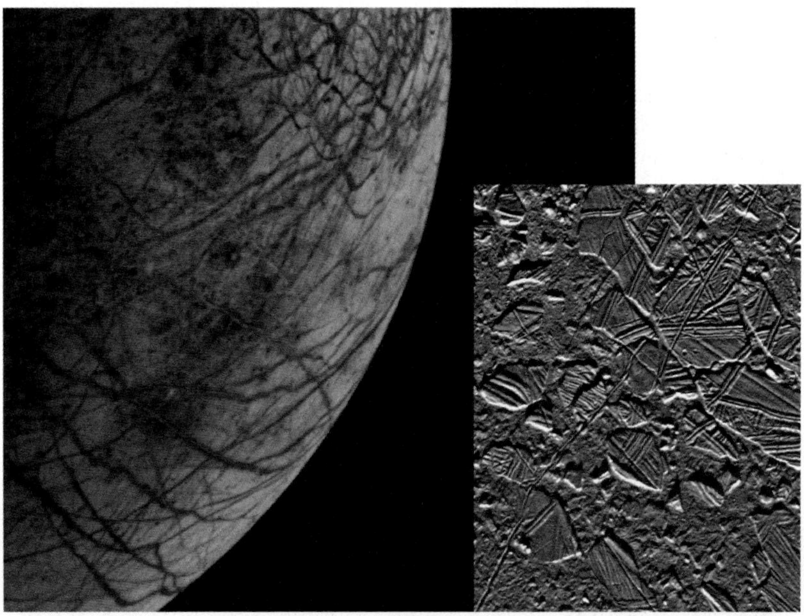

Fig. 11.1. The Jovian moon Europa as observed by Voyager 2 in 1979 (Smith et al., 1979b). Very few impact features are seen on the water-ice surface. Extensive streaks and blushes indicate that an outer crust has been fractured, filled in from the interior, and re-frozen many times. Inset: Detailed image of the surface of Europa taken by the Galileo spacecraft in 1997. The area shown is approximately 34 by 42 kilometers in size, with a resolution of 54 meters. Crustal plates of ice are seen, which have broken and then "rafted" together into positions resembling those of terrestrial pack-ice. The size and shape of these plates has suggested a fluid or slush-like environment close to the surface during its breakup (Greeley et al., 1998b).

The surface of this moon is not only almost devoid of impact craters (indicating a young age), but is criss-crossed by a remarkable network of shallow ripples, crack-like features, plate-like features, and distinct blemishes of the surface water ice (e.g. Greeley et al., 1998a). Galileo's Near-Infrared Mapping Spectrometer (NIMS) detected evidence for hydrated salts at various surface locations – suggesting evaporation from a globally mixed water layer (McCord et al., 1999). Detailed imaging also indicates regions consistent with low viscosity surface flows (now frozen), and anomalously shallow impact craters (i.e. filled in from the interior, Moore et al., 1998). While all of this evidence points towards the presence of extensive subsurface liquid, the most compelling result comes from the detection of an *induced* magnetic field (Kivelson et al., 2000; Zimmer et al., 2000). The magnitude and response of this field (to Europa's position relative to the powerful Jovian magnetosphere) indicates the presence of a near-surface, global, conducting layer – consistent with a salty water ocean of at least 10 km thickness. Figure 11.2 illustrates the possible internal structure of Europa in this case.

Cassen et al. (1979) predicted that Europa could be experiencing tidal heating sufficient to maintain a subsurface ocean. The tides arise from its eccentric orbit ($e \simeq 0.01$) produced by the mean-motion 4 : 2 : 1 Laplacian resonance between Io, Europa, and Ganymede. Heating results from flexure between periapsis and apoapsis around Jupiter (Sect. 11.3.1). Following the Voyager data there were significant investigations of the implications of tidal heating for Europa (e.g. Squyres et al., 1983; Melosh et al., 2004), and the implications for habitability within subsurface oceans and episodic fracture zones enabling photosynthetic (Greenberg et

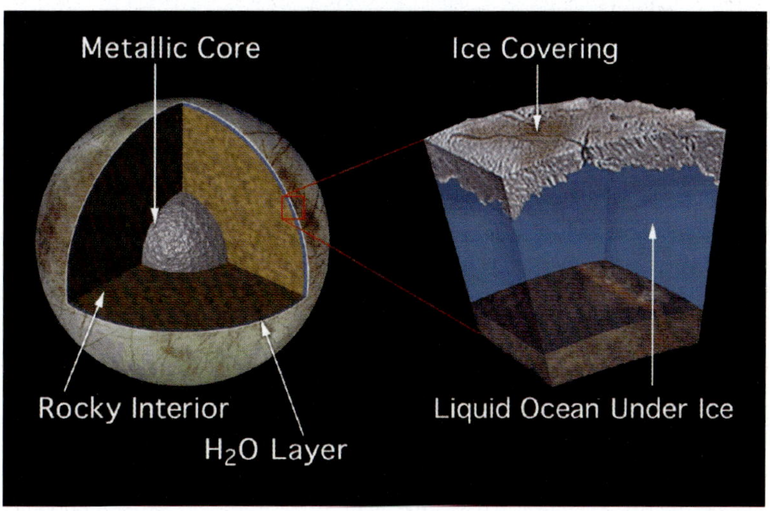

Fig. 11.2. Cutaway illustration of the possible internal structure of Europa (NASA/JPL). The presence of a metallic core and rocky interior is inferred from Europa's mean density, radius, and measurement of its gravitational field during spacecraft flybys. The presence of a liquid water ocean is strongly suggested but not yet confirmed.

al., 2000) or non-photosynthetic biospheres to exist (Chyba, 2000). These investigations demonstrated that moons could represent an entirely new class of habitable environment – potentially less dependent on stellar insolation and driven largely by dynamical energy dissipation – which could furthermore influence climate in cases where atmospheres exist (Reynolds et al., 1987). The dynamically dense nature of moon systems, and their evolution due to moon-planet tides, appears to lend itself to situations of moon-moon orbital resonances. External perturbations could also be significant (see below). Furthermore, if the core-accretion model for planet formation is correct (e.g. Lissauer, 1993), then giant planets will form preferentially beyond the so-called "snow line" in a system Hayashi (1981). This may lead to moons naturally accumulating significant icy mantles, thereby circumventing many of the present uncertainties in "dry" versus "wet" formation for terrestrial planets (e.g. Raymond et al., 2006)).

Other moons in our Solar System, which would otherwise be inert, also show evidence for what may be (at least partially) dynamically driven heating. For example, the recent detection by *Cassini* of water "geysers" on Enceladus in the Saturnian system (e.g. Porco et al., 2006) points towards a remarkably active geology, even on such a small moon, only 500 km in diameter (Fig. 11.3). The precise origin of this activity is currently unresolved (see however Nimmo et al., 2007; Hurford et al., 2007), but is of enormous interest since it may indicate that there are reservoirs of subsurface liquid water. Even the small Uranian moon Ariel shows some evidence for tidally induced heating and cryovolcanism (Melosh et al., 1989).

Fig. 11.3. The Saturnian moon Enceladus back-lit by the Sun as imaged by the *Cassini* orbiter. A series of discrete, fountain-like sprays (or "geysers") are seen above the southern polar region of the moon. It is likely that these are erupting from subsurface, pressurized pockets or reservoirs of water at temperatures above 273 K (Image credit: NASA/JPL/Space Science Institute, 2005).

Some moons are large enough to exhibit characteristics that are generally considered to be the province of planets. For example, Titan harbors a thick atmosphere dominated by nitrogen (98.4%) and methane (1.6%) (with a 1.5 atmosphere pressure at its surface), and a rich and complex low-temperature (90-95 K) hydrocarbon surface chemistry. Such an environment may represent conditions that not only were at one time present on the early Earth (although Titan is far colder), but also those which could conceivably allow for alternative forms of life (Lunine et al., 1998). Finally, Ganymede evidently has an extensive icy mantle (although it is slightly larger than Mercury it is only half the mass) and has an *internally generated* magnetic field (Kivelson et al., 1998), indicating that it may have a molten iron or iron-sulfur core.

11.1.1 Habitable Zones and Exoplanets

Given the intriguing nature of the population of moons in our Solar System it is a logical extension to ask whether moons elsewhere could represent potential habitats for life in the Universe. There are several specific motivations for this: (1) The formation of moons around giant planets seems likely to be a generic phenomenon (see Sect. 11.2) and the remarkable diversity seen in exoplanet systems (e.g. Marcy et al., 2006) could be matched by the diversity of moons. (2) The propensity of ice-mantled moons to be heated by tidal effects and thereby sustain surface or subsurface liquid water environments, and even volcanic/tectonic activity in their silicate interiors. (3) The possibility that the host planets of moons orbit within the classical circumstellar habitable zone, or have a habitable surface environment maintained by a combination of stellar and tidal input (Reynolds et al., 1987; Scharf, 2006). (4) Cold, but chemically rich and active moons, such as Titan, could represent the best current example of an environment for entirely alternative life mechanisms.

In the broader context, such potential habitats are of enormous interest in seeking both the origins of life, and the capacity for life to survive through unfavorable circumstances. For example, not only could moons like Europa offer potential "incubators" for life, they might – through forward contamination (e.g. Gladman et al., 2006) – offer relatively safe haven for microbial life set adrift due to cataclysmic events on inner, terrestrial-type, worlds.

However, there are a number of possible difficulties with the idea of giant planet moons as habitats: (1) Large bodies of liquid water have not been conclusively proven on any moon in our Solar System (e.g. the water "geysers" from Enceladus could originate in a variety of scenarios, not all of which indicate persistent or extensive liquid water reservoirs (Porco et al., 2006)). (2) Radiation environments within gas giant magnetospheres can be highly active and potentially detrimental to surface life through both biochemical damage and atmospheric modification and sputtering (Chyba, 2000; Williams et al., 1997). (3) Although tidal heating can originate from a variety of dynamical situations it may not always be sufficiently long-lived to enable the development of a biosphere (e.g. Scharf, 2006).

11.2 Moon Formation

The precise nature of the formation history of moon systems is still highly uncertain. Most modern theories assume that following the initial formation of the giant planet (via core-accretion or dynamical instability), and the collapse of its gaseous envelope into a denser, cooling atmosphere, a circumplanetary nebula, or sub-disk, forms around it. This hypothesis appears to be supported by hydrodynamical models (Lubow et al., 1999; Lubow & D'Angelo, 2006). This sub-disk of gas and solids then forms moons in an analogous fashion to the coagulation of the proto-planetary disk itself (in contrast to a system such as the Earth-Moon, Canup, 2004). The sub-disk is an actively changing environment, which continues to both receive material from, and return it to, the surrounding environment. This differs from earlier models which assumed a minimum mass sub-solar nebula (e.g. Peale, 1999). Consequently, the moons that form will reflect the environmental circumstances of their host planet (e.g. Alibert et al., 2005; Canup & Ward, 2006).

Once at this stage, further growth of the planet is primarily through the accretion of material via the sub-disk (e.g. Alibert et al., 2005). This appears to be a reasonable consequence of the core-accretion model of planet formation, and probably also of the gas instability model (Boss, 2005). It also qualitatively agrees with modern efforts to accelerate core-accretion through turbulence driven 'vortices', which concentrate material (Klahr & Bodenheimer, 2006). Most models then make a further distinction between a young sub-disk, which is actively fed at its edges by the circumstellar/proto-planetary disk and via vertical infall, and a 'late' sub-disk (Canup & Ward, 2002; Alibert et al., 2005; Canup & Ward, 2006). Other models argue for an entirely gas-poor 'late-late' formation of moons (Estrada & Mosqueira, 2006).

In the late sub-disk, the circumstellar disk has largely dissipated, and the sub-disk is on its way to dissipating. During the young sub-disk stage, any moons that form may migrate inwards (in an analogous way to fast Type-I planet migration, possibly sometimes to slower Type-II – disk gap migration). Thus, these early moons are likely accreted by the planet, and do not survive. The consequence is that the surviving moons are typically those which are the last generation to form, during the late sub-disk stage. As the sub-disk itself dissipates, these objects cease migration and remain as the final moon population. Such models suggest that there may be a "universal" scaling law between the host planet mass and the total mass of its moons. Canup & Ward (2006) find that their models consistently produce moons totalling a few 10^{-4} of the host planet mass. This agrees well with the observed relation for Jupiter and Saturn, and even Neptune, where they make the argument that as a captured body, Triton's mass is necessarily similar to whatever objects it usurped during its initial encounter. This scaling clearly does not hold for the Earth's Moon, which formed through a late-time impact event (Canup, 2004).

It remains to be seen whether this is universally correct. No models attempt to include magneto-hydrodynamics in the planet-moon-disk interactions. For planet formation this may be a critical ingredient (e.g. Oishi et al., 2007) and can fundamentally alter the characteristics and rates of migration. For moon formation it may also play a role given the powerful magnetosphere of giant planets. Nonetheless, it does seem reasonable to assume that large moons (i.e. Mars or Earth sized) will be more likely around planets more massive than Jupiter.

Further issues must come into play in considering both the moons in our own Solar System, and particularly those in exoplanetary systems. These include the effects of host planet migration and orbital configuration (e.g. does planet orbital eccentricity drive eccentricity in moon orbits?), and circumstellar disk thermal and chemical structure. These are complex questions which relate directly to the potential habitability of moons. The accumulated volatile (water) component of moons is particularly important, as is the potential for tidal heating. The host planet orbit also determines the stellar insolation of the moons, its time variability, and the long-term stability of a moon system. This latter point is a crucial one, sometimes overlooked. While the simple Hill-Sphere radius indicates the outermost stable satellite orbit, it is insufficient to correctly predict the planetary orbital radius within which moon systems lack long-term stability. More detailed analyses (Ward & Reid, 1973; Barnes & O'Brien, 2002) have shown that around stars of mass $> 0.15\ M_\odot$, moon systems are only truly long-term stable for planetary orbits > 0.6 AU. Within this radius stellar tides may act to remove moons over timescales less than some 5 Gyr. Although this is a conservative evaluation (in the sense that some planets within 0.6 AU may still have quite long lived moon systems), it does eliminate the so-called hot Jupiters (within some 0.1 AU of their parent stars, Marcy et al., 2006) as candidate hosts for significant moons.

Thus, while it seems reasonable to assume that familiar moon systems will exist around many giant exoplanets there is still much work to be done.

11.3 Environmental Conditions of Moons

Much as with planets, the amount of stellar radiation received by a moon will help dictate some of its fundamental characteristics and potential habitability. The additional complication with moons lies in their range of masses, from tiny (e.g. Enceladus) to large (e.g. Ganymede or Titan) and how this relates to the retention of volatiles – which we discuss below.

In order to make an initial estimate of the orbital radius at which stellar insolation produces a given surface temperature, the classical prescription for estimating the equilibrium surface temperature of a fast-rotating body can be applied, namely:

$$T_{eq} = \left(\frac{(1 - A_B)L_*}{16\pi\epsilon\sigma d^2} \right)^{1/4} \tag{11.1}$$

where A_B is the Bond albedo, L_* the parent stellar luminosity, and d the distance from the parent star. The factor ϵ is a crude, first-order, correction in the case where an atmosphere is assumed (for zero-atmosphere $\epsilon = 1$). It incorporates the infrared optical depth, and for a present-day Earth-type atmosphere $\epsilon \simeq 0.62$. In Fig. 11.4 this expression is used to estimate the time-averaged surface temperature

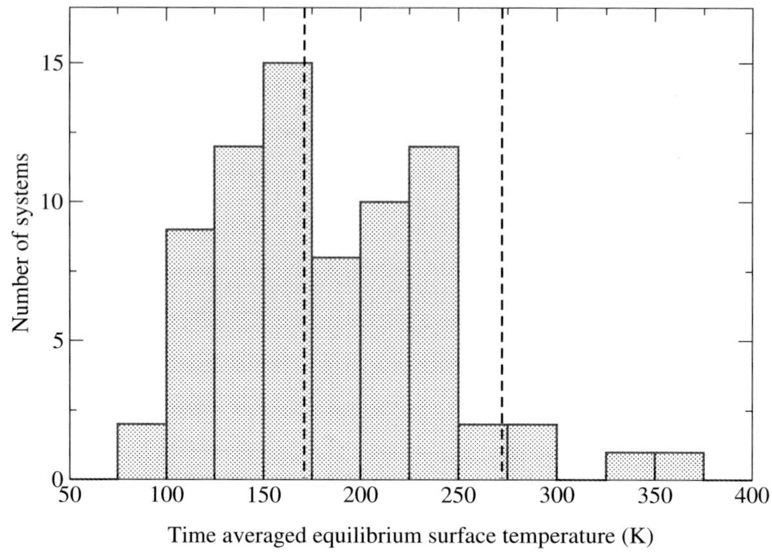

Fig. 11.4. Time averaged surface temperatures for hypothetical icy-moons (albedo 0.68 commensurate with Europa, no atmospheres) around a subset of 74 known exoplanets with orbital semi-major axes greater than 0.6 AU. Vertical dashed lines correspond to the water-ice sublimation temperature in a vacuum (170 K), and the water ice/liquid transition above the triple-point pressure (273 K).

for atmosphere-free water-ice mantled moons around a subset of known exoplanets – excluding those within 0.6 AU of their parent stars (Sect. 11.4). A significant fraction occupy the temperature range above 170 K – the sublimation temperature for pure water ice in a vacuum.

The rate of water sublimation for an ice mantled body can be estimated by considering the water vapor pressure over ice (e.g. Spencer, 1987). This is plotted in Fig. 11.5 as a function of temperature.

This plot assumes *only* sublimation, and does not account for re-deposition of material, which at $\sim$ 170 K may be at rates approximately equal to those of sublimation – depending on local surface temperature variations (e.g. latitudinal variation on Galilean satellites, Spencer, 1987). For $T_{eq} > 170$ K sublimation rates are extremely fast – with 100km depth of water ice sublimating in only a few 10^6 years at 170 K – if there is no re-deposition.

The escape velocity from the surface of a Europa mass moon ($\sim 0.0082 M_{\oplus}$) is ~ 2 km s^{-1}, compared to a mean velocity of a water molecule in a gas at 170 K of ~ 0.4 km s^{-1}. Applying the thermal (Jeans) escape methodology (e.g. Lammer et al., 2004) then the typical flux of escaping gas particles at these temperatures is at least a factor 10^8 lower than that for gas in the exosphere of a large moon with 1000 K temps (e.g. similar to the Martian exosphere). Thus, pure thermal escape appears unlikely to be a dominant mechanism for material loss in cold moons beyond the sublimation line, and even up to the ice-line at 273 K. However, for moons harboring

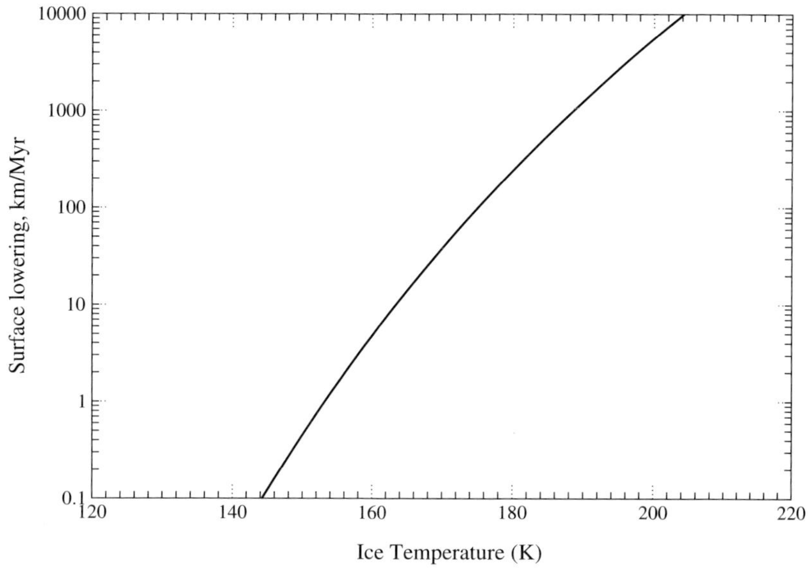

Fig. 11.5. Water ice sublimation surface lowering rates (km per Myr) as a function of temperature, following (Spencer, 1987).

atmospheres, dissociation of molecular species can promote thermal loss for moons of mass $< 0.12M_\oplus$ (Williams et al., 1997).

Sputtering by charged particles trapped within the magnetosphere of the giant planet host is likely to be a highly efficient atmospheric/volatile loss mechanism for moons of all masses (Williams et al., 1997). This is true unless a moon possesses an intrinsic magnetic field. A field such as that measured for Ganymede ($0.03M_\oplus$) (Kivelson et al., 1998) could prevent rapid particle loss. Finally, to represent a classical habitable environment a moon probably requires a climate stabilizing system such as the geophysical carbon-cycle on Earth (Kasting et al., 1993). In the absence of tidal heating (Sect. 11.3.1) Williams et al. (1997) estimate that a moon must exceed some $0.23M_\oplus$ in order to sustain plate tectonics.

Small ($< 0.12M_\oplus$) icy moons with mean surface temperatures in the 170-273 K range may therefore be likely to lose sublimated material and eventually all surface volatiles over relatively short timescales. By extension (based on Fig. 11.4) it appears that 15-27% (albedo ranging from 0.3 to 0.68) of *all* currently known exoplanets (i.e. including those within 0.6 AU of their parent star) might be capable of harboring small, ice-mantled moons with the potential for tidally heated subsurface oceans. The implications of volatile retention are also compared schematically to the theoretical population of exoplanets in Fig. 11.6. The criteria applied to produce the shaded zone described in Fig. 11.6 are likely overly conservative, but nonetheless there are 4 currently known exoplanet systems that fall within this zone – all with G-type parent stars (GJ 3021b, HD 80606b, HD 104985b, and 70 Vir b).

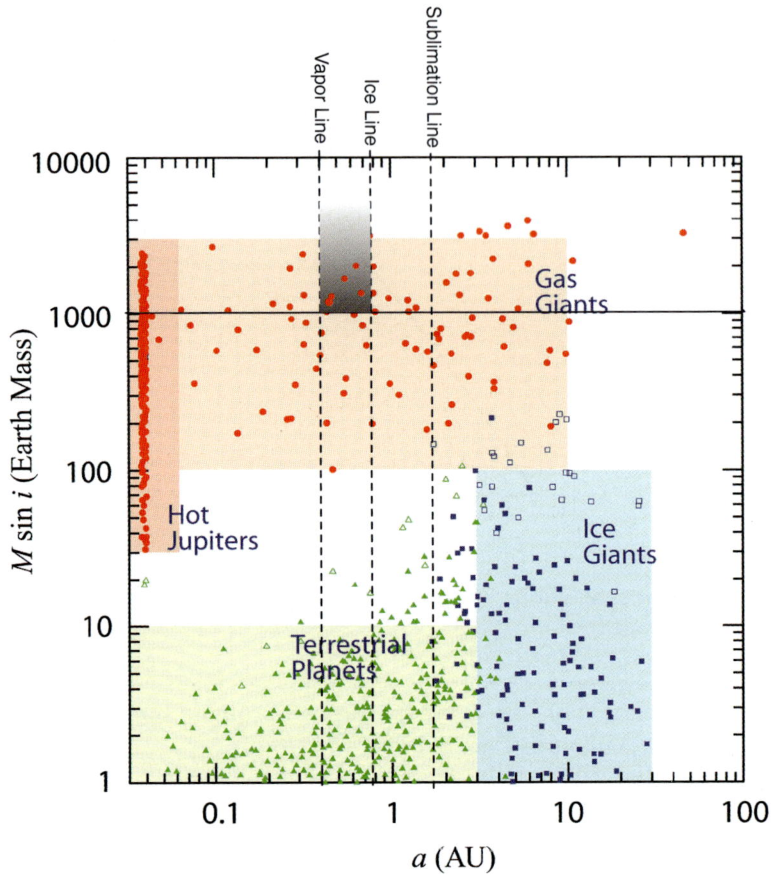

Fig. 11.6. The results of 3000 planet formation models are shown (adapted from Ida & Lin, 2004). Planet mass is plotted versus semi-major axis. The major planetary families are labeled. Vertical dashed lines correspond to (left to right): The vapor line – the distance from a 1 $L_\odot$ star at which a planetary body of albedo 0.68 (commensurate with a reflective, icy, moon) and a terrestrial-type atmosphere and greenhouse effect could attain a surface equilibrium temperature of 273 K. The ice line corresponds to the same physical model, but for a surface temperature of 373 K. The sublimation line corresponds to an atmosphere-free body which attains a surface temperature of 170 K – corresponding to mean water ice sublimation in a vacuum. The horizontal solid line at 1000 $M_\oplus$ corresponds to the giant planet mass that could yield a Mars sized (0.1 $M_\oplus$) moon according to the scaling suggested by Canup & Ward (2006) (Sect. 11.2). The shaded zone above this line between the vapor and ice lines therefore corresponds to the most probable region where an atmosphere retaining, "habitable", moon could be found (e.g. Williams et al., 1997)

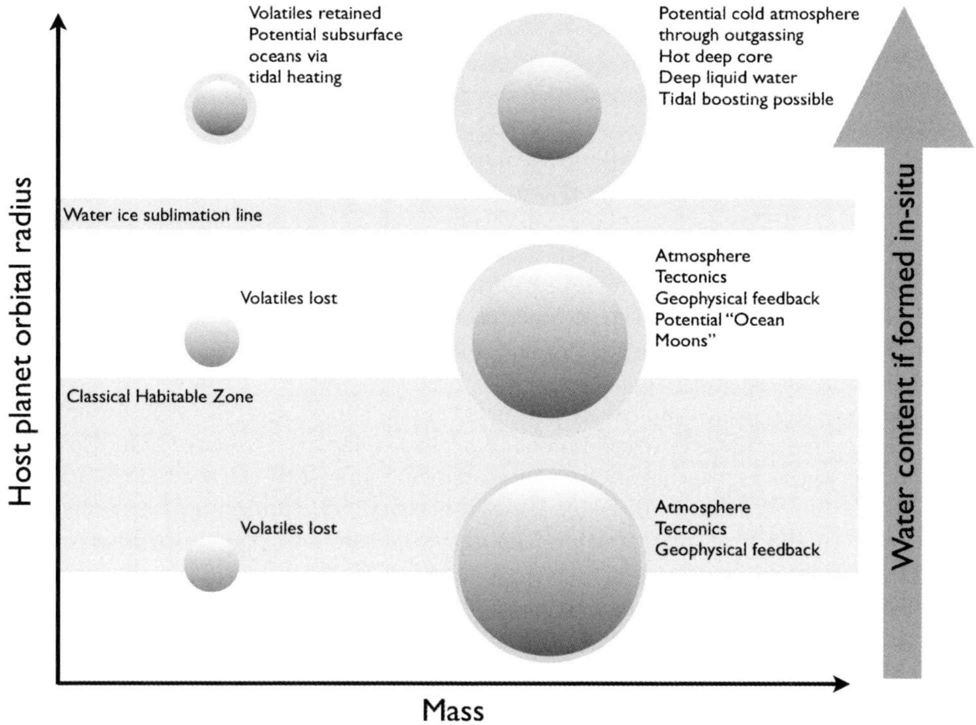

Fig. 11.7. Possible classes of moons as a function of mass and distance from the parent star (see Sect. 11.3)

In Fig. 11.7 a schematic is used to summarize some of the variations in moon characteristics that might be expected. The initial volatile composition of moons is assumed to increase with distance from the parent star (therefore host planet migration and sub-disk temperature structure is ignored). Low mass moons will have difficulty in retaining their initial complement of volatiles unless they are beyond at least the water ice sublimation line in a system. If they do retain volatiles then they are good candidates for tidally heated environments containing liquid water. More massive moons (i.e. those of at least 0.12-0.23 $M_{\oplus}$) may both retain volatiles and an atmosphere, as well as exhibit active tectonics (potentially boosted by tidal heating) which can provide climate stabilizing feedback (e.g. Kasting et al., 1993). Large moons that experience stellar irradiation commensurate with the classical circumstellar habitable zone may range from relatively "dry" to "wet". With tidal heating the potential for very wet, "ocean" moons is increased since these may form further from the star and therefore have a larger intrinsic volatile content. The most massive moons at large radii may resemble a class of objects akin to Titan, but with a potentially wide range in actual surface conditions.

11.3.1 Tidal Heating and Boosted Temperatures

If tidal heating plays a role on moons it can provide both benefits and obstacles to life beyond the simple maintenance of mean temperatures or liquid water oceans. Smaller bodies cool down more rapidly and, as likely in the case of Mars, geological activity also ceases more rapidly (this may also be due to the loss of water as a geophysical lubricant and driver of tectonic plate subduction). In this case, tidal effects may act to maintain geological activity (as seen in an extreme case on Io, Smith et al., 1979b), and therefore mechanisms of climate stabilization, such as the carbon-silicate cycle (Kasting et al., 1993). The negative result may be that tidal effects overdo it (again, Io is the obvious example), and on smaller moons can cause the loss of volatiles, and on larger moons may render the surface environment too unstable.

There are numerous routes to tidal heating in moon systems. These include; mean motion resonances between moons (e.g. the Laplacian resonance of Io, Europa, Ganymede, or the Enceladus-Dione 1:2 resonance), spin-orbit librational resonances (e.g. Enceladus, Wisdom, 2004), and moon eccentricity "pumping" through tidal effects due to the host planet orbital eccentricity (akin to planets in binary star systems). In order to gauge the relative magnitude of tidal heating, a surface heat flow has generally been used (e.g. Reynolds et al., 1987; Scharf, 2006).

For a satellite or moon in a non-zero eccentricity orbit the rate of tidal dissipation ($\dot{E}$), assuming Keplerian motion, synchronous rotation, and zero obliquity, may be written in terms of the surface heat flow (H_T) for a moon of mass M_s:

$$H_T = \frac{21}{38}\frac{G^{5/2}}{\mu Q}\left(\frac{3}{4\pi}\right)^{5/3} e_s^2 \frac{\rho_s^{1/3}}{a_s^{15/2}} M_s^{5/3} M_p^{5/2} \tag{11.2}$$

where μ is the satellite elastic rigidity (assumed uniform), $1/Q$ is the satellite specific dissipation function, R_s the satellite radius and e_s the satellite orbital eccentricity, which is assumed to be small (see e.g., Peale et al., 1980).

To examine the combination of stellar insolation and tidal heating (first postulated by Reynolds et al., 1987) a very naive assumption can be made that the equilibrium temperature of an object's surface – defined as some *ad hoc* layer of outer material – is that of a pure black body receiving both an input radiation flux *and* an input flux from tidal heating. This condition can be written as a form of zero-order energy-balance equation;

$$\epsilon\sigma T_{eq}^4 = \frac{(1-A_B)}{4} f_{rad} + H_T \tag{11.3}$$

where f_{rad} and H_T are the stellar flux and tidal surface heat flow respectively. An implicit assumption is that the tidal surface heat flow acts exactly like a radiation field. In other words the energy flux is assumed to be entirely thermalized by the object's surface, and does not allow for non-thermal energy dissipation (e.g. the tidal surface energy flow could equally power the bulk rearrangement of an object's surface).

Applying this produces encouraging results. A Mars-sized (0.1 $M_\oplus$) moon retaining a terrestrial-like atmosphere with an orbital radius and eccentricity similar to that of Europa around any of the known exoplanets capable of retaining moons (Fig. 11.4), could readily attain a habitable surface temperature (i.e. 273 K$< T <$ 373 K). The tidal energy flow required at such a moon's surface would range from 0.1 to 100 Wm^{-2}, compared to the actual 0.1 Wm^{-2} for Europa and 1.5 Wm^{-2} for Io. The total spin energy of Jupiter is some 10^{33} J, and hence such a hypothetical moon could, in principle, sustain this level of tidal dissipation for several Gyr – if the driving conditions (e.g. orbital resonance) are maintained.

11.4 Moon Detection

Moons can both perturb the orbital phase of their host planet and add to the optical blocking during stellar transits. Both effects are of course small. Nonetheless, current studies suggest that they are in a regime that will become accessible with future (and possibly existing) instruments. Doyle & Deeg (2004) (see also Sartoretti & Schneider, 1999) have shown that the transit timing (i.e. the ingress or egress time) of a Titan-Saturn like system has a characteristic variation due to Titan on the order of 30 seconds. A Jupiter-Europa like system has a variation at the 6 second level. Szabó et al. (2006) have discussed the effect of moons on transit lightcurve shapes and depths, and find the effect to be at the level of a few 0.01 milli-magnitudes. The transit timing method may therefore be the most tractable. Indeed, Doyle & Deeg (2004) argue that both the COROT (CNES) and Kepler (NASA) space-based transit missions are sensitive to Titan-Saturn systems (for COROT a 9th magnitude star is required, for Kepler the star can be as faint as 12th magnitude). Kepler may even be sensitive to Jupiter-Europa systems.

Ehrenreich et al. (2006) have further suggested that future large-aperture instruments (such as GMT, TMT, and OWL) may have sufficient sensitivity to detect the ensemble signature of spectroscopic absorption during transit due to moon atmospheres. This may sound extraordinarily unlikely. Only Titan in our Solar System retains a substantial atmosphere. However, this is unlikely to be truly representative. For masses around 0.12-0.23 $M_{\oplus}$ at habitable temperatures, a body is likely to not only retain an atmosphere (even within the magnetosphere of a giant planet), but also sustain tectonic activity vital for atmospheric re-cycling and resupply (Sect. 11.3). Furthermore, giant planets appear to exist within zones of stellar irradiation well suited to 'temperate' surface conditions for any moons they harbor (Fig. 11.4). Thus, a wide range of atmospheric types may be possible.

Rather remarkably, the net projected disk area of the Galilean moons and the other regular Jovian satellites amounts to approximately 1% of the area of Jupiter's disk. This is roughly equivalent to the ratio of an Earth-sized planet's disk compared to that of Jupiter. If moon populations do indeed scale with the host planet size (as postulated by Canup & Ward (2006)) then this disk area scaling will also continue to hold to first order. Although this presents a very small perturbation to the net reflected light of a giant planet (and an equally small perturbation to infrared emission, via moon-planet transit in particular) phase curve photometry could potentially offer a route to seek the modulation effects of moon systems – due to both direct contributions to the net luminosity of a planet, and the shadowing of the planetary surface by moons (and ring structures) (Cabrera & Schneider, 2007).

It appears likely that in the near future the existence (or absence) of exomoon systems will be established – first through transit-timing experiments and secondly through ultra-high precision photometric studies. Assuming moons are indeed found around some giant exoplanets we will be able to begin to evaluate their potential habitability. In the longer term, very large-aperture instruments will open up exciting avenues for transit spectroscopy capable of detecting moon atmospheres.

11.5 Life on Exomoons

Although entirely in the realm of speculation it is nonetheless interesting to consider what life might have to deal with on exomoons. For example, tidal heating may be episodic on geological timescales if driven by moons wandering in and out of resonance conditions. If a moon has a sufficiently massive silicate/metal core then in certain situations radiogenic heating, combined with the insulating properties of the outer ice crust, could maintain a partial sub-surface liquid water environment, close to the core. Upon re-entering a period of tidal heating this liquid zone would expand until a new equilibrium is reached. Biota might be capable of dealing with these extended "deep-freezes".

If a combination of tidal and radiogenic heating maintain a heat flow between a silicate/metal core and a body of liquid water then the opportunity for hydrothermal systems akin to those on Earth arises (Lowell & DuBose (2005), and see Fig. 11.8). Biota associated with such systems do not rely on photosynthetic chemical pathways (such as considered for the upper ocean environment on Europa, Sect. 11.1) – although there is intriguing evidence that some terrestrial species can indeed harvest the infrared photon flux from deep ocean vents (Beatty et al., 2005). Given the possibility that even on Earth these systems could play a key role in the origin of life (e.g. Gold, 1992) it is reasonable to speculate that they could be of central importance to life on moons.

It is also important to remember that moons inhabit a rather unique orbital architecture. Based on our own Solar System it appears that they will often share their environment with other nearby satellites. This raises some interesting possibilities for systems containing multiple habitable objects. There is likely to be (especially during early epochs) significant transfer of material between moons due to spallation-like impacts (Melosh, 1984). The short dynamical timescale of the system can result in quite rapid transfer, and the possibility of extensive exchange of both organic chemistry and even active biological material.

Moons will in general also be subject to diurnal total shadowing, or eclipse, events by their host planet. Although such periods would often be short (depending on orbital geometry), they could act to constantly perturb any surface climate. This is particularly relevant for high mass moons that could best represent a terrestrial-like habitable environment. While some limited studies have been made on the climatic effect of strong annual variations in stellar flux (i.e. for planets on eccentric

Fig. 11.8. A terrestrial hydrothermal vent system located at an ocean depth of approximately 2km on the Endeavor Ridge in the Pacific Ocean. Superheated (300°) water, rich in minerals, is being forced up from the ocean floor. On cooling, mineral deposition forms the "chimney" structures seen in the center of this picture. Surrounding these is an oasis of life, from extremophilic microbial species to the remarkable tube-worm colony on the left (Photo: V. J. Tunnicliffe, University of Victoria).

orbits, Williams & Pollard, 2002), little is known about the impact of rapid "on/off" insolation.

The magnetosphere of a giant planet may represent a difficult radiation environment for life on moons. However, it may also serve to alter the internal dynamics of a moon with consequences for life. The case of Ganymede's intrinsic magnetic field is intriguing in this respect, and it has been suggested that the moon's liquid core dynamo could in fact have been "spun up" by Jupiter's magnetosphere (Kivelson et al., 1998). Here then is another route to maintaining, or encouraging, a molten interior and all that follows in terms of potential habitability. It is also true that a giant planet's magnetic field may vary considerably. Saturn's field strength is some 30 times weaker than that of Jupiter (although still 580 times that of the Earth). The degree to which charged particles are trapped and accelerated may therefore vary significantly from planet to planet.

11.6 Summary

Are moons of exoplanets likely habitats for life? Certainly within our own Solar System there are at least two moons (Europa and Enceladus) that appear to meet some minimum criteria to encourage our interest in searching for evidence of life, and there may be more (Ganymede and Titan). Our expanding view of the diversity of life on Earth has created a plethora of variations on the classical ideas of habitability – from hot to cold, and chemically reactive to inert environments. On the basis of these facts alone it would seem reasonable to consider that moons elsewhere could be equally as interesting as terrestrial-type planets. The added temptation is that moons are likely to exist in a multitude of different physical, chemical, and thermodynamical configurations, including those resembling that of the Earth – potentially all within a *single* planetary system. It would be prudent to allow for the possibility that our terrestrial-centric view of life needs yet another revision.

Acknowledgments
Support from a NASA Astrobiology: Exobiology and Evolutionary Biology, Planetary Protection research grant (number NNG05GO79G) is acknowledged. Support was also provided by Columbia University through its Initiatives in Science and Engineering program and the Columbia Astrobiology Center.

References

Alibert, Y., Mousis, O., & Benz, W. : *Modeling the Jovian subnebula. I. Thermodynamic conditions and migration of proto-satellites*,Astronomy & Astrophysics, **439**, 1205 (2005)

Barnes, J. W., & O'Brien, D. P. : *Stability of Satellites around Close-in Extrasolar Giant Planets*, Astrophysical Journal, **575**, 1087 (2002)

Beatty, J. T. et al. : *An obligately photosynthetic bacterial anaerobe from a deep sea hydrothermal vent*,Proceedings of Natl. Acad. of Sci. **102**, 9306 (2005)

Boss, A. P. : *Evolution of the Solar Nebula. VII. Formation and Survival of Protoplanets Formed by Disk Instability*,Astrophysical Journal, **629**, 535 (2005)

Cabrera, J., & Schneider, J. : *Detecting companions to extrasolar planets using mutual events*, Astron. & Astrophys., **464**, 1133 (2007)

Canup, R. M. : *Dynamics of Lunar Formation*,Ann. Rev. Astron. & Astrophys., **42**, 441 (2004)

Canup, R. M., & Ward, W. R. :*Formation of the Galilean Satellites: Conditions of Accretion*, Astronomical Journal, **124**, 3404 (2002)

Canup, R. M., & Ward, W. R. : *A common mass scaling for satellite systems of gaseous planets*, Nature, **441**, 834 (2006)

Cassen, P., Reynolds, R. T., & Peale, S. J. : *Is there liquid water on Europa?*, Geophys. Res. Lett., **6**, 731 (1979)

Chyba, C. F. : *Energy for microbial life on Europa*, Nature, **403**, 381 (2000)

Doyle, L. R., & Deeg, H.-J. : *Timing Detection of Eclipsing Binary Planets and Transiting Extrasolar Moons*, IAU Symposium, **213**, 80 (2004)

Ehrenreich, D., Tinetti, G., Lecavelier Des Etangs, A., Vidal-Madjar, A., & Selsis, F. : *The transmission spectrum of Earth-size transiting planets*, Astron. & Astrophys. **448**, 379 (2006)

Estrada, P. R., & Mosqueira, I. : *A gas-poor planetesimal capture model for the formation of giant planet satellite systems*, Icarus, **181**, 486 (2006)

Gladman, B., Dones, L., Levison, H., Burns, J., & Gallant, J. : *Meteoroid Transfer to Europa and Titan*, 37th Annual Lunar and Planetary Science Conference, **37**, 2165 (2006)

Gold, T. : *The Deep, Hot Biosphere*, Proceedings of Natl. Acad. of Sci. **89**, 6045 (1992)

Greeley, R., et al. : *Europa: Initial Galileo Geological Observations*, Icarus, **135**, 4 (1998a)

Greeley, R., et al. : *Terrestrial Sea Ice Morphology: Considerations for Europa*, Icarus, **135**, 25 (1998b)

Greenberg, R., Geissler, P., Tufts, B. R., & Hoppa, G. V. : *Habitability of Europa's crust: The role of tidal-tectonic processes*, Journal of Geophysical Research, **105**, 17551 (2000)

Hayashi, C. : *Structure of the Solar Nebula, Growth and Decay of Magnetic Fields and Effects of Magnetic and Turbulent Viscosities on the Nebula*, Progress of Theoretical Physics Supplement, **70**, 35 (1981)

Hurford, T. A., Helfenstein, P., Hoppa, G. V., Greenberg, R., Bills, B. G. : *Eruptions arising from tidally controlled periodic openings of rifts on Enceladus*, Nature, **447**, 292, (2007)

C. Huygens: *Cosmotheoros*, (1695)
online translation http://www.phys.uu.nl/~huygens/cosmotheoros_en.htm

Ida, S., & Lin, D. N. C. : *Toward a Deterministic Model of Planetary Formation. I. A Desert in the Mass and Semimajor Axis Distributions of Extrasolar Planets*, Astrophysical Journal, **604**, 388 (2004)

Kasting, J. F., Whitmire, D. P., & Reynolds, R. T. : *Habitable Zones around Main Sequence Stars*, Icarus, **101**, 108 (1993)

Klahr, H., & Bodenheimer, P. : *Formation of Giant Planets by Concurrent Accretion of Solids and Gas inside an Anticyclonic Vortex*, Astrophysical Journal, **639**, 432 (2006)

Kivelson, M. G., et al. : *Ganymede's magnetosphere: Magnetometer overview*, Journal of Geophysical Research, **103**, 19963 (1998)

Kivelson, M. G., et al. : *Galileo Magnetometer Measurements: A Stronger Case for a Subsurface Ocean at Europa*, Science, **289**, 1340 (2000)

G. P. Kuiper: *Infrared observations of planets and satellites*, AJ, **62**, 295 (1957)

Lammer, H., Selsis, F., Ribas, I., Guinan, E. F., Bauer, S. J., & Weiss, W. W. : *Hydrodynamic escape of exo-planetary atmospheres*, In: Second Eddington Workshop: Stellar structure and habitable planet finding. Edited by F. Favata, S. Aigrain and A. Wilson, ESA SP-538, 339 (2004)

Lissauer, J. J. : *Planet formation*, Ann. Rev. Astron. & Astrophys., **31**, 129 (1993)

Lowell, R. P., & DuBose, M. : *Hydrothermal systems on Europa*, Geophysical Research Letters, **32**, 5202 (2005)

Lubow, S. H., Seibert, M., & Artymowicz, P. : *Disk Accretion onto High-Mass Planets*, Astrophysical Journal, **526**, 1001 (1999)

Lubow, S. H., & D'Angelo, G. : *Gas Flow across Gaps in Protoplanetary Disks*, Astrophysical Journal, **641**, 526 (2006)

Lunine, J. I., Lorenz, R. D., & Hartmann, W. K. : *Some speculations on Titans past, present and future*, Planetary and Space Science, **46**, 1099 (1998)

Marcy, G., Fischer, D. A., Butler, R. P., & Vogt, S. S. : *Properties of exoplanets: a Doppler study of 1330 stars*, Planet Formation, 179, Edited by Hubert Klahr and Wolfgang Brandner. Cambridge, UK (2006)

McCord, T. B., et al. : *Hydrated salt minerals on Europa's surface from the Galileo near-infrared mapping spectrometer (NIMS) investigation*, Journal of Geophysical Research, **104**, 11827 (1999)

Melosh, H. J. : *Impact ejection, spallation, and the origin of meteorites*, Icarus, **59**, 234 (1984)

Melosh, H. J., Janes, D. M., Jankowski, D. G., & Squyres, S. W. : *Ice Volcanism on Ariel*, Science, **245**, 194 (1989)

Melosh, H. J., Ekholm, A. G., Showman, A. P., & Lorenz, R. D. : *The temperature of Europa's subsurface water ocean*, Icarus, **168**, 498 (2004)

Moore, J. M., et al. : *Large Impact Features on Europa: Results of the Galileo Nominal Mission*, Icarus, **135**, 127 (1998)

Nimmo, F., Spencer, J. R., Pappalardo, R. T., Mullen, M. E.: *Shear heating as the origin of the plumes and heat flux on Enceladus*, Nature, **447**, 289 (2007)

Oishi, J. S., Mac Low, M.-M., & Menou, K. : *Turbulent Torques on Protoplanets in a Dead Zone*, ArXiv Astrophysics e-prints, arXiv:astro-ph/0702549 (2007)

Peale, S. J., Cassen, P., & Reynolds, R. T. : *Tidal dissipation, orbital evolution, and the nature of Saturn's inner satellites*, Icarus, **43**, 65 (1980)

Peale, S. J. : *Origin and Evolution of the Natural Satellites*, Ann. Rev. Astron & Astrophys., **37**, 533 (1999)

Porco, C. C., et al. : *Cassini Observes the Active South Pole of Enceladus*, Science, **311**, 1393 (2006)

Proctor, R. A. : *Other worlds than ours : The plurality of Worlds studied under the light of recent scientific researches*, London : Longman Green (1870).

Raymond, S. N., Mandell, A. M., & Sigurdsson, S. : *Exotic Earths: Forming Habitable Worlds with Giant Planet Migration*, Science, **313**, 1413 (2006)

Reynolds, R. T., Squyres, S. W., Colburn, D. S., & McKay, C. P. : *On the habitability of Europa*, Icarus, **56**, 246 (1983)

Reynolds, R. T., McKay, C. P., & Kasting, J. F. : *Europa, tidally heated oceans, and habitable zones around giant planets*, Advances in Space Research, **7**, 125 (1987)

Sartoretti, P., & Schneider, J. : *On the detection of satellites of extrasolar planets with the method of transits*, Astron. & Astrophys. Suppl., **134**, 553 (1999)

Scharf, C. A. : *The Potential for Tidally Heated Icy and Temperate Moons around Exoplanets*, Astrophysical Journal, **648**, 1196 (2006)

Smith, B. A., et al.: *The Jupiter system through the eyes of Voyager 1*, Science, **204**, 951 (1979a)

Smith, B. A., et al.: *The Galilean satellites and Jupiter – Voyager 2 imaging science results*, Science, **206**, 927 (1979b)

Spencer, J. R. : *Thermal segregation of water ice on the Galilean satellites*, Icarus, **69**, 297 (1987)

Squyres, S. W., Reynolds, R. T., & Cassen, P. M. : *Liquid water and active resurfacing on Europa*, Nature, **301**, 225 (1983)

Szabó, G. M., Szatmáry, K., Divéki, Z., & Simon, A. : *Possibility of a photometric detection of "exomoons"*, Astron. & Astrophys. **450**, 395 (2006)

Ward, W. R., & Reid, M. J. : *Solar tidal friction and satellite loss*, Mon. Notices Royal Astr. Soc., **164**, 21 (1973)

Williams, D. M., Kasting, J. F., & Wade, R. A. : *Habitable moons around extrasolar giant planets*, Nature, **385**, 234 (1997)

Williams, D. M., & Pollard, D. : *Earth-like worlds on eccentric orbits: excursions beyond the habitable zone*, International Journal of Astrobiology, **1**, 61 (2002)

Wisdom, J. : *Spin-Orbit Secondary Resonance Dynamics of Enceladus*, Astronomical Journal, **128**, 484 (2004)

Zimmer, C., Khurana, K. K., & Kivelson, M. G. : *Subsurface Oceans on Europa and Callisto: Constraints from Galileo Magnetometer Observations*, Icarus, **147**, 329 (2000)

Index

Printing: Mercedes-Druck, Berlin
Binding: Stein + Lehmann, Berlin